MULTISCALE ANALYSIS
OF COMPLEX TIME SERIES

THE WILEY BICENTENNIAL–KNOWLEDGE FOR GENERATIONS

*E*ach generation has its unique needs and aspirations. When Charles Wiley first opened his small printing shop in lower Manhattan in 1807, it was a generation of boundless potential searching for an identity. And we were there, helping to define a new American literary tradition. Over half a century later, in the midst of the Second Industrial Revolution, it was a generation focused on building the future. Once again, we were there, supplying the critical scientific, technical, and engineering knowledge that helped frame the world. Throughout the 20th Century, and into the new millennium, nations began to reach out beyond their own borders and a new international community was born. Wiley was there, expanding its operations around the world to enable a global exchange of ideas, opinions, and know-how.

For 200 years, Wiley has been an integral part of each generation's journey, enabling the flow of information and understanding necessary to meet their needs and fulfill their aspirations. Today, bold new technologies are changing the way we live and learn. Wiley will be there, providing you the must-have knowledge you need to imagine new worlds, new possibilities, and new opportunities.

Generations come and go, but you can always count on Wiley to provide you the knowledge you need, when and where you need it!

WILLIAM J. PESCE
PRESIDENT AND CHIEF EXECUTIVE OFFICER

PETER BOOTH WILEY
CHAIRMAN OF THE BOARD

MULTISCALE ANALYSIS
OF COMPLEX TIME SERIES
Integration of Chaos and Random Fractal Theory, and Beyond

Jianbo Gao
Department of Electrical and Computer Engineering
University of Florida

Yinhe Cao
BioSieve

Wen-wen Tung
Department of Earth and Atmospheric Sciences
Purdue University

Jing Hu
Department of Electrical and Computer Engineering
University of Florida

WILEY-INTERSCIENCE
A John Wiley & Sons, Inc., Publication

Published by John Wiley & Sons, Inc., Hoboken, New Jersey.
Published simultaneously in Canada.

For general information on our other products and services or for technical support, please contact our Customer Care Department within the United States at (800) 762-2974, outside the United States at (317) 572-3993 or fax (317) 572-4002.

Wiley also publishes its books in a variety of electronic formats. Some content that appears in print may not be available in electronic format. For information about Wiley products, visit our web site at www.wiley.com.

Wiley Bicentennial Logo: Richard J. Pacifico

Library of Congress Cataloging-in-Publication Data:

Multiscale analysis of complex time series : integration of chaos and random
    fractal theory, and beyond / Jianbo Gao . . . [et al.].
        p. cm.
    Includes index.
    ISBN 978-0-471-65470-4 (cloth)
    1. Time series analysis.   2. Chaotic behavior in systems.   3. Fractals.   I.
Gao, Jianbo, 1966–
    QA280.M85 2007
    519.5'5—dc22   2007019072

To our teachers,
Xianyi Zhou of Zhejiang
University,
Yilong Bai and Zhemin
Zheng of Chinese Academy
of Sciences,
Michio Yanai of UCLA,
and to our families

CONTENTS

Preface   xiii

1 Introduction   **1**

    1.1   Examples of multiscale phenomena   4

    1.2   Examples of challenging problems to be pursued   9

    1.3   Outline of the book   12

    1.4   Bibliographic notes   14

2 Overview of fractal and chaos theories   **15**

    2.1   Prelude to fractal geometry   15

    2.2   Prelude to chaos theory   18

    2.3   Bibliographic notes   23

    2.4   Warmup exercises   23

3 Basics of probability theory and stochastic processes   **25**

    3.1   Basic elements of probability theory   25

       3.1.1   Probability system   25

       3.1.2   Random variables   27

       3.1.3   Expectation   30

3.1.4   Characteristic function, moment generating function, Laplace
transform, and probability generating function   32
3.2   Commonly used distributions   34
3.3   Stochastic processes   41
3.3.1   Basic definitions   41
3.3.2   Markov processes   43
3.4   Special topic: How to find relevant information for a new field quickly   49
3.5   Bibliographic notes   51
3.6   Exercises   51

4   Fourier analysis and wavelet multiresolution analysis   53

4.1   Fourier analysis   54
4.1.1   Continuous-time (CT) signals   54
4.1.2   Discrete-time (DT) signals   55
4.1.3   Sampling theorem   57
4.1.4   Discrete Fourier transform   58
4.1.5   Fourier analysis of real data   58
4.2   Wavelet multiresolution analysis   62
4.3   Bibliographic notes   67
4.4   Exercises   67

5   Basics of fractal geometry   69

5.1   The notion of dimension   69
5.2   Geometrical fractals   71
5.2.1   Cantor sets   71
5.2.2   Von Koch curves   74
5.3   Power law and perception of self-similarity   75
5.4   Bibliographic notes   76
5.5   Exercises   76

6   Self-similar stochastic processes   79

6.1   General definition   79
6.2   Brownian motion (Bm)   81
6.3   Fractional Brownian motion (fBm)   84
6.4   Dimensions of Bm and fBm processes   87
6.5   Wavelet representation of fBm processes   89
6.6   Synthesis of fBm processes   90
6.7   Applications   93
6.7.1   Network traffic modeling   93
6.7.2   Modeling of rough surfaces   97
6.8   Bibliographic notes   97

| | 6.9 | Exercises | 98 |

7   **Stable laws and Levy motions**   **99**

	7.1	Stable distributions	100
	7.2	Summation of strictly stable random variables	103
	7.3	Tail probabilities and extreme events	104
	7.4	Generalized central limit theorem	107
	7.5	Levy motions	108
	7.6	Simulation of stable random variables	109
	7.7	Bibliographic notes	111
	7.8	Exercises	112

8   **Long memory processes and structure-function–based multifractal analysis**   **115**

	8.1	Long memory: basic definitions	115
	8.2	Estimation of the Hurst parameter	118
	8.3	Random walk representation and structure-function–based multifractal analysis	119
		8.3.1 Random walk representation	119
		8.3.2 Structure-function–based multifractal analysis	120
		8.3.3 Understanding the Hurst parameter through multifractal analysis	121
	8.4	Other random walk–based scaling parameter estimation	124
	8.5	Other formulations of multifractal analysis	124
	8.6	The notion of finite scaling and consistency of H estimators	126
	8.7	Correlation structure of ON/OFF intermittency and Levy motions	130
		8.7.1 Correlation structure of ON/OFF intermittency	130
		8.7.2 Correlation structure of Levy motions	131
	8.8	Dimension reduction of fractal processes using principal component analysis	132
	8.9	Broad applications	137
		8.9.1 Detection of low observable targets within sea clutter	137
		8.9.2 Deciphering the causal relation between neural inputs and movements by analyzing neuronal firings	139
		8.9.3 Protein coding region identification	147
	8.10	Bibliographic notes	149
	8.11	Exercises	151

9   **Multiplicative multifractals**   **153**

	9.1	Definition	153
	9.2	Construction of multiplicative multifractals	154
	9.3	Properties of multiplicative multifractals	157

9.4   Intermittency in fully developed turbulence   163
    9.4.1   Extended self-similarity   165
    9.4.2   The log-normal model   167
    9.4.3   The log-stable model   168
    9.4.4   The β-model   168
    9.4.5   The random β-model   168
    9.4.6   The p model   169
    9.4.7   The SL model and log-Poisson statistics of turbulence   169
9.5   Applications   171
    9.5.1   Target detection within sea clutter   173
    9.5.2   Modeling and discrimination of human neuronal activity   173
    9.5.3   Analysis and modeling of network traffic   176
9.6   Bibliographic notes   178
9.7   Exercises   179

10 Stage-dependent multiplicative processes   **181**

10.1   Description of the model   181
10.2   Cascade representation of $1/f^\beta$ processes   184
10.3   Application: Modeling heterogeneous Internet traffic   189
    10.3.1   General considerations   189
    10.3.2   An example   191
10.4   Bibliographic notes   193
10.5   Exercises   193

11 Models of power-law-type behavior   **195**

11.1   Models for heavy-tailed distribution   195
    11.1.1   Power law through queuing   195
    11.1.2   Power law through approximation by log-normal distribution   196
    11.1.3   Power law through transformation of exponential distribution   197
    11.1.4   Power law through maximization of Tsallis nonextensive entropy   200
    11.1.5   Power law through optimization   202
11.2   Models for $1/f^\beta$ processes   203
    11.2.1 $1/f^\beta$ processes from superposition of relaxation processes   203
    11.2.2 $1/f^\beta$ processes modeled by ON/OFF trains   205
    11.2.3 $1/f^\beta$ processes modeled by self-organized criticality   206
11.3   Applications   207
    11.3.1   Mechanism for long-range-dependent network traffic   207
    11.3.2   Distributional analysis of sea clutter   209
11.4   Bibliographic notes   210
11.5   Exercises   211

12 Bifurcation theory   213

   12.1 Bifurcations from a steady solution in continuous time systems   213
       12.1.1 General considerations   214
       12.1.2 Saddle-node bifurcation   215
       12.1.3 Transcritical bifurcation   215
       12.1.4 Pitchfork bifurcation   215
   12.2 Bifurcations from a steady solution in discrete maps   217
   12.3 Bifurcations in high-dimensional space   218
   12.4 Bifurcations and fundamental error bounds for fault-tolerant computations 218
       12.4.1 Error threshold values for arbitrary K-input NAND gates 219
       12.4.2 Noisy majority gate   222
       12.4.3 Analysis of von Neumann's multiplexing system   226
   12.5 Bibliographic notes   233
   12.6 Exercises   233

13 Chaotic time series analysis   235

   13.1 Phase space reconstruction by time delay embedding   236
       13.1.1 General considerations   236
       13.1.2 Defending against network intrusions and worms   237
       13.1.3 Optimal embedding   240
   13.2 Characterization of chaotic attractors   243
       13.2.1 Dimension   244
       13.2.2 Lyapunov exponents   246
       13.2.3 Entropy   251
   13.3 Test for low-dimensional chaos   254
   13.4 The importance of the concept of scale   258
   13.5 Bibliographic notes   258
   13.6 Exercises   259

14 Power-law sensitivity to initial conditions (PSIC)   261

   14.1 Extending exponential sensitivity to initial conditions to PSIC   262
   14.2 Characterizing random fractals by PSIC   263
       14.2.1 Characterizing $1/f^{\beta}$ processes by PSIC   264
       14.2.2 Characterizing Levy processes by PSIC   265
   14.3 Characterizing the edge of chaos by PSIC   266
   14.4 Bibliographic notes   268

15 Multiscale analysis by the scale-dependent
Lyapunov exponent (SDLE)   271

   15.1 Basic theory   271
   15.2 Classification of complex motions   274

15.2.1   Chaos, noisy chaos, and noise-induced chaos   274
15.2.2 $1/f^\beta$ processes   276
15.2.3   Levy flights   277
15.2.4   SDLE for processes defined by PSIC   279
15.2.5   Stochastic oscillations   279
15.2.6   Complex motions with multiple scaling behaviors   280
15.3   Distinguishing chaos from noise   283
15.3.1   General considerations   283
15.3.2   A practical solution   284
15.4   Characterizing hidden frequencies   286
15.5   Coping with nonstationarity   290
15.6   Relation between SDLE and other complexity measures   291
15.7   Broad applications   297
15.7.1   EEG analysis   297
15.7.2   HRV analysis   298
15.7.3   Economic time series analysis   300
15.7.4   Sea clutter modeling   303
15.8   Bibliographic notes   304

Appendix A: Description of data   **307**

A.1   Network traffic data   307
A.2   Sea clutter data   308
A.3   Neuronal firing data   309
A.4   Other data and program listings   309

**Appendix B: Principal Component Analysis (PCA), Singular Value
Decomposition (SVD), and Karhunen-Loève (KL) expansion**   **311**

Appendix C: Complexity measures   **313**

C.1   FSLE   314
C.2   LZ complexity   315
C.3   PE   317

References   319

Index   347

PREFACE

Complex interconnected systems, including the Internet, stock markets, and human heart or brain, are usually comprised of multiple subsystems that exhibit highly nonlinear deterministic as well as stochastic characteristics and are regulated hierarchically. They generate signals that exhibit complex characteristics such as nonlinearity, sensitive dependence on small disturbances, long memory, extreme variations, and nonstationarity. A complex system usually cannot be studied by decomposing the system into its constituent subsystems, but rather by measuring certain signals generated by the system and analyzing the signals to gain insights into the behavior of the system. In this endeavor, data analysis is a crucial step. Chaos theory and random fractal theory are two of the most important theories developed for data analysis. Unfortunately, no single book has been available to present all the basic concepts necessary for researchers to fully understand the ever-expanding literature and apply novel methods to effectively solve their signal processing problems. This book attempts to meet this pressing need by presenting chaos theory and random fractal theory in a unified way.

Integrating chaos theory and random fractal theory and going beyond them has proven to be much harder than we had thought, because the foundations for chaos theory and random fractal theory are entirely different. Chaos theory is mainly concerned about apparently irregular behaviors in a complex system that are generated by nonlinear deterministic interactions of only a few numbers of degrees of freedom, where noise or intrinsic randomness does not play an important role, while random fractal theory assumes that the dynamics of the system are inherently random. After postponing delivery of the book for more than two and half years, we are finally satisfied. The book now contains many new results in Chapters 8–15 that have not been published elsewhere, culminating in the development of a multiscale complexity measure that is computable from short, noisy time series. As shown in Chapter 15, the measure can readily classify major types of complex motions, effectively deal with nonstationarity, and simultaneously characterize the behaviors of complex signals on a wide range of scales, including complex irregular behaviors on small scales and orderly behaviors, such as oscillatory motions, on large scales.

This book has adopted a data-driven approach. To help readers better understand and appreciate the power of the materials in the book, nearly every significant concept or approach presented is illustrated by applying it to effectively solve real problems, sometimes with unprecedented accuracy. Furthermore, source codes, written in various languages, including Java, Fortran, C, and Matlab, for many methods are provided in a dedicated book website, together with some simulated and experimental data (see Sec. A.4 in Appendix A).

This book contains enough material for a one-year graduate-level course. It is useful for students with various majors, including electrical engineering, computer science, civil and environmental engineering, mechanical engineering, chemical engineering, medicine, chemistry, physics, geophysics, mathematics, finance, and population ecology. It is also useful for researchers working in relevant fields and practitioners who have to solve their own signal processing problems.

We thank Drs. Vince Billock, Gijs Bosman, Yenn-Ru Chen, Yuguang Fang, Jose Fortes, John Harris, Hsiao-ming Hsu, Sheng-Kwang Hwang, Mark Law, Jian Li, Johnny Lin, Jiamin Liu, Mitch Moncrieff, Jose Principe, Vladimir Protopopescu, Nageswara Rao, Ronn Ritke, Vwani Roychowdhury, Izhak Rubin, Chris Sackellares, Zhen-Su She, Yuch-Ning Shieh, Peter Stoica, Martin Uman, Kung Yao, and Keith White for many useful discussions. Drs. Jon Harbor, Andy Majda, and Robert Nowack have read part of Chapter 15, while Dr. Alexandre Chorin has read a number of chapters. We are grateful for their many useful suggestions and encouragement. One of the authors, Jianbo Gao, taught a one-year course entitled "Signal Processing with Chaos and Fractals" at the University of Florida, in the fall of 2002 and the spring of 2003. Students' enthusiasm has been instrumental in driving us to finish the book. He would particularly thank his former and current students Jing Ai, Ung Sik Kim, Jaemin Lee, Yan Qi, Dongming Xu, and Yi Zheng for their contributions to the many topics presented here. We would like to thank the editors at Wiley, Helen Greenberg, Whitney Lesch, Val Moliere, Christine Punzo, and George Telecki, for their patience and encouragement. Finally, we thank IPAM at UCLA and MBI at the Ohio State University for generously supporting us to attend a number of interesting workshops organized by the two institutions.

CHAPTER 1

INTRODUCTION

Complex systems are usually comprised of multiple subsystems that exhibit both highly nonlinear deterministic and stochastic characteristics and are regulated hierarchically. These systems generate signals that exhibit complex characteristics such as sensitive dependence on small disturbances, long memory, extreme variations, and nonstationarity. A stock market, for example, is strongly influenced by multilayered decisions made by market makers, as well as by interactions of heterogeneous traders, including intraday traders, short-period traders, and long-period traders, and thus gives rise to highly irregular stock prices. The Internet, as another example, has been designed in a fundamentally decentralized fashion and consists of a complex web of servers and routers that cannot be effectively controlled or analyzed by traditional tools of queuing theory or control theory and give rise to highly bursty and multiscale traffic with extremely high variance, as well as complex dynamics with both deterministic and stochastic components. Similarly, biological systems, being heterogeneous, massively distributed, and highly complicated, often generate nonstationary and multiscale signals. With the rapid accumulation of complex data in health sciences, systems biology, nano-sciences, information systems, and physical sciences, it has become increasingly important to be able to analyze multiscale and nonstationary data.

1

Multiscale signals behave differently, depending upon the scale at which the data are examined. How can the behaviors of such signals on a wide range of scales be simultaneously characterized? One strategy we envision is to use existing theories synergistically instead of individually. To make this possible, appropriate scale ranges where each theory is most pertinent need to be identified. This is a difficult task, however, since different theories may have entirely different foundations. For example, chaos theory is mainly concerned about apparently irregular behaviors in a complex system that are generated by nonlinear deterministic interactions with only a few degrees of freedom, where noise or intrinsic randomness does not play an important role. Random fractal theory, on the other hand, assumes that the dynamics of the system are inherently random. Therefore, to make this strategy work, different theories need to be integrated and even generalized.

The second vital strategy we envision is to develop measures that explicitly incorporate the concept of scale so that different behaviors of the data on varying scales can be simultaneously characterized by the same scale-dependent measure. In the most ideal scenario, a scale-dependent measure can readily classify different types of motions based on analysis of short, noisy data. In this case, one can readily see that the measure will be able not only to identify appropriate scale ranges where different theories, including information theory, chaos theory, and random fractal theory, are applicable, but also to automatically characterize the behaviors of the data on those scale ranges.

The vision presented above dictates the style and the scope of this book, as depicted in Fig. 1.1. Specifically, we aim to build an effective arsenal by synergistically integrating approaches based on chaos and random fractal theory, and going beyond this, to complement conventional approaches such as spectral analysis and machine learning techniques. To make such an integration possible, four important efforts are made:

1. Wavelet representation of fractal models as well as wavelet estimation of fractal scaling parameters will be carefully developed. Furthermore, a new fractal model will be developed. The model provides a new means of characterizing long-range correlations in time series and a convenient way of modeling non-Gaussian statistics. More importantly, it ties together different approaches in the vast field of random fractal theory (represented by the four small boxes under the "Random Fractal" box in Fig. 1.1).

2. Fractal scaling break and truncation of power-law behavior are related to specific features of real data so that scale-free fractal behavior as well as structures defined by specific scales can be simultaneously characterized.

3. A new theoretical framework for signal processing — power-law sensitivity to initial conditions (PSIC) — will be developed, to provide chaos and random fractal theory a common foundation so that they can be better integrated.

4. The scale-dependent Lyapunov exponent (SDLE), which is a variant of the finite-size Lyapunov exponent (FSLE), is an excellent multiscale measure. We shall develop a highly efficient algorithm for calculating it and show that it can readily classify different types of motions, aptly characterize complex behaviors of real-world multiscale signals on a wide range of scales, and, therefore, naturally solve the classic problem of distinguishing low-dimensional chaos from noise. Furthermore, we shall show that the SDLE can effectively deal with nonstationarity and that existing complexity measures can be related to the value of the SDLE on specific scales.

To help readers better understand and appreciate the power of the materials in this book, nearly every significant concept or approach presented will be illustrated by applying it to effectively solve real problems, sometimes with unprecedented accuracy. Furthermore, source codes, written in various languages, including Fortran, C, and Matlab, for many methods are provided together with some simulated and experimental data.

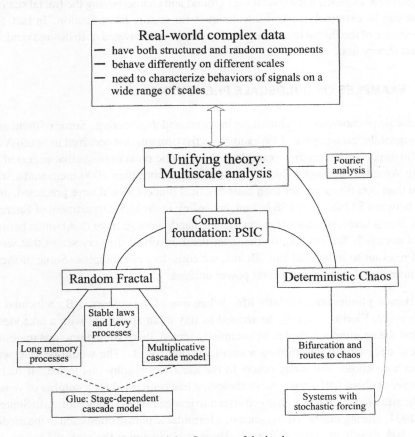

Figure 1.1. Structure of the book.

In the rest of this chapter, we give a few examples of multiscale phenomena so that readers can better appreciate the difficulties and excitement of multiscale signal processing. We then highlight a number of multiscale signal processing problems that will be discussed in depth later in the book. Finally, we outline the structure of the book.

Before proceeding, we note a subtle but important distinction between fractal (and wavelet)-based multiscale analysis methods and multiscale phenomena. As we shall discuss in more detail in Chapter 2, fractal phenomena are situations where a part of an object or phenomenon is exactly or statistically similar to another part or to the whole. Because of this, no specific scales can be defined. Fractal property can thus be considered as a common feature across vastly different scales; such a feature can be considered as a background. While this can be viewed as a multiscale phenomenon, it is in fact one of the simplest. More complicated situations can be easily envisioned. For example, a fractal scaling may be broken at certain spatio-temporal scales determined by the periodicities. Here, the periodicities are part of the multiscale phenomenon but not part of the fractal phenomenon. As we shall see in Chapter 8, exploiting the fractal background and characterizing the fractal scaling break can be extremely powerful techniques for feature identification. In fact, the importance of the fractal feature as a background has motivated us to discuss random fractal theory first.

1.1 EXAMPLES OF MULTISCALE PHENOMENA

Multiscale phenomena are ubiquitous in nature and engineering. Some of them are, unfortunately, catastrophic. One example is the tsunami that occurred in South Asia at Christmas 2004. Another example is the gigantic power outage that occurred in North America on August 14, 2003. It affected more than 4000 megawatts, was more than 300 times greater than mathematical models would have predicted, and cost between $4 billion and $6 billion, according to the U.S. Department of Energy. Both events involved scales that, on the one hand, were so huge that human beings could not easily fathom and, on the other hand, involved the very scales that were most relevant to individual life. Below, we consider six examples. Some of them are more joyful than tsunamis and power outages.

Multiscale phenomena in daily life. When one of the authors, J.B., relocated to Gainesville, Florida, in 2002, he wanted to stay in an apartment with a lake view. Behind the apartment complex he eventually leased, there were numerous majestic cypress trees closely resembling a small wetland forest. The would-be lake was in fact a sinkhole, like many others in the karst topography of Florida. It had a diameter of about half a mile. Nevertheless, it had been dried for a number of years.

The situation completely changed after hurricane Frances struck Florida in September, 2004. During that hurricane season, a formidable phrase often used in the media was "dark clouds the size of Texas." Texas is about twice the size of Florida, so the ratio between the size of the sinkhole behind J.B.'s apartment and the clouds associated with the tropical storm system is on the order of 10^{-3}. Obviously, the sizes of the sinkhole and the cloud system define two very different scales.

During the passage of hurricane Frances, within only three days, the water level in the center of the sinkhole rose 4–5 meters. By the spring of 2005, the sinkhole had fully developed into a beautiful lake ecosystem: wetland plants and trees blossomed; after a few rains, the water swarmed with tiny fishes; each day at around sunset, hundreds of egrets flew back to the lake, calling; dozens of ducks constantly played on the water, generating beautiful water waves; turtles appeared; even alligators came – one day one of them was seen to be killing a snake for food. All these activities occurred on a scale comparable to or much smaller than that of the lake, and therefore much smaller than the size of the clouds accompanying hurricane Frances. In spite of causing devastating destructions to the east coast of Florida, hurricane Frances also replenished and diversified ecosystems in its path. Therefore, although a rather rare and extreme event, hurricane Frances can never be ignored because it made a huge impact on lives long after its passing. An important lesson to learn from this example is that a rare event may not simply be treated as an outlier and ignored.

One of the most useful parameters for characterizing water level change in a river or lake is the Hurst parameter, named after the distinguished hydrologist Hurst, who monitored the water level changes in Niles for decades. The Hurst parameter measures the persistence of correlations. Intuitively, this corresponds to the situation of the sinkhole behind J.B.'s apartment: when it is dry, it can stay dry for years, but with its current water level, it is unlikely to become dry again any time soon. In the past decade, researchers have found that persistent correlation is a prevailing feature of network traffic. Can this feature be ignored when designing or examining the quality of service of a network? The answer is no. We shall have much to say about the Hurst parameter in general and the impact of persistent correlation on network traffic modeling as an example of an application in this book.

Multiscale phenomena in genomic DNA sequences. DNA is a large molecule composed of four basic units called nucleotides. Each nucleotide contains phosphate, sugar, and one of the four bases: adenine, guanine, cytosine, and thymine (usually denoted A, G, C, and T). The structure of DNA is described as a double helix. The two helices are held together by hydrogen bonds. Within the DNA double helix, A and G form two and three hydrogen bonds with T and C on the opposite strand, respectively. The total length of the human DNA is estimated to be 3.2×10^9 base pairs. The most up-to-date estimate of the number of genes in humans is about 20,000 – 25,000, which is comparable to the number of genes in many other species. Typically, a gene is several hundred bases long. This is about $10^{-7} - 10^{-6}$ of the total length of the human genome, comparable to the ratio between the size of an alligator and the size of the clouds accompanying hurricane Frances. There are other, shorter functional units, such as promoters, enhancers,

Figure 1.2. Turbulent flow by Da Vinci.

prohibitors, and so on. Some genes or functional units could repeat, exactly or with slight modifications (called mutations), after tens of thousands of bases. Therefore, genomic DNA sequences are full of (largely static) multiscale phenomena.

Multiscale modeling of fluid motions. Fascinated by the complex phenomena of water flowing and mixing, Leonardo da Vinci made an exquisite portrait of turbulent flow of fluid, involving vortices within vortices over an ever-decreasing scale. See Fig. 1.2 and
http://www.efluids.com/efluids/gallery/gallery_pages/da_vinci_page.htm.
The central theme in multiscale modeling of fluid motions is the determination of what information on the finer scale is needed to formulate an equation for the "effective" behavior on the coarser scale. In fact, this is also the central theme of multiscale modeling in many other fields, such as cloud-resolving modeling for studying atmospheric phenomena on scales much larger than individual clouds, and modeling in materials science and biochemistry, where one strives to relate the functionality of the material or organism to its fundamental constituents, their chemical nature and geometric arrangement.

Multiscale phenomena in computer networks. Large-scale communications networks, especially the Internet, are among the most complicated systems that man has ever made, with many multiscale aspects. Intuitively speaking, these multiscales come from the hierarchical design of a protocol stack, the hierarchical topological architecture, and the multipurpose and heterogeneous nature of the Internet. More precisely, there are multiscales in (1) time, manifested by the prevailing fractal, multifractal, and long-range-dependent properties in traffic, (2) space, essentially due to topology and geography and again manifested by scale-free properties, (3) state, e.g., queues and windows, and (4) size, e.g., number of nodes and number of users. Also, it has been observed that the failure of a single router may trigger routing instability, which may be severe enough to instigate a route flap storm. Furthermore, packets may be delivered out of order or even get dropped, and packet reordering is not a pathological network behavior. As the next-generation Internet applications such as remote instrument control and computational steering are being developed, another facet of complex multiscale behavior is beginning to surface in terms of transport dynamics. The networking requirements for these next-generation applications belong to (at least) two broad classes *involving vastly disparate time scales*: (1) high bandwidths, typically multiples of 10 Gbps, to support bulk data transfers, and (2) stable bandwidths, typically at much lower bandwidths such as 10–100 Mbps, to support interactive, steering, and control operations.

Is there any difference between this example and examples 2 and 3? The answer is yes. In multiscale modeling of fluid motions, the basic equation, the Navier-Stokes equation, is known. Therefore, the dynamics of fluid motions can be systematically studied through a combined approach of theoretical modeling, numerical simulation, and experimental study. However, there is no fundamental equation to describe a DNA sequence or a computer network.

Multiscale phenomena in sea clutter. Sea clutter is the radar backscatter from a patch of ocean surface. The complexity of the signals comes from two sources: the rough sea surface, sometimes oscillatory, sometimes turbulent, and the multipath propagation of the radar backscatter. This can be well appreciated by imagining radar pulses massively reflecting from the wavetip of Fig. 1.3. To be quantitative, in Fig. 1.4, two 0.1 s duration sea clutter signals, sampled with a frequency of 1 KHz, are plotted in (a,b), a 2 s duration signal is plotted in (c), and an even longer signal (about 130 s) is plotted in (d). It is clear that the signal is not purely random, since the waveform can be fairly smooth on short time scales (Fig. 1.4(a)). However, the signal is highly nonstationary, since the frequency of the signal (Fig. 1.4(a,b)) as well as the randomness of the signal (Fig. 1.4(c,d)) change over time drastically. Therefore, naive Fourier analysis or deterministic chaotic analysis of sea clutter may not be very useful. From Fig. 1.4(e), where $X_t^{(m)}$ is the nonoverlapping running mean of X over block size m and X is the sea clutter amplitude data, it can be further concluded that neither autoregressive (AR) models nor textbook fractal models can describe the data. This is because AR modeling requires exponentially decaying autocorrelation (which amounts to $Var(X_t^{(m)}) \sim m^{-1}$, or a Hurst parameter of 1/2; see Chapters 6 and 8), while fractal modeling requires the variation between $Var(X_t^{(m)})$ and m to follow a power law. However, neither behavior is observed in Fig. 1.4(e). Indeed, although extensive work has been done on sea clutter, its nature is still poorly understood. As a result, the important problem of target detection

Figure 1.3.   Schematic of a great wave (a tsunami, woodblock print from the 19th century Japanese artisit Hokusai). Suppose that our field of observation includes the wavetip of length scale of a few meters. It is then clear that the complexity of sea clutter is mainly due to massive reflection of radar pulses from a wavy and even turbulent ocean surface.

within sea clutter remains a tremendous challenge. We shall return to sea clutter later.

Multiscale and nonstationary phenomena in heart rate variability (HRV). HRV is an important dynamical variable in cardiovascular function. Its most salient feature is spontaneous fluctuation, even if the environmental parameters are maintained constant and no perturbing influences can be identified. It has been observed that HRV is related to various cardiovascular disorders. Therefore, analysis of HRV is very important in medicine. However, this task is very difficult, since HRV data are highly complicated. An example for a normal young subject is shown in Fig. 1.5. Evidently, the signal is highly nonstationary and multiscaled, appearing oscillatory for some period of time (Figs. 1.5(b,d)), and then varying as a power law for another period of time (Figs. 1.5(c,e)). The latter is an example of the so-called $1/f$ processes, which will be discussed in depth in later chapters. While the multiscale nature of such signals cannot be fully characterized by existing methods, the nonstationarity of the data is even more troublesome, since it requires the data to be properly segmented before further analysis by methods derived from spectral analysis, chaos theory, or random fractal theory. However, automated segmentation of complex biological signals to remove undesired components is itself a significant open problem, since it is closely related to, for example, the challenging task of

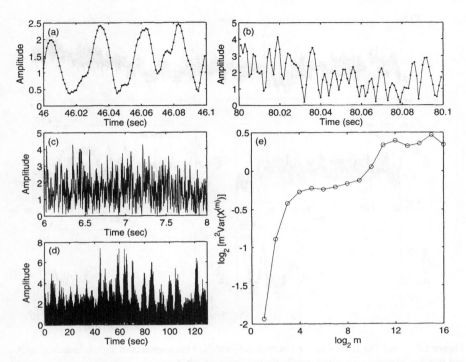

Figure 1.4.   (a,b) Two 0.1 s duration sea clutter signals; (c) a 2 s duration sea clutter signal; (d) the entire sea clutter signal (of about 130 s); and (e) $\log_2[m^2 Var(X^{(m)})]$ vs. $\log_2 m$, where $X^{(m)} = \{X_t^{(m)} : X_t^{(m)} = (X_{tm-m+1} + \cdots + X_{tm})/m, \ t = 1, 2, \cdots\}$ is the non-overlapping running mean of $X = \{X_t : t = 1, 2, \cdots\}$ over block size m and X is the sea clutter amplitude data. To better see the variation of $Var(X_t^{(m)})$ with m, $Var(X_t^{(m)})$ is multiplied by m^2. When the autocorrelation of the data decays exponentially fast (such as modeled by an AR process), $Var(X_t^{(m)}) \sim m^{-1}$. Here $Var(X_t^{(m)})$ decays much faster. A fractal process would have $m^2 Var(X_t^{(m)}) \sim m^\beta$. However, this is not the case. Therefore, neither AR modeling nor ideal textbook fractal theory can be readily applied here.

accurately detecting transitions from normal to abnormal states in physiological data.

1.2 EXAMPLES OF CHALLENGING PROBLEMS TO BE PURSUED

In this book, a wide range of important problems will be discussed in depth. As a prelude, we describe 10 of them in this section.

P1: Can economic time series be modeled by low-dimensional noisy chaos?
Since late 1980s, considerable efforts have been made to determine whether irregular economic time series are chaotic or random. By analyzing real as well as simulated economic time series using the neural network–based Lyapunov expo-

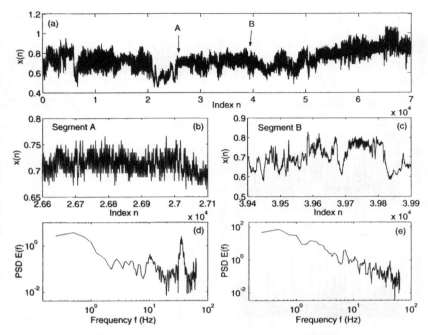

Figure 1.5.   (a) The HRV data for a normal subject; (b,c) the segments of signals indicated as A and B in (a); (d,e) power spectral density for the signals shown in (b,c).

nent estimator, a number of recent studies have suggested that the world economy may not be characterized by low-dimensional chaos, since the largest Lyapunov exponent is negative. As will be explained in Chapter 13, the sum of the positive Lyapunov exponents gives a tight upper bound of Kolmogorov-Sinai (KS) entropy. KS entropy characterizes the rate of creation of information in a dynamical system. It is zero, positive, and infinite for regular, chaotic, and random motions, respectively. Therefore, a negative largest Lyapunov exponent in economic time series amounts to negative KS entropy and thus implies regular economic dynamics. But economy is anything but simple! How may we resolve this dilemma? An answer will be provided in Chapter 15.

P2: Network traffic modeling
Data transfer across a network is a very complicated process owing to interactions among a huge number of correlated or uncorrelated users, congestion, routing instability, packet reordering, and many other factors. Traditionally, network traffic is modeled by Poisson or Markovian models. Recently, it has been found that long-range dependence (LRD) is a prevailing feature of real network traffic. However, it is still being debated which type of model, Poisson or Markovian type, LRD, or multifractal, should be used to evaluate the performance of a network. It would be very desirable to develop a general framework that not only includes all the traffic

models developed so far as special cases, but also accurately and parsimoniously models real network traffic. Can such a goal be reached? A surprisingly simple answer will be given in Chapter 10.

P3: Network intrusion and worm detection
Enterprise networks are facing ever-increasing security threats from various types of intrusions and worms. May the concepts and methods developed in this book be useful for protecting against these problems? A surprisingly simple and effective solution will be presented in Sec. 13.1.2.

P4: Sea clutter modeling and target detection within sea clutter
Accurate modeling of sea clutter and robust detection of low observable targets within sea clutter are important problems in remote sensing and radar signal processing applications. For example, they may be helpful in improving navigation safety and facilitating environmental monitoring. As we have seen in Fig. 1.4, sea clutter data are highly complicated. Can some concept or method developed in this book provide better models for sea clutter? The answer is yes. In fact, we will find that many of the new concepts and methods developed in the book are useful for this difficult task.

P5: Fundamental error bounds for nano- and fault-tolerant computations
In the emerging nanotechnologies, faulty components may be an integral part of a system. For the system to be reliable, the error of the building blocks has to be smaller than a threshold. Therefore, finding exact error thresholds for noisy gates is one of the most challenging problems in fault-tolerant computations. We will show in Chapter 12 that bifurcation theory offers an amazingly effective approach to solve this problem.

P6: Neural information processing
Mankind's desire to understand neural information processing has been extremely useful in developing modern computing machines. While such a desire will certainly motivate the development of new bio-inspired computations, understanding neural information processing has become increasingly pressing owing to the recent great interest in brain-machine interfaces and deep brain stimulation. In Secs. 8.9.2 and 9.5.2, we will show how the various types of fractal analysis methods developed in the book can help understand neuronal firing patterns.

P7: Protein coding sequence identification in genomic DNA sequences
Gene finding is one of the most important tasks in the study of genomes. Indices that can discriminate DNA sequences' coding and noncoding regions are crucial elements of a successful gene identification algorithm. Can multiscale analysis of genome sequences help construct novel codon indices? The answer is yes, as we shall see in Chapter 8.

P8: Analysis of HRV

We have seen in Fig. 1.5 that HRV data are nonstationary and multiscaled. Can multiscale complexity measures readily deal with nonstationarity in HRV data, find the hidden differences in HRV data under healthy and disease conditions, and shed new light on the dynamics of the cardiovascular system? An elegant answer will be given in Chapter 15.

P9: EEG analysis

Electroencephalographic (EEG) signals provide a wealth of information about brain dynamics, especially related to cognitive processes and pathologies of the brain such as epileptic seizures. To understand the nature of brain dynamics as well as to develop novel methods for the diagnosis of brain pathologies, a number of complexity measures from information theory, chaos theory, and random fractal theory have been used to analyze EEG signals. Since these three theories have different foundations, it has not been easy to compare studies based on different complexity measures. Can multiscale complexity measures offer a unifying framework to overcome this difficulty and, more importantly, to offer new and more effective means of providing a warning about pathological states such as epileptic seizures? Again, an elegant answer will be given in Chapter 15.

P10: Modeling of turbulence

Turbulence is a prevailing phenomenon in geophysics, astrophysics, plasma physics, chemical engineering, and environmental engineering. It is perhaps the greatest unsolved problem in classical physics. Multifractal models, pioneered by B. Mandelbrot, are among the most successful in describing intermittency in turbulence. In Sec. 9.4, we will present a number of multifractal models for the intermittency phenomenon of turbulence in a coherent way.

1.3   OUTLINE OF THE BOOK

We have discussed the purpose and the basic structure of the book in Fig. 1.1. To facilitate our discussions, Chapter 2 is a conceptual chapter consisting of two sections describing fractal and chaos theories. Since fractal theory will be treated formally starting with Chapter 5, the section on fractal theory in Chapter 2 will be quite brief; the section on chaos theory, however, will be fairly detailed because this theory will not be treated in depth until Chapter 13. The rest of the structure of the book is largely determined by our own experience in analyzing complicated time series arising from fields as diverse as device physics, radar engineering, fluid mechanics, geophysics, physiology, neuroscience, vision science, and bioinformatics, among many others. Our view is that random fractals, when used properly, can be tremendously helpful, especially in forming new hypotheses. Therefore, we shall spend a lot of time discussing signal processing techniques using random fractal theory.

To facilitate the discussion on random fractal theory, Chapter 3 reviews the basics of probability theory and stochastic processes. The correlation structure of the latter will be emphasized so that comparisons between stochastic processes with short memory and fractal processes with long memory can be made in later chapters. Chapter 4 briefly discusses Fourier transform and wavelet multiresolution analysis, with the hope that after this treatment, readers will find complicated signals to be friendly. Chapter 5 briefly resumes the discussion of Sec. 2.1 and discusses the basics of fractal geometry. Chapter 6 discusses self-similar stochastic processes, in particular the fractional Brownian motion (fBm) processes. Chapter 7 discusses a different type of fractal processes, the Levy motions, which are memoryless but have heavy-tailed features. Beyond Chapter 7, we focus on various techniques of signal processing: Chapter 8 discusses structure-function–based multifractal technique and various methods for assessing long memories from a time series together with a number of applications. The latter include topics as diverse as network traffic modeling, detection of low observable objects within sea clutter radar returns, gene finding from DNA sequences, and so on. Chapter 9 discusses a different type of multifractal, the multiplicative cascade multifractal. Fractal analysis culminates in Chapter 10, where a new model is presented. This is a wonderful model, since it "glues" together the structure-function–based fractal analysis and the cascade model. When one is not sure which type of fractal model should be used to analyze data, this model may greatly simplify the job. Chapter 11 discusses models for generating heavy-tailed distributions and long-range correlations. In Chapter 12, we switch to a completely different topic — bifurcation theory — and apply it to solve the difficult problem of finding exact error threshold values for fault-tolerant computations. In Chapter 13, we discuss the basics of chaos theory and chaotic time series analysis. In Chapter 14, we extend the discussion in Chapter 13 and consider a new theoretical framework, the power-law sensitivity to initial conditions, to provide chaos theory and random fractal theory with a common foundation. Finally, in Chapter 15, we discuss an excellent multiscale measure — the scale-dependent Lyapunov exponent — and its numerous applications.

Complex time series analysis is a very diverse field. For ease of illustrating the practical use of many concepts and methods, we have included many of our own works, some published, some appearing here for the first time, as examples. This, however, does not mean that our own contribution to this field is very significant. We must emphasize that even though the topics covered in this book are very broad, they are still just a subset of the many interesting issues and methods developed for complex time series analysis. For example, fascinating topics such as chaos control and chaos-based noise reduction and prediction are not touched on at all. Furthermore, we may not have covered all aspects of the chosen topics. Some of the omissions are intentional, since this book is designed to be useful not only for people in the field but also as an introduction to the field — especially to be used as a textbook. Of course, some of the unintentional omissions are due to our ignorance. While we apologize if some of important works are not properly reported here, we hope that readers will search for relevant literature on interesting topics using the literature search method discussed in Chapter 3. Finally, readers are strongly encouraged to work on the homework problems in the end of each chapter, especially those related to computer simulations. Only by doing this can they truly appreciate the beauty as well as the limitation of an unfamiliar new concept.

1.4 BIBLIOGRAPHIC NOTES

In the past two decades, a number of excellent books on chaos and fractal theories have been published. An incomplete list includes [33, 36, 130, 198, 294, 387] on fractal theory, [381,455] on stable laws, [25,45,106,206,250,330,404,409,415,475] on chaos theory, and [1, 249] on chaotic time series analysis. In particular, [249] can be considered complementary to this book. A collection of significant early papers can be found in [331].

In Sec. 1.3, we stated that topics including chaos control, synchronization in chaotic systems, chaos-based noise reduction and prediction would not be touched on in the book. Readers interested in chaos control are referred to the comprehensive book by Ott [330], those interested in chaos synchronization are referred to [338], and those interested in prediction from data are referred to [249] for chaos-based approaches and to [213] for the Kalman filtering approach. For prediction of a dynamical system with known equations but only partial knowledge of initial conditions, we refer readers to [77]. Finally, readers wishing to pursue the physical mechanisms of self-similarity and incomplete self-similarity are strongly encouraged to read the exquisite book by Barenblatt [31].

CHAPTER 2

OVERVIEW OF FRACTAL AND CHAOS THEORIES

The formal treatment of fractal theory will not begin until Chapter 5, and chaos theory will not be treated in depth until Chapter 13. In this chapter, we devote two sections to fractal and chaos theories to provide readers with a general overview of these theories without getting into the technical details.

2.1 PRELUDE TO FRACTAL GEOMETRY

Euclidean geometry is about lines, planes, triangles, squares, cones, spheres, etc. The common feature of these different objects is regularity: none of them is irregular. Now let us ask a question: Are clouds spheres, mountains cones, and islands circles? The answer is obviously no. In pursuing answers to such questions, Mandelbrot has created a new branch of science — fractal geometry.

For now, we shall be satisfied with an intuitive definition of a fractal: a set that shows irregular but self-similar features on many or all scales. Self-similarity means that part of an object is similar to other parts or to the whole. That is, if we view an irregular object with a microscope, whether we enlarge the object by 10 times or by 100 times or even by 1000 times, we always find similar objects. To understand this better, let us imagine that we were observing a patch of white cloud drifting away in the sky. Our eyes were rather motionless: we were staring more or less in the same direction. After a while, the part of the cloud we saw drifted away, and we were viewing a different part of the cloud. Nevertheless, our feeling remained more or less the same.

Mathematically, a fractal is characterized by a power-law relation, which translates into a linear relation in the log-log scale. In Chapter 5, we shall explain why a power-law relation implies self-similarity. For now, let us again resort to imagination. We are walking down a wild, jagged mountain trail or coastline. We would like to know the distance covered by our route. Suppose our ruler has a length of ϵ, which could be our step size, and different hikers may have different step sizes — a person riding a horse has a huge step size, while a group of people with a little child must have a tiny step size. The length of our route is

$$L = N(\epsilon) \cdot \epsilon , \qquad (2.1)$$

where $N(\epsilon)$ is the number of intervals needed to cover our route. It is most remarkable that typically $N(\epsilon)$ scales with ϵ in a power-law manner,

$$N(\epsilon) \sim \epsilon^{-D}, \quad \epsilon \to 0 , \qquad (2.2)$$

with D being a noninteger, $1 < D < 2$. Such a nonintegral D is often called the fractal dimension to emphasize the fragmented and irregular characteristics of the object under study.

Let us now try to understand the meaning of the nonintegral D. For this purpose, let us consider how length, area, and volume are measured. A common method of measuring a length, a surface area, or a volume is to cover it with intervals, squares, or cubes whose length, area, or volume is taken as the unit of measurement. These unit intervals, squares, and volumes are called unit boxes. Suppose, for instance, that we have a line whose length is 1. We want to cover it by intervals (boxes) whose length is ϵ. It is clear that we need $N(\epsilon) \sim \epsilon^{-1}$ boxes to completely cover the line. Similarly, if we want to cover an area or volume by boxes with linear length ϵ, we would need $N(\epsilon) \sim \epsilon^{-2}$ to cover the area, or $N(\epsilon) \sim \epsilon^{-3}$ boxes for the volume. Such D is called the topological dimension and takes on a value of 1 for a line, 2 for an area, and 3 for a volume. For isolated points, D is zero. That is why a point, a line, an area, and a volume are called $0 - D$, $1 - D$, $2 - D$, and $3 - D$ objects, respectively.

Now let us examine the consequence of $1 < D < 2$ for a jagged mountain trail. It is clear that the length of our route increases as ϵ becomes smaller, i.e., when $\epsilon \to 0$, $L \to \infty$. To be more concrete, let us visualize a race between the hare and the tortoise on a fractal trail with $D = 1.25$. Assume that the length of the average step taken by the hare is 16 times that taken by the tortoise. Then we have

$$L_{\text{hare}} = \frac{1}{2} L_{\text{tortoise}} .$$

That is, the tortoise has to run twice the distance of the hare! Put differently, if you were walking along a wild mountain trail or coastline and tired, slowing down

(a)   (b)

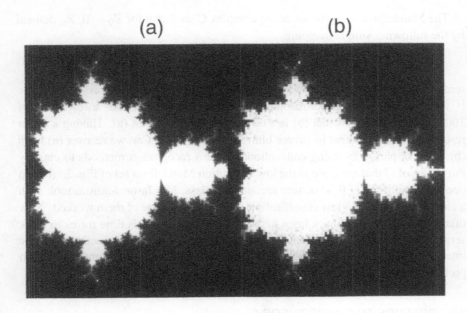

Figure 2.1.   Mandelbrot set

your pace and shrinking your steps, then you were in trouble. It certainly would be worse if you also got lost.

To better appreciate the open-endedness of the concept of a fractal, let us say a few words about the path associated with one's desire to achieve a certain nontrivial goal. This can be viewed as an extension of the above discussion.

A highly motivated person trying to achieve a grand goal often desires to be a lucky person such that the route to success is smooth and full of excitement. Typically, however, the opposite is true: the route is full of frustrations and failures, and one could then lose confidence and even become depressed. That is, the path might be a bumpy fractal one. A better strategy to achieve one's goal is perhaps to stop whining, accept the fact that often the most wanted help or environment may not be there, and take steps as large as possible to finish the journey.

Now, a word for outside observers: if you could not help but offer advice, expect your own fractal-like path of frustration. This is because advice is often drawn from past painful struggles, and is thus incompatible with a mind seeking lesser complexity or difficulty, or simply cheers or fun.

Now back to signal processing. In practice, random fractals are more useful than fractal geometry. We shall devote a lot of time to a discussion of random fractal theories. For the moment, to stimulate readers' curiosity as well as to echo Mandelbrot's observation that clouds are not spheres, nor mountains cones, nor islands circles, we design a simple pattern recognition problem related to the sensational Mandelbrot set.

The Mandelbrot set is the set of the complex C such that for $Z_0 = 0$, Z_n defined by the following simple mapping,

$$Z_{n+1} = Z_n^2 + C, \qquad (2.3)$$

remains bounded. Fig. 2.1 shows an example of C in the square region $[-0.5, 1.5] \times [-1.2, 1.2]$, where (a) has a resolution of 800×800 and (b) has a resolution of 100×100. In other words, (b) is a downsampled version of (a). Having a lower resolution, some features in (b) are blurred. *Now we ask*: Can we recover (a) from (b), fully or partly, by using conventional pattern recognition methods to classify the values of C that are close to the low-resolution Mandelbrot set of Fig. 2.1(b)? In one of the author's (J.B.'s) pattern recognition class, Mr. Jason Johnson took such a challenge and tried a few classification methods. But none of them worked. Now that fractal phenomena have been found to be ubiquitous, it is time for us to think seriously about how conventional stochastic pattern recognition methods can be integrated with fractal-based methods so that broader classes of pattern recognition problems can be tackled with higher accuracy.

2.2 PRELUDE TO CHAOS THEORY

Imagine that we are observing an aperiodic, highly irregular time series. Can a signal arise from a deterministic system that can be characterized by only a very few state variables instead of a random system with infinite numbers of degrees of freedom? A chaotic system is capable of just that. This discovery has such far-reaching implications in science and engineering that sometimes chaos theory is considered one of the three most revolutionary scientific theories of the twentieth century, along with relativity and quantum mechanics.

At the center of chaos theory is the concept of sensitive dependence on initial conditions: a very minor disturbance in initial conditions leads to entirely different outcomes. An often used metaphor illustrating this point is that sunny weather in New York could be replaced by rainy weather sometime in the future after a butterfly flaps its wings in Boston. Such a feature contrasts sharply with the traditional view, largely based on our experience with linear systems, that small disturbances (or causes) can only generate proportional effects, and that in order for the degree of randomness to increase, the number of degrees of freedom has to be infinite.

Mathematically, the property of sensitive dependence on initial conditions can be characterized by an exponential divergence between nearby trajectories in the phase space. Let $d(0)$ be the small separation between two arbitrary trajectories at time 0, and let $d(t)$ be the separation between them at time t. Then, for true low-dimensional deterministic chaos, we have

$$d(t) \sim d(0)e^{\lambda_1 t} \qquad (2.4)$$

where λ_1 is called the largest positive Lyapunov exponent.

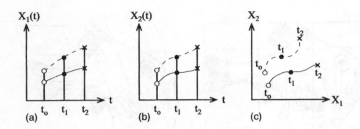

Figure 2.2.   Schematic illustrating the concept of phase space.

Another fundamental property of a chaotic attractor is that it is an attractor —
the trajectories in the phase space are bounded. The incessant stretching due to
exponential divergence between nearby trajectories, and folding due to boundedness
of the attractor, then cause the chaotic attractor to be a fractal, characterized by
Eq. (2.2), where $N(\epsilon)$ represents the (minimal) number of boxes, of linear length
not greater than ϵ, needed to cover the attractor in the phase space. Typically, D is
a nonintegral number called the box-counting dimension of the attractor.

At this point, it is worthwhile to take time to discuss a phase space (or state space)
and transformation in the phase space. Let us assume that a system is fully characterized by two state variables, X_1 and X_2. When monitoring the motion of the system,
one can plot out the waveforms for $X_1(t)$ and $X_2(t)$, as shown in Figs. 2.2(a,b).
Alternatively, one can monitor the trajectory defined by $(X_1(t), X_2(t))$, where t
appears as an implicit parameter (Fig. 2.2(c)). The space spanned by X_1 and X_2 is
called the phase space (or state space). It could represent position and velocity, for
example.

The introduction of phase space enables one to study the dynamics of a complicated system geometrically. For example, a globally stable fixed point solution is
represented as a single point in the phase space; solutions with arbitrary initial conditions will all converge to it sooner or later. Similarly, a globally stable limit cycle
is represented as a closed loop in the phase space; again, solutions with arbitrary
initial conditions will all converge to it.

To make readers unfamiliar with the concept of phase space even more comfort-
able, we note that this concept is frequently used in daily life. As an example, let
us imagine that we were driving to our friend's home for a party. On our way, there
was a traffic jam, and our car got stuck. Afraid of being late, we decided to call our
friend. How would we describe our situation to her? Usually, we would tell her
where we got stuck and how quickly or slowly we were driving, but not the signal
waveforms shown in Figs. 2.2(a,b).

We can now talk about the transformations in phase space. For this, we ask readers
to imagine how a patch of dust would be swept across the sky on a very windy day:
if the dust originally resembled the face of a person, after being swept the face would
get twisted badly. As another example, let us consider the fish transformation shown

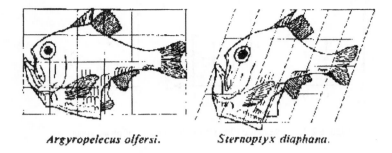

Argyropelecus olfersi.   *Sternoptyx diaphana.*

Figure 2.3.   To appreciate a chaotic transformation, one can imagine that the head and tail of the fish gets mixed up.

in Fig. 2.3. In his famous book *Growth and Form*, Sir D'Arcy Thompson found that simple plane transformations can bring two different types of fishes together. With awesome intuition, he then concluded that different fishes have the same origin and that this is a vivid demonstration of Darwin's theory of evolution. To appreciate a chaotic transformation, imagine that the head and tail of the fish get mixed up, just as a "dust face" gets twisted by a gusty wind. In exercise 3, we shall make these ideas more concrete.

To appreciate more concretely the concept of sensitive dependence on initial conditions, let us consider the map on a circle,

$$x_{n+1} = 2x_n \mod 1 , \tag{2.5}$$

where x is positive, and mod 1 means that only the fractional part of $2x_n$ will be retained as x_{n+1}. This map can also be viewed as a Bernoulli shift, or binary shift. Suppose that we represent an initial condition x_0 in binary

$$x_0 = 0.a_1a_2a_3\cdots = \sum_{j=1}^{\infty} 2^{-j}a_j, \tag{2.6}$$

where each of the digits a_j is either 1 or 0. Then

$$x_1 = 0.a_2a_3a_4\cdots ,$$

$$x_2 = 0.a_3a_4a_5\cdots ,$$

and so on. Thus, a digit that is initially far to the right of the decimal point, say the 40th digit (corresponding to $2^{-40} \approx 10^{-12}$), and hence has only a very minor role in determining the initial value of x_0, eventually becomes the first and the most important digit. In other words, a small change in the initial condition makes a large change in x_n.

Since chaotic as well as simple, regular motions have been observed almost everywhere, we have to ask a fundamental question: Can a chaotic motion arise from a regular one and vice versa? The answer is yes, and lies in the study of bifurcations and routes to chaos. Here, the key concept is that the dynamics of a system are controlled by one or a few parameters. When the parameters are changed, the behavior of the system may undergo qualitative changes. The parameter values where such qualitative changes occur are called bifurcation points.

To understand the idea better, let us reflect on graduate student life. In the initial spinning period, which could last for a few years, a student may be learning relevant materials diligently, but has few new ideas and thus is not productive at all. At some point, the student suddenly feels a call to work out something interesting. Beyond that point, he or she becomes quite productive, and the professor can now relax a bit. Here, the parameter we could vaguely identify would be the maturity of the student.

As another example, let us consider cooperation between two companies. When the profit is small, the companies cooperate well and both make profits. After a while, the market has grown so considerably that they become competitors. At a certain point, in order to dominate the market, they decide to sacrifice profit. After a while, the loss makes them seek cooperation again, and so on. It is clear that such a process is highly nonlinear.

To be more concrete, let us now consider the logistic map

$$x_{n+1} = f(x_n) = rx_n(1 - x_n), \quad 0 \leq x_n \leq 1, \tag{2.7}$$

where $0 \leq r \leq 4$ is the control parameter. Let us now fix r to be 2 and iterate the map starting from $x_0 = 0.3$. With a calculator or computer, we quickly find that x_n becomes arbitrarily close to 0.5 for large n. Now if we start from $x_0 = 0.5$, we find that $x_1 = x_2 = \cdots = 0.5$. This means that 0.5 is a stable fixed point solution. For now, let us not bother about the formal analysis of the stability of fixed point solutions and simple bifurcations (which can be found in Chapter 12), but instead adopt a simple simulation approach: For any allowable fixed r, we arbitrarily choose an initial value for x_0, and iterate Eq. (2.7). After throwing away sufficiently many iterations so that the solution of the map has converged to some attractor, we retain, say, 100 iterations and plot those 100 points against each r. When the map has a globally attracting fixed point solution, then after the transients die out, the recorded values of x_n all become the same. One then only observes a single point for that specific r. When the solution is periodic with period m, then one observes m distinct points for that specific r. When the motion becomes chaotic, one observes as many distinct points as one records (100 in our example). Figure 2.4(a) shows that the logistic map undergoes a period-doubling bifurcation to chaos.

In fact, Fig. 2.4(a) contains more structures than one could comprehend at a casual glance. For example, if one magnifies the small rectangular region in Fig. 2.4(a), one obtains Fig. 2.4(b). To have hands-on experience with such self-similar features, readers new to chaos theory are strongly encouraged to write a simple code to reproduce Figs. 2.4(a) and 2.4(b).

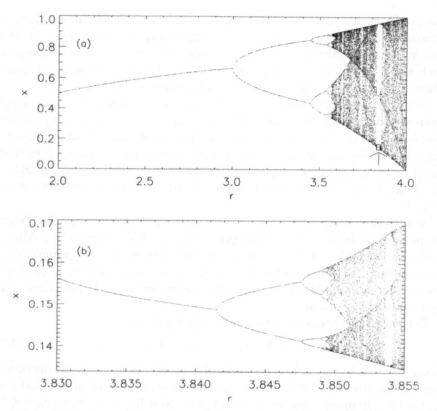

Figure 2.4.   Bifurcation diagram for the logistic map; (b) is an enlargement of the little rectangular box indicated by the arrow in (a).

While one might think that the first period-2 bifurcation in Fig. 2.4(a) is too simple to be interesting, we note that the worst case operation of a noisy NAND gate can be described by a simple map, which upon transformation can be made equivalent to the logistic map and that the first period-2 bifurcation point gives the error threshold value for the noisy NAND gate to function reliably. We shall have more to say on this in Chapter 12.

Period-doubling bifurcation to chaos is one of the most famous routes to chaos identified so far. It has been observed in many different fields. An important universal quantity associated with this route is the limit of the ratio of the differences between successive bifurcation parameters,

$$\delta = \lim_{k\to\infty} \frac{r_k - r_{k-1}}{r_{k+1} - r_k} = 4.669201\cdots.$$

Other well-known routes to chaos include the quasi-periodicity route and the intermittency route. The quasi-periodicity route to chaos occurs when a critical parameter is varied, the motion becomes periodic with one basic periodicity, quasi-periodic with two or more basic periodicities, and suddenly the motion becomes chaotic. Recently, it has been found that this route underlies the complicated Internet transport dynamics. The third route to chaos, intermittency, refers to the state of a system operating between smooth and erratic modes, depending on the variation of a key parameter. This route to chaos may also be very relevant to transportations on the Internet.

2.3 BIBLIOGRAPHIC NOTES

Motivated newcomers to the field of chaos and fractal theories should find James Gleick's best-seller *Chaos* [192] and Mandelbrot's classic book *The Fractal Geometry of Nature* [294] entertaining and inspirational. Sir D'Arcy Thompson's book is [434]. For the three routes to chaos, we refer to [131, 351, 379] as well as to the comprehensive book by Edward Ott [330]. Readers interested in chaotic Internet transportations are referred to [163, 164, 363]. An interesting paper on route flap storms is [274], while [41] is on pathological network behavior. Finally, [114] is a classic book on pattern classification.

2.4 WARMUP EXERCISES

After finishing the following three problems, you will be enlightened: "Aha, they are much simpler than I thought." With this, all intimidations shall be gone.

1. Write a simple code (say, using Matlab) to generate the Mandelbrot set.

2. Reproduce the bifurcation diagram for the logistic map and explore self-similarities.

3. We now resume the discussion of the distorted face. To be concrete, let us consider the chaotic Henon map,

$$
\begin{aligned}
x_{n+1} &= 1 - ax_n^2 + y_n, \\
y_{n+1} &= bx_n,
\end{aligned}
\tag{2.8}
$$

with $a = 1.4$ and $b = 0.3$. Now sample a unit circle centered at the origin by, say, 1000 or 10,000 points, treat them as initial conditions, and iterate the Henon map. Plot out images of the circle at, say, iteration 10, 100, 1000, etc. Can you still see a "trace" of the circle when the number of iterations increases?

You could perform the same operation on a "face" represented by five circles: one big, representing the head, and four small, representing two eyes, the nose, and the mouth.

A densely sampled circle should soon converge to the famous Henon attractor. What do you get for the Henon attractor?

Now assume that we start from just one arbitrary initial condition and iterate the map many times. After throwing away sufficiently many points (called transient points), we record, say, 1000 or 10,000 points. Plot out those points as dots (i.e., do not connect them by lines). Do you obtain the same pattern that you got earlier? Can you explain why they are the same?

CHAPTER 3

BASICS OF PROBABILITY THEORY AND STOCHASTIC PROCESSES

In this chapter we first review the most important definitions and results of elementary probability theory. Then we describe a number of commonly used distributions and discuss their main properties. To facilitate applications, we shall also explain how random variables of a specific distribution can be simulated. Finally, we briefly discuss random processes. Our main purpose here is to equip readers with an intuitive understanding. Therefore, occasionally, mathematical rigor will be sacrificed.

Most readers might think that the materials covered in this chapter have been so well developed in many excellent textbooks that they might not have much to do with research of considerable current interest. To counter such a belief, when appropriate, we will comment on where such elementary materials have shed new light on complex problems. We particularly remind research-oriented readers of the literature search using the concept of the power-law network, described in Sec. 3.4.

3.1 BASIC ELEMENTS OF PROBABILITY THEORY

3.1.1 Probability system

Probability theory is concerned with the description of random events occurring in the real world. To better appreciate the basic elements of the theory, let us first examine the main features of an experiment with random outcomes. There are three main features:

1. A set of possible experimental outcomes (obtained under precisely controlled identical experimental conditions).

2. A grouping of these outcomes into classes called results.

3. The relative frequency of these results in many independent trials of the experiment.

The relative frequency f_c of a result is merely the number of times the result is observed divided by the number of times the experiment is performed; as the number of experimental trials increases, we expect f_c to be close to a constant value. As an example, suppose a fair coin is tossed a thousand times; then the relative frequency for either head or tail to be observed will be close to 1/2. As another example, if a fair die is thrown many times, the relative frequency with which one of the numbers $\{1, 2, 3, 4, 5, 6\}$ is observed will be close to 1/6. There are situations in which one may want to group the six numbers into only two classes: odd numbers $\{1, 3, 5\}$ and even numbers $\{2, 4, 6\}$ or small numbers $\{1, 2, 3\}$ and large numbers $\{4, 5, 6\}$. Then there are only two classes or results of the die-throwing experiment. One can also construct complicated experiments by combining simple ones, such as first tossing a coin, then throwing a die.

Abstracting from the above considerations, one obtains three basic elements that constitute the probability theory:

1. A sample space S, which is a collection of objects, corresponding to the set of mutually exclusive, exhaustive outcomes of an experiment. Each point ω in S is called a sample point.

2. A family of events E, denoted as $\{A, B, C, \cdots\}$, in which each event is a set of sample points $\{\omega\}$. An event corresponds to a class or result of a real-world experiment.

3. A probability measure P which assigns each event, say A, a real nonnegative number $P(A)$, which corresponds to the relative frequency in the experimental situation. This assignment must satisfy three properties (axioms):

   - For any event A, $0 \leq P(A) \leq 1$.
   - $P(S) = 1$.
   - If A and B are mutually exclusive events, then
     $P(A \cup B) = P(A) + P(B)$.
     More generally, for any sequence of mutually exclusive events $E_1, E_2, \cdots$
     (that is, events satisfying $E_i E_j = \Phi$, $i \neq j$, where Φ is the empty set),

$$P\left(\bigcup_{i=1}^{\infty} E_i\right) = \sum_{i=1}^{\infty} P(E_i).$$

The triplet (S, E, P), along with the three axioms, forms a probability system.

Let us now present two other definitions. The first is the conditional probability of event A given that event B occurred, $P(A|B)$. It is defined as

$$P(A|B) = \frac{P(AB)}{P(B)} \tag{3.1}$$

whenever $P(B) \neq 0$. The introduction of the conditional event B forces us to shift attention from the original sample space S to a new sample space defined by the event B. In terms of real-world applications, this amounts to renormalizing the problem by magnifying the probabilities associated with the conditional events by dividing by the term $P(B)$ as given above.

The second important notion is that of statistical independence of events. Two events A and B are said to be statistically independent if and only if

$$P(AB) = P(A)P(B). \tag{3.2}$$

For three events A, B, C, we require that each pair of them satisfies Eq. (3.2) and in addition

$$P(ABC) = P(A)P(B)P(C). \tag{3.3}$$

It is worth emphasizing that Eq. (3.3) alone is not sufficient to define statistical independence of three events A, B, C. To see this, let us consider a situation schematically shown in Fig. 3.1, where

$$P(A) = P(B) = P(C) = 1/5,$$

and the events A, B, C intersect with each other on a common subset O,

$$P(AB) = P(BC) = P(CA) = P(ABC) = P(O) = p.$$

If one chooses $p = 1/25$, then Eq. (3.2) is satisfied but Eq. (3.3) is not. If one chooses $p = 1/125$, then Eq. (3.3) is satisfied but Eq. (3.2) is not. In fact, Eqs. (3.2) and (3.3) cannot be simultaneously satisfied in this situation.

It is easy to see that the notion of statistical independence can be readily extended to any $n > 2$ events.

3.1.2 Random variables

Up to now, we have associated probabilities directly with random events. Often it is more convenient to represent a random event by a number, called a random variable, and talk about the probability of a random variable. In fact, in throwing a die, we naturally have numbers as outcomes. When tossing a coin, we can denote head by 1 and tail by -1. Such a mapping is particularly convenient, noticing that the summation of a sequence of 1 and -1 gives the number of net wins (e.g., head) or losses (e.g., tail) in a total of N experiments. The summation of a sequence of random variables is called a random walk process. We will talk more about it later.

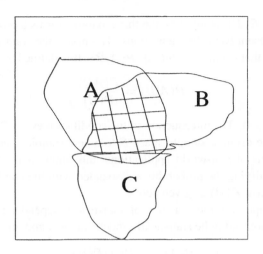

Figure 3.1.   A situation where $AB = BC = AC = ABC$.

Let us denote a random variable by X, whose value depends on the outcome, ω, of a random experiment. The event $(X \leq x)$ is thus equivalent to $\{\omega : X(\omega) \leq x\}$. Let us denote the probability of $(X \leq x)$ by $F_X(x)$, which is called the cumulative distribution function (CDF),

$$F_X(x) = P(X \leq x). \tag{3.4}$$

The important properties of $F_X(x)$ include

$$F_X(x) \geq 0,$$

$$F_X(\infty) = 1,$$

$$F_X(-\infty) = 0,$$

$$P(a < X \leq b) = F_X(b) - F_X(a) \geq 0 \quad \text{for} \quad a \leq b.$$

Therefore, $F_X(x)$ is a nonnegative, monotonically nondecreasing function with limits 0 and 1 at $-\infty$ and ∞, respectively. Conventionally, $F_X(x)$ is assumed to be continuous from the right. When the derivative of $F_X(x)$ exists, it is often more convenient to work with its derivative, called the probability density function (PDF),

$$f_X(x) = \frac{dF_X(x)}{dx}. \tag{3.5}$$

"Inverting" Eq. (3.5) yields

$$F_X(x) = \int_{-\infty}^{x} f_X(y)dy. \tag{3.6}$$

Therefore,

$$P(a < X \le b) = \int_a^b f_X(y)dy.$$

Since probability has to be nonnegative, letting $a \to b$, we see that the last equation implies that $f_X(x) \ge 0$.

Given a probability system (S, E, P), we can also define many random variables in the same sample space. In real-world applications, such a situation amounts to describing a random event by a high-dimensional vector instead of a scalar. Such an extension is necessary if the outcome of an experiment cannot be characterized by just one number. For ease of illustration below, let us mainly consider the case of two random variables X and Y defined for some probability system (S, E, P). The joint CDF is then defined by

$$F_{XY}(x, y) = P(X \le x, Y \le y), \tag{3.7}$$

which is merely the probability that X takes on a value not larger than x and at the same time Y takes on a value not larger than y; that is, it is the sum of the probabilities associated with all sample points in the intersection of the two events $\{\omega : X(\omega) \le x\}$ and $\{\omega : Y(\omega) \le y\}$. Associated with this function is a joint PDF defined as

$$f_{XY}(x, y) = \frac{\partial^2 F_{XY}(x, y)}{\partial x \partial y}. \tag{3.8}$$

Given a joint PDF, the "marginal" density function for one of the variables is given by integrating over all possible values of the second variable. For example,

$$f_X(x) = \int_{y=-\infty}^{\infty} f_{XY}(x, y)dy. \tag{3.9}$$

We are now in a position to define the notion of independence between random variables. Two random variables X and Y are said to be independent if and only if their joint PDF factors into the product of the one-dimensional PDFs:

$$f_{XY}(x, y) = f_X(x)f_Y(y).$$

This is very much like the definition for two independent events as given in Eq. (3.2). However, for three or more random variables, the definition is essentially the same as for two, namely, $X_1, X, \cdots, X_n$ are said to be independent random variables if and only if

$$f_{X_1 X_2 \cdots X_n}(x_1, x_2, \cdots, x_n) = f_{X_1}(x_1)f_{X_2}(x_2) \cdots f_{X_n}(x_n). \tag{3.10}$$

This is simpler than the definition required for multiple events to be independent. Loosely speaking, this simplification comes from the fact that PDFs characterize "elementary" events of an experiment with random outcomes. Mathematically, this is because Eq. (3.10) implies that the following equation is true:

$$f_{X_{i1} X_{i2} \cdots X_{ik}}(x_{i1}, x_{i2}, \cdots, x_{ik}) = f_{X_{i1}}(x_{i1})f_{X_{i2}}(x_{i2}) \cdots f_{X_{ik}}(x_{ik}),$$

where $k \leq n$ and $i1, i2, \cdots , ik$ are all different. This is easily proven by resorting to the definition of a marginal distribution and integrating both sides of Eq. (3.10) with respect to the random variables one wishes to remove.

With more than one random variable, we can now define the conditional distributions and densities. For example, we can ask for the CDF of the random variable X conditioned on some given value of the random variable Y, which is merely the probability $P[X \leq x | Y = y]$. Similarly, the conditional PDF on X, given Y, is defined as

$$f_{X|Y}(x|y) = \frac{d}{dx}P[X \leq x | Y = y] = \frac{f_{XY}(x,y)}{f_Y(y)},$$

much as the definition for the conditional probability of events.

We can also define one random variable Y in terms of a second random variable X. In this case, Y is referred to as a function of the random variable X. In its most general form, we have

$$Y = g(X),$$

where $g(\cdot)$ is some given function of its argument. Thus, once the value of X is determined, the value of Y can be computed. Since the value of X depends upon the sample point ω, we can write $Y = g(X(\omega)) = Y(\omega)$. Therefore,

$$F_Y(y) = P(Y \leq y) = P[\{\omega : g(X(\omega)) \leq y\}].$$

In general, the computation of this last equation may be complicated. It is clear that a random variable may be a function of many random variables rather than just one. For example,

$$Y = \sum_{i=1}^{n} X_i.$$

This simple equation describes a random walk process, which we will discuss in depth later.

Before ending this section, let us comment on statistical independence and vector description of a real-world problem. A complicated experiment or situation typically has multiple features; each of them could be described by a number. If the features identified are all different, then the dependence among the random variables describing each feature would be quite minor. This is a desired situation. If, however, the dependence among the random variables is quite high, then the number of independent features identified would be considerably smaller than the number of variables designed for the problem. It could be that the problem under study might be less complicated than we thought or that some important features have not been identified.

3.1.3   Expectation

An important class of measures associated with the CDF and the PDF for a random variable is expectations or mean values. They deal with a special type of integral of the PDF. Since PDF is the derivative of CDF, when CDF is not differentiable, a difficulty arises when one tries to define PDF. This difficulty can be resolved by the use of impulse functions. Alternatively (and preferably), one resorts to Stieltjes integrals. A Stieltjes integral is defined in terms of a nondecreasing function $F(x)$ and a continuous function $\phi(x)$; in addition, two sets of points $\{t_k\}$ and $\{\zeta_k\}$ such that $t_{k-1} < \zeta_k \le t_k$ are defined and a limit is considered where max $|t_k - t_{k-1}| \to 0$. With these, consider the sum

$$\sum_k \phi(\zeta_k)[F(t_k) - F(t_{k-1})].$$

This sum tends to a limit as the intervals shrink to zero independent of the sets $\{t_k\}$ and $\{\zeta_k\}$, and the limit is referred to as the Stieltjes integral of ϕ with respect to F. This integral is written as

$$\int \phi(x)dF(x).$$

When $F(x)$ is differentiable or if impulse functions are allowed in PDF, we can write

$$dF(x) = f_X(x)dx.$$

Therefore,

$$\int \phi(x)dF(x) = \int \phi(x)f_X(x)dx.$$

We are now ready to define expectations.

The mean or average of X is given by

$$E[X] = \overline{X} = \int_{-\infty}^{\infty} x\, dF(x) = \int_{-\infty}^{\infty} x f_X(x)dx.$$

The nth moment of X is defined as

$$E(X^n) = \int_{-\infty}^{\infty} x^n dF_X(x) = \int_{-\infty}^{\infty} x^n f_X(x)dx.$$

Similarly, the nth central moment of X is defined as

$$E[(x - \overline{X})^n] = \int_{-\infty}^{\infty} (x - \overline{X})^n f_X(x)dx.$$

The variance of X is defined by

$$Var(X) = \sigma_X^2 = E(X^2) - [E(X)]^2,$$

which is the second central moment. Its square root σ_X is called the standard deviation, and the coefficient of variation is defined by

$$C_X = \sigma_X / \overline{X}.$$

Let us now consider the case where the random variable Y is a function of the random variable X, $Y = g(X)$. We have

$$E_Y[Y] = \int_{-\infty}^{\infty} y f_Y(y) dy.$$

Since $f_Y(y) dy$ and $f_X(x) dx$ describe the same probability for the set of sample points of an experiment, they have to be equal; hence,

$$E_Y[Y] = \int_{-\infty}^{\infty} g(x) f_X(x) dx.$$

We can similarly define expectations for multiple random variables. For example,

$$E[X + Y] = \int_{-\infty}^{\infty} \int_{-\infty}^{\infty} (x + y) f_{XY}(x, y) dx dy,$$

$$E[XY] = \int_{-\infty}^{\infty} \int_{-\infty}^{\infty} xy f_{XY}(x, y) dx dy.$$

It is easy to prove that

$$E[X + Y] = E[X] + E[Y]$$

without any condition. On the other hand, in order to have

$$E[XY] = E[X]E[Y],$$

the random variables X and Y have to be independent. Extending the above two equations, we have

$$E[X_1 + X_2 + \cdots + X_n] = E[X_1] + E[X_2] + \cdots + E[X_n]$$

without any condition and

$$E[g(X)h(Y)] = E[g(X)]E[h(Y)]$$

when X and Y are independent.

3.1.4   Characteristic function, moment generating function, Laplace transform, and probability generating function

In computations using probability theory, one has to evaluate integrals of PDFs. To simplify computations, one can resort to a number of functions. Here we review four closely related functions, called the characteristic function, the moment generating function, the Laplace transform, and the probability generating function. Their usefulness will be illustrated in the next section when we talk about various types of distributions.

The characteristic function of a random variable X, denoted by $\Phi_X(u)$, is given by

$$\Phi_X(u) = E[e^{juX}] = \int_{-\infty}^{\infty} e^{jux} f_X(x)dx,$$

where $j = \sqrt{-1}$ and u is an arbitrary real variable. Clearly,

$$|\Phi_X(u)| \le \int_{-\infty}^{\infty} |e^{jux}||f_X(x)dx| = 1.$$

The equality is achieved at $u = 0$. Expanding e^{jux} in terms of its power series, we have

$$e^{jux} = 1 + jux + \frac{(jux)^2}{2!} + \cdots.$$

We thus have

$$\Phi_X(u) = 1 + ju\overline{X} + \frac{(ju)^2}{2!}\overline{X^2} + \cdots.$$

Differentiating the above equation n times on both sides with respect to u, we have

$$\left.\frac{d^n\Phi_X(u)}{du^n}\right|_{u=0} = j^n\overline{X^n}.$$

For simplicity, we shall define

$$g^{(n)}(x_0) = \left.\frac{d^n g(x)}{dx^n}\right|_{x=x_0}.$$

Therefore,

$$\Phi_X^{(n)}(0) = j^n\overline{X^n}.$$

The moment generating function $M_X(v)$ is obtained by replacing ju by a real variable v in the characteristic function,

$$M_X(v) = E[e^{vX}] = \int_{-\infty}^{\infty} e^{vx} f_X(x)dx.$$

Similarly, we have

$$M_X^{(n)}(0) = \overline{X^n}.$$

The Laplace transform $L_X(s)$ of the PDF is defined as

$$L_X(s) = E[e^{-sX}] = \int_{-\infty}^{\infty} e^{-sx} f_X(x)dx.$$

Similarly, we can prove that

$$L_X^{(n)}(0) = (-1)^n\overline{X^n}.$$

It is clear that the three functions $\Phi_X(u), M_X(v), L_X(s)$ are closely related. In particular, we have

$$\Phi_X(js) = M_X(-s) = L_X(s).$$

In the case of a discrete random variable defined by

$$g_k = P[X = k],$$

we can define the probability generating function $G(z)$ as follows:

$$G(z) = E[z^X] = \sum_k z^k g_k,$$

where z is a complex variable. In fact, $G(z)$ is just the z-transform of the discrete sequence g_k. When $|z| \leq 1$, we have, as with the continuous transforms,

$$|G(z)| \leq \sum_k |z^k||g_k| \leq 1.$$

It is easy to see that the first derivative evaluated at $z = 1$ yields the mean of X

$$G^{(1)}(1) = \overline{X},$$

and the second derivative yields

$$G^{(2)}(1) = \overline{X^2} - \overline{X}.$$

Alternatively, one may obtain $\overline{X^2}$ by differentiating $zG^{(1)}(z)$ and then taking $z = 1$:

$$\overline{X^2} = \frac{d}{dz} zG^{(1)}(z)\Big|_{z=1}.$$

The foregoing discussions make it clear that in many situations, these functions may greatly simplify the calculations of expectations of a given distribution. This will be illustrated in the next section. Another important application of these functions is related to the following property. Let X and Y be independent random variables with characteristic functions $\Phi_X(u)$ and $\Phi_Y(u)$, respectively. Let $Z = X + Y$. Then the characteristic function for Z is

$$\Phi_Z(u) = \Phi_X(u) \cdot \Phi_Y(u).$$

This property is of crucial importance for the discussion of stable laws and Levy motions in Chapter 7. The proof is left as an exercise.

3.2 COMMONLY USED DISTRIBUTIONS

In this section, we describe a number of commonly used distributions and discuss the relations among them.

(1) Exponential and related distributions. The CDF for the exponential distribution is given by

$$F_X(x) = 1 - e^{-\lambda x}, \quad x \geq 0. \tag{3.11}$$

Its PDF is

$$f_X(x) = \lambda e^{-\lambda x}, \quad x \geq 0. \tag{3.12}$$

The exponential distribution has the remarkable memoryless property, which is the defining property of the Markov process. It can be shown as follows:

$$
\begin{aligned}
P(X \leq x + x_0 | X > x_0) &= \frac{P(x_0 < X \leq x + x_0)}{P(X > x_0)} \\
&= \frac{P(X \leq x + x_0) - P(X \leq x_0)}{P(X > x_0)}.
\end{aligned}
$$

Using Eq. (3.11) we then have

$$P(X \leq x + x_0 | X > x_0) = \frac{1 - e^{-\lambda(x+x_0)} - (1 - e^{-\lambda x_0})}{1 - (1 - e^{-\lambda x_0})} = 1 - e^{-\lambda x}.$$

Since the conditional distribution does not depend on x_0, the distribution is memoryless. To understand this property, let us imagine that we are waiting for a bus to come to a bus stop. Assume that the time interval between two successive arrivals follows an exponential distribution, with a mean, say, of 15 min. Then, even if we have waited for 20 minutes, on average, we will have to wait another 15 min to get on a bus; the time that we have waited does not have any effect. In Sec. 3.3.2, when discussing continuous-time Markov chains, we will prove that the exponential distribution is the only continuous distribution that has the memoryless property.

The mean and variance of an exponentially distributed random variable can be directly calculated using the definition. Instead of doing so, here let us find its Laplace transform:

$$L_X(s) = \int_0^\infty e^{-sx} \lambda e^{-\lambda x} dx = \frac{\lambda}{s + \lambda}. \tag{3.13}$$

From the Laplace transform, we find

$$\overline{X} = -L_X^{(1)}(0) = \frac{1}{\lambda},$$

$$\overline{X^2} = L_X^{(2)}(0) = \frac{2}{\lambda^2},$$

$$\sigma_X^2 = \overline{X^2} - (\overline{X})^2 = \frac{1}{\lambda^2}.$$

Therefore, its mean and standard deviation are both $1/\lambda$.

Now let us consider a sum of k exponentially distributed random variables

$$Y = X_1 + X_2 + \cdots X_k,$$

where X_i are independent. Usually $X_1, \cdots, X_k$ are called identically and independently distributed (iid) random variables. The PDF for Y is given by the convolution of the densities on each of the X_i's, since they are independently distributed. The Laplace transform of the PDF for Y is therefore given by

$$L_Y(s) = [L_{X_i}(s)]^k = \left(\frac{\lambda}{s+\lambda}\right)^k.$$

Inversion of the above equation gives the Erlang distribution:

$$f_Y(y) = \frac{\lambda(\lambda y)^{k-1}}{(k-1)!}e^{-\lambda y}, \quad y \geq 0, \tag{3.14}$$

which is a special case of the gamma distribution:

$$f(y) = \frac{1}{\Gamma(t)}\lambda^t y^{t-1}e^{-\lambda y}, \quad y \geq 0, \tag{3.15}$$

where parameters $\lambda, t > 0$, and $\Gamma(t)$ is the gamma function:

$$\Gamma(t) = \int_0^\infty y^{t-1}e^{-y}dy.$$

The relations $\Gamma(t) = (t-1)\Gamma(t-1)$ for any t, $\Gamma(1/2) = \sqrt{\pi}$ and $\Gamma(n) = (n-1)!$, where n is an integer, can be used to evaluate the gamma distribution.

The discrete version of the exponential distribution is called the geometrical distribution, given by

$$P(X = k) = f(k) = p(1-p)^{k-1}, \tag{3.16}$$

where $0 < p < 1$. Often $f(k)$ is called the mass function. This distribution arises in the following way. Suppose that independent Bernoulli trials (with parameter p) are performed at times $1, 2, \cdots$. Let X be the time that elapses before the first success; X is called the waiting time. Then $P(X > k) = (1-p)^k$ and thus

$$P(X = k) = P(X > k - 1) - P(X > k) = p(1-p)^{k-1}.$$

It is easy to prove that the mean and variance are $1/p$ and $(1-p)/p^2$, respectively. Readers are encouraged to prove, preferably at this point, that the geometrical distribution is the unique discrete distribution that has the memoryless property.

(2) Normal and related distributions. The normal distribution is perhaps the most important continuous distribution. It has two parameters, μ and σ^2. Its density function is

$$f(x) = \frac{1}{\sqrt{2\pi\sigma^2}}e^{-\frac{(x-\mu)^2}{2\sigma^2}}, \quad -\infty < X < \infty. \tag{3.17}$$

It is denoted by $N(\mu, \sigma^2)$. If $\mu = 0$ and $\sigma^2 = 1$, then

$$f(x) = \frac{1}{\sqrt{2\pi}}e^{-\frac{x^2}{2}}, \quad -\infty < X < \infty$$

is the density of the standard normal distribution. The characteristic function for $N(\mu, \sigma^2)$ is

$$\Phi_X(u) = e^{j\mu u - \frac{\sigma^2 u^2}{2}}. \tag{3.18}$$

The normal distribution arises in many ways. In particular, it can be obtained as a continuous limit of the binomial distribution, as will be discussed shortly. This result is a special case of the central limit theorem, which says that in many cases the sum of a large number of independent (or at least not too dependent) random variables is approximately normally distributed.

Now let us consider the distribution for $x = \sum_{i=1}^{d} x_i^2$, where $x_i \sim N(0, 1)$ and x_i's are independent. The distribution for x is called the chi-squared distribution, with degree d. It can be obtained by setting $\lambda = 1/2$ and $t = d/2$ in the gamma distribution (Eq. 3.15)).

Next let us consider $X = \sqrt{X_1^2 + X_2^2}$, where $X_i \sim N(0, \sigma^2)$, $i = 1, 2$, and X_1 and X_2 are independent. The distribution for X is called the Rayleigh distribution given by

$$f(x) = \frac{x}{\sigma^2} e^{-\frac{x^2}{2\sigma^2}}, \quad x \geq 0. \tag{3.19}$$

Finally, let

$$Y = e^X, \quad \text{where } X \sim N(\mu, \sigma^2).$$

The distribution for Y is called the log-normal distribution, given by

$$f(y) = \frac{1}{\sigma y \sqrt{2\pi}} e^{-\frac{(\ln y - \mu)^2}{2\sigma^2}}. \tag{3.20}$$

The log-normal distribution has been found to accurately describe, among many other things, switching times in multistable visual perceptions and many different types of network traffic processes, including variable bit rate (VBR) video traffic, the call duration and dwell time in a wireless network, data connections and messages, and page size, page request frequency, and user's think time in the World Wide Web (WWW) traffic. One mechanism for the log-normal distribution is that the random variable Y is a multiplication of a sequence of iid random variables, $Y = X_1 X_2 \cdots X_n$, where each X_i may represent one event in a complex chain of events. This could correspond to propagation of a signal in a complex system. By taking the logarithm and then applying the central limit theorem, one readily sees that Y follows the log-normal distribution. We shall have more to say on this when discussing multiplicative cascade multifractal models.

(3) Binomial and related distributions. Consider the Bernoulli trial, where a coin is tossed. Denote $Y = 1$ when a head turns up and $Y = 0$ when a tail turns up. Assume the probability for a head to turn up to be $0 < p < 1$. Toss the coin n times, and let $X = Y_1 + Y_2 + \cdots + Y_n$. The total number X of heads takes values in the set $\{0, 1, 2, \cdots, n\}$ and is a discrete random variable. Then X has the binomial distribution

$$P(X = k) = \binom{n}{k} p^k q^{n-k}, \tag{3.21}$$

where $q = 1 - p$. It is easy to prove that

$$E(X) = np, \quad \text{var}(X) = npq.$$

Now let us consider two special cases:

- $n \to \infty$ and $p \to 0$ in such a way that $E(X) = np$ approaches a nonzero constant λ. Then, for $k = 0, 1, 2, \cdots$,

$$P(X = k) = \binom{n}{k} p^k (1 - p)^{n-k} \sim \frac{1}{k!}\left(\frac{np}{1-p}\right)^k \left(1 - \frac{np}{n}\right)^n \to \frac{\lambda^k}{k!}e^{-\lambda},$$

which is the Poisson distribution.

The probability generating function for the Poisson distribution is

$$G(z) = E[z^K] = \sum_{k=0}^{\infty} z^k P(X = k) = \sum_{k=0}^{\infty} z^k e^{-\lambda}\frac{\lambda^k}{k!} = e^{\lambda(z-1)}.$$

Therefore,

$$E[X] = G^{(1)}(1) = \lambda,$$

$$\sigma_X^2 = G^{(2)}(1) + G^{(1)}(1) - [G^{(1)}(1)]^2 = \lambda.$$

These results can, of course, also be obtained from the mean and variance for the binomial distribution, noting that $np = \lambda$, $npq \to np = \lambda$.

- If $npq \gg 1$, then

$$P(X = k) = \binom{n}{k} p^k q^{n-k} \approx \frac{1}{\sqrt{2\pi npq}}e^{-\frac{(k-np)^2}{2npq}}$$

for k in the $\sqrt{npq}$ neighborhood of np. This result is called the DeMoivre-Laplace theorem. It says that when n is large, X follows a normal distribution with mean np and variance npq. This result can be understood by noting that X is the sum of n Bernoulli random variables and then applying the central limit theorem.

One can ask a different question: what is the distribution when k is far away from the mean? The answer is given by Bernstein's inequality:

$$P\left(\frac{1}{n}X \geq p + \epsilon\right) \leq e^{-\frac{1}{4}n\epsilon^2} \quad for \quad \epsilon > 0.$$

Such an inequality is known as large-deviation estimation.

(4) Heavy-tailed distributions. The distributions discussed up to now all have finite variance and are thus called thin-tailed. Do distributions with infinite variance exist? The answer is, of course, yes. As an example, let us derive the Cauchy distribution.

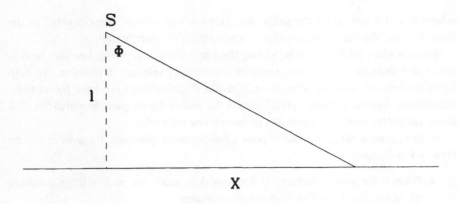

Figure 3.2.   Schematic for deriving the Cauchy distribution.

Suppose there is a point light source S emitting photons. S is at the distance l away from a plane. See Fig. 3.2. Let the angle Φ be uniformly distributed on $(-\pi/2, \pi/2)$. Due to symmetry, we only wish to know the distribution for the random variable X. Since X and Φ are related through the relation $x = l \tan \phi$, using equality $f(x) = f(\phi)\left|\frac{d\phi}{dx}\right|$, we obtain the distribution for X:

$$f(x) = \frac{l}{\pi(l^2 + x^2)}. \tag{3.22}$$

This is the Cauchy distribution. It does not have any finite moments. For large x, the distribution can be written as

$$f(x) \sim x^{-2}, \quad x \to \infty.$$

This is a special case of the so-called heavy-tailed distribution, commonly expressed as

$$f(x) \sim x^{-\alpha-1}, \quad x \to \infty.$$

Equivalently, one may write

$$P[X \geq x] \sim x^{-\alpha}, \quad x \to \infty. \tag{3.23}$$

This expression is in fact more popular, since it emphasizes the tail of the distribution. It is easy to prove that when $\alpha < 2$, the variance and all moments higher than second-order moments do not exist. Furthermore, when $\alpha \leq 1$, the mean also diverges.

When the power-law relation extends to the entire range of the allowable x, we have the Pareto distribution:

$$P[X \geq x] = \left(\frac{b}{x}\right)^{\alpha}, \quad x \geq b > 0, \quad \alpha > 0, \tag{3.24}$$

where α and b are called the shape and location parameters, respectively. In the discrete case, the Pareto distribution is called the Zipf distribution.

Some readers might find it mind-boggling to believe that power-law-like heavy-tailed distributions with infinite variance can be very relevant in practice. To help those readers, we note that Pareto- or Zipf-like distributions have been found to be ubiquitous. See the Bibliographic Notes at the end of the chapter. See also Sec. 3.4 about an application of the concept of power-law networks.

At this point, it is appropriate to pose a fundamental question. Its answer will be revealed in Chapter 7.

- What is the proper theoretical framework to study the sum of large numbers of iid random variables with infinite variance?

(5) Simulation of random variables. Numerically simulating random variables of various distributions is an important exercise and can provide great help with one's research. Here we explain how to obtain different distributions from uniform $[0, 1]$ random variables.

As we explained earlier, the value of a random variable U represents an outcome of a random experiment. Such an outcome can certainly be represented by the value of another random variable X. The probability of an event of the experiment is then either $dF_U(u) = f_U(u)du$ or $dF_X(x) = f_X(x)dx$, where $F_U(u)$ and $F_X(x)$ are the CDFs for the U and X, while $f_U(u)$ and $f_X(x)$ are the PDFs. Now suppose U is a uniform $[0, 1]$ random variable, while X is a random variable defined on the interval $[a, b]$, and a can be $-\infty$, while b can be ∞. Then we have

$$\int_a^X dF_X(x) = \int_0^U du.$$

Since $F_X(x)$ is monotonically nondecreasing, its inverse function exists. We then have

$$X = F_X^{-1}(U). \tag{3.25}$$

When there is no closed-form formula for F_X^{-1}, we can solve the problem numerically. Below we consider a few simple examples.

Example 1: Exponential distribution with parameter λ. Since $F_X(x) = 1 - e^{-\lambda x}$,

$$X = -\frac{1}{\lambda}\ln(1 - U) = -\frac{1}{\lambda}\ln U', \tag{3.26}$$

where $U' = 1 - U$. It is easy to see that U' is also a uniform $[0, 1]$ random variable.

Example 2: Rayleigh distribution. In this case, $F_X(x) = 1 - e^{-\frac{x^2}{2\sigma^2}}$. Therefore,

$$X = \sqrt{-2\sigma^2 \ln U}. \tag{3.27}$$

Example 3: Standard normal distribution. Several methods are available for generating normal random variables. One simple (but not very efficient) way is to use the central limit theorem. That is, we form

$$X = U_1 + U_2 + \cdots + U_n,$$

where $U_i, i = 1, 2, \cdots, n$ are all uniform $[0, 1]$ random variables. When n is large, X is approximately normally distributed. In practice, $n = 12$ is already quite acceptable. Larger n gives more accurate results, but the simulation takes longer. The following method is more efficient.

First, we note (see exercise 3) that if x and y are $N(0, \sigma)$ and independent, if we form

$$x \cos(\omega t) + y \sin(\omega t) = r \cos(\omega t - \varphi), \quad |\varphi| < \pi,$$

then the random variables r and φ are independent, φ is uniform in the interval $[-\pi, \pi]$, and r has a Rayleigh distribution. Now it is easy to see that X_1 and X_2 given by the following equations are $N(0, \sigma)$:

$$
\begin{aligned}
X_1 &= \sqrt{-2\sigma^2 \ln U} \cos(2U_1 - 1)\pi, \\
X_2 &= \sqrt{-2\sigma^2 \ln U} \sin(2U_2 - 1)\pi,
\end{aligned}
\tag{3.28}
$$

where U, U_1, and U_2 are all uniform $[0, 1]$ random variables. This is called the Box-Muller method.

Example 4: Pareto distribution. From Eq. (3.24), we have

$$F_X(x) = 1 - \left(\frac{b}{x}\right)^\alpha, \quad 0 < b \le x.$$

Therefore,

$$X = bU^{-\frac{1}{\alpha}}. \tag{3.29}$$

3.3 STOCHASTIC PROCESSES

In this section, we first present basic definitions of stochastic processes. Then we discuss Markov processes, with an emphasis on their correlation structures, to facilitate comparison between Markov processes and fractal processes with long-range correlations in later chapters.

3.3.1 Basic definitions

Given a probability system (S, E, P), which consists of a sample space S, a set E of events $\{A, B, \cdots\}$, and a probability measure P, a stochastic process is defined as follows: For each sample point $\omega \in S$, we assign a time function $X(t, \omega)$. This family of functions forms a stochastic process. Note that for each allowable parameter t (often interpreted as a time index), $X(t, \omega)$ is a random variable. When ω is fixed, $X(t, \omega)$ is a function of t. Examples of stochastic processes are abundant. The closing price of a given security on the New York Stock Exchange and motions of dust in sky are familiar examples of stochastic processes. A stochastic process is also called a random process. For simplicity, we denote $X(t, \omega)$ by $X(t)$.

How can a random process be specified? For this purpose, we define, for each allowable t, a CDF $F_X(x, t)$, which is given by

$$F_X(x, t) = P[X(t) \leq x].$$

Further, we define for each of n allowable t, $\{t_1, t_2, \cdots, t_n\}$, a joint CDF, given by

$$F_{X_1 X_2 \cdots X_n}(x_1, x_2, \cdots, x_n; \, t_1, t_2, \cdots, t_n)$$
$$= P[X(t_1) \leq x_1, X(t_2) \leq x_2, \cdots, X(t_n) \leq x_n].$$

For simplicity, we use the vector notation $F_{\mathbf{X}}(\mathbf{x}, \mathbf{t})$ to denote this function. Of course, in general, specifying the joint CDF is a formidable task.

A stochastic process is said to be stationary if all $F_{\mathbf{X}}(\mathbf{x}, \mathbf{t})$ are invariant to shifts in time; that is, for any given constant τ, the following holds:

$$F_{\mathbf{X}}(\mathbf{x}, \mathbf{t} + \tau) = F_{\mathbf{X}}(\mathbf{x}, \mathbf{t}),$$

where $\mathbf{t} + \tau$ denotes the vector $(t_1 + \tau, t_2 + \tau, \cdots, t_n + \tau)$.

Next, let us define the PDF for a stochastic process. It is given by

$$f_{\mathbf{X}}(\mathbf{x}, \mathbf{t}) = \frac{\partial F_{\mathbf{X}}(\mathbf{x}, \mathbf{t})}{\partial \mathbf{X}} = \frac{\partial^n F_{X_1 X_2 \cdots X_n}(x_1, x_2, \cdots, x_n; \, t_1, t_2, \cdots, t_n)}{\partial x_1 \partial x_2 \cdots \partial x_n}.$$

Using the first-order PDF, we can define the mean of a stochastic process:

$$\overline{X(t)} = E[X(t)] = \int_{-\infty}^{\infty} x f_X(x; \, t) dx.$$

Using the second-order PDF, we can define the autocorrelation of $X(t)$,

$$R_{XX}(t_1, t_2) = E[X(t_1)X(t_2)] = \int_{-\infty}^{\infty} \int_{-\infty}^{\infty} x_1 x_2 f_{X_1 X_2}(x_1, x_2; \, t_1, t_2) dx_1 dx_2,$$

and the covariance function of $X(t)$,

$$C_{XX}(t_1, t_2) = R_{XX}(t_1, t_2) - \overline{X(t_1)X(t_2)}.$$

In the case of a stationary random process, we have

$$\overline{X(t)} = \overline{X} = \text{constant}, \tag{3.30}$$

$$R_{XX}(t_1, t_2) = R_{XX}(t_2 - t_1). \tag{3.31}$$

That is, R_{XX} is a function only of the time difference $\tau = t_2 - t_1$. $R_{XX}(\tau)$ is an even function. It attains the maximum value at $\tau = 0$, which is $E\{x(t)^2\}$. A random process is said to be wide-sense stationary if Eqs. (3.30) and (3.31) hold.

Clearly, all stationary processes are wide-sense stationary, but the converse is not true.

The concept of autocorrelation can be readily extended to the concept of cross-correlation, $R_{uv}(t_1, t_2)$, between two random processes $u(t)$ and $v(t)$. It is defined by

$$R_{uv}(t_1, t_2) = E\{u(t_1)v(t_2)\}. \tag{3.32}$$

It is often convenient to study a random process in the frequency domain. However, we shall postpone treatment of this material until Chapter 4.

3.3.2 Markov processes

A Markov process is a random process whose past has no influence on the future if its present is specified. This means the following: If $t_{n-1} < t_n$, then

$$p[x(t_n) \leq x_n | x(t), t \leq t_{n-1}] = p[x(t_n) \leq x_n | x(t_{n-1})].$$

From this, it follows that if

$$t_1 < t_2 < \cdots < t_n,$$

then

$$p[x(t_n) \leq x_n | x(t_{n-1}), \cdots, x(t_1)] = p[x(t_n) \leq x_n | x(t_{n-1})]. \tag{3.33}$$

From Eq. (3.33), it follows that

$$f(x_n | x_{n-1}, \cdots, x_1) = f(x_n | x_{n-1}).$$

Applying the chain rule, we thus obtain

$$f(x_1, \cdots, x_n) = f(x_n | x_{n-1}) f(x_{n-1} | x_{n-2}) \cdots f(x_2 | x_1) f(x_1). \tag{3.34}$$

Example 1: If x_n satisfies the recursion equation

$$x_{n+1} - g(x_n, n) = \eta_n,$$

where $g(x_n, n)$ is a function of x_n and n, η_n is a white noise of zero mean. Then x_n is a Markov process, since x_{n+1} is determined solely by x_n and η_n, and hence is independent of x_k for $k < n$.

Example 2: Let us consider a first-order autoregressive process denoted by AR(1). It is a special case of example 1 and is given by

$$x_{n+1} - \overline{x} = a(x_n - \overline{x}) + \eta_n, \tag{3.35}$$

where the constant coefficient a satisfies $0 \neq |a| < 1$.

We first find the variance σ^2 of the process. Taking the square of both sides of Eq. (3.35) and then taking the expectation, we have

$$R(0) = \sigma^2 = \sigma_n^2 / (1 - a^2),$$

where σ_n^2 is the variance of the noise η_n. Next, we compute the autocorrelation function. We have

$$
\begin{aligned}
R(1) = E[(x_{n+1} - \overline{x})(x_n - \overline{x})] &= E[(x_n - \overline{x})(a(x_n - \overline{x}) + \eta_n)] \\
&= aE[(x_n - \overline{x})^2] = a\sigma^2 = aR(0).
\end{aligned}
$$

Similarly, we can obtain

$$
R(2) = a^2\sigma^2 = a^2 R(0).
$$

Similar expressions can be obtained for negative time lag m. By induction, we then see that the autocorrelation function of x_n is given by

$$
R(m) = \sigma^2 a^{|m|}. \tag{3.36}
$$

It decays exponentially.

Example 3: In example 1, if $x_0 = 0$ and $g(x_n, n) = x_n$, then we have the random walk process given by

$$
x_n = x_{n-1} + \eta_n = \eta_1 + \eta_2 + \cdots + \eta_n.
$$

Since the variance of x_n is proportional to n, this process is not stationary. In general, a Markov process may not be stationary.

Below, we shall only consider homogeneous Markov processes where the conditional density $f(x_n|x_{n-1})$ does not depend on n. However, the first-order density $f(x_n)$ may depend on n. In general, a homogeneous process is not stationary. In many cases, it tends toward a stationary process as $n \to \infty$. To further simplify the matter, we shall only consider Markov processes with a finite or countably infinite number of states. Such processes are usually called Markov chains.

3.3.2.1 Homogeneous discrete-time (DT) Markov chains

Let the N states be denoted by $E_1, E_2, \cdots, E_N$, or simply $1, 2, \cdots, N$. A DT Markov chain is specified by the transition probability matrix $\mathbf{P} = [p_{ij}]$, where

$$
p_{ij} = P[X_n = j | X_{n-1} = i], \tag{3.37}
$$

and its initial probability vector is

$$
\vec{\pi}_0 = [\pi_0(1), \pi_0(2), \cdots, \pi_0(N)], \tag{3.38}
$$

where the subscript 0 denotes the time step 0. Being a probability, p_{ij} has to be nonnegative. Furthermore, by the axioms of probability, the sum of each row of the matrix $\mathbf{P}$ has to be 1:

$$
\sum_{j=1}^{N} p_{ij} = 1.
$$

Now we can compute the probabilities of the system at any time step. In particular, at the time step 1, we have

$$\vec{\pi}_1 = \vec{\pi}_0 \mathbf{P}.$$

In general, we have

$$\vec{\pi}_k = \vec{\pi}_{k-1} \mathbf{P}.$$

When $\vec{\pi}_*$ satisfies

$$\vec{\pi}_* = \vec{\pi}_* \mathbf{P}$$

it is called the stationary distribution of the Markov chain.

Recall that when a vector $\vec{B}$ satisfies

$$\vec{B} \mathbf{P} = \lambda \vec{B}$$

it is called a left eigenvector of the matrix $\mathbf{P}$, while the number λ is called the eigenvalue. Therefore, $\vec{\pi}_*$ is the left eigenvector of $\mathbf{P}$ associated with the eigenvalue 1. Now let us assume that $\mathbf{P}$ has N distinct eigenvalues denoted by $1, \lambda_2, \lambda_3, \cdots, \lambda_N$. Their corresponding left eigenvectors are

$$\vec{\pi}_*, \vec{B}_2, \vec{B}_3, \cdots, \vec{B}_N.$$

These eigenvectors are linearly independent (but not necessarily orthogonal, since the matrix $\mathbf{P}$ is typically not symmetric). Expressing $\vec{\pi}_0$ as a linear combination of these eigenvectors,

$$\vec{\pi}_0 = \alpha_1 \vec{\pi}_* + \alpha_2 \vec{B}_2 + \cdots + \alpha_N \vec{B}_N,$$

where α_i are coefficients, we then have

$$\vec{\pi}_n = \vec{\pi}_0 \mathbf{P}^n = \alpha_1 \vec{\pi}_* + \alpha_2 \lambda_2^n \vec{B}_2 + \cdots + \alpha_N \lambda_N^n \vec{B}_N. \tag{3.39}$$

We assume that when $n \to \infty$, all the exponential terms die out, and

$$\vec{\pi}_n \to \vec{\pi}_*.$$

Therefore, the coefficient α_1 has to be 1. The speed of convergence to the stationary distribution, of course, depends on the magnitudes of the eigenvalues $\lambda_2, \lambda_3, \cdots, \lambda_N$. Eq. (3.39) makes it clear that in general, a Markov process is not stationary. However, when $n \to \infty$, it may become stationary.

Example 4: Consider a two-state Markov chain. Its transition probability matrix is given by

$$\mathbf{P} = \begin{bmatrix} 1-p & p \\ q & 1-q \end{bmatrix}$$

and the initial probability vector is given by

$$\vec{\pi}_0 = [\pi_0(0), 1 - \pi_0(0)].$$

It is easy to find that the two eigenvalues are given by

$$\lambda_1 = 1, \quad \lambda_2 = 1 - p - q$$

and their corresponding eigenvectors are

$$\left[\frac{q}{p+q}, \frac{p}{p+q}\right], \quad [1, -1].$$

We have normalized the first eigenvector, since it corresponds to the stationary distribution $\vec{\pi}_*$. Also notice that the two eigenvectors are linearly independent but not orthogonal. Decomposing $\vec{\pi}_0$ in terms of the two eigenvectors, we have

$$[\pi_0(0), 1 - \pi_0(0)] = \left[\frac{q}{p+q}, \frac{p}{p+q}\right] + c[1, -1],$$

where $c = \pi_0(0) - q/(p+q)$. Therefore,

$$\vec{\pi}_n = \vec{\pi}_* + (\vec{\pi}_0 - \vec{\pi}_*)(1 - p - q)^n$$

when $p + q \neq 2$. When $p + q = 2$, the binary Markov chain is simply a deterministic system with period 2.

Now let us examine the time the system spends in a given state. We assert that the number of time units the system spends in the same state is geometrically distributed. To prove this, let us focus on an arbitrary state E_i. Assume that at time step k the system is at E_i. The system will remain in this state at the next time step with probability p_{ii}; similarly, it will leave this state at the next time step with probability $1 - p_{ii}$. If indeed it does remain in this state at the next time step, then the probability of its remaining for an additional time step is again p_{ii}, while the probability of leaving at this second time step is $1 - p_{ii}$. Since these probabilities are independent, we thus have

- P[system remains in E_i for exactly m additional time units given that it has entered E_i] $= (1 - p_{ii})p_{ii}^m$.

This is the geometrical distribution, as we have claimed. Note that the geometrical distribution is the unique discrete memoryless distribution.

3.3.2.2 Homogeneous continuous-time (CT) Markov chains
Simplifying the Markov property to a state space with a finite or infinitely countable number of states, we obtain the definition for a homogeneous CT Markov chain:

Definition: The random process $X(t)$ forms a CT Markov chain if for all integers n and for any sequence $t_1, t_2, \cdots, t_{n+1}$ such that $t_1 < t_2 < \cdots < t_{n+1}$ we have

$$\begin{aligned}
P[X(t_{n+1}) &= j | X(t_1) = i_1, X(t_2) = i_2, \cdots, X(t_n) = i_n] \\
&= P[X(t_{n+1}) = j | X(t_n) = i_n].
\end{aligned} \tag{3.40}$$

The main body of a homogeneous CT Markov chain is described by differential equations. In developing signal processing tools based on random fractals, we do not have much use for these equations. Hence, we will not present the general theory here. Below, we shall discuss the memoryless property of CT Markov chains and then describe the Poisson process.

We have proven that a DT Markov chain has geometrically distributed state times (also called sojourn times). We will now prove that a CT Markov chain has exponentially distributed sojourn times. Let τ_i be a random variable representing the time the process spends in state E_i. Since the influences of the past trajectory of the process on its future development are completely specified by giving the current state of the process, we need not specify how *long* the process has been in the current state. This means that the remaining time in E_i must have a distribution that depends only upon i and not upon how long the process has been in E_i. We may write this statement as

$$P(\tau_i > s + t | \tau_i > s) = h(t),$$

where $h(t)$ is a function only of the additional time t (and not of the expended time s). The above equation can be rewritten as

$$P(\tau_i > s + t | \tau_i > s) = \frac{P(\tau_i > s + t, \ \tau_i > s)}{P(\tau_i > s)} = \frac{P(\tau_i > s + t)}{P(\tau_i > s)}.$$

The last step follows since the event $\tau_i > s + t$ implies the event $\tau_i > s$. We can rewrite the last equation as

$$P(\tau_i > s + t) = P(\tau_i > s)\, h(t). \tag{3.41}$$

Setting $s = 0$ and observing that $P(\tau_i > 0) = 1$, we immediately have

$$h(t) = P(\tau_i > t).$$

Substituting this last equation in Eq. (3.41), we obtain

$$P(\tau_i > s + t) = P(\tau_i > s)\, P(\tau_i > t) \tag{3.42}$$

for $s, t \geq 0$. We now show that the only continuous distribution that satisfies Eq. (3.42) is the exponential distribution. First, we have, by definition, the following general relationship:

$$\frac{d}{dt} P(\tau_i > t) = \frac{d}{dt}[1 - P(\tau_i \leq t)] = -f_{\tau_i}(t), \tag{3.43}$$

where $f_{\tau_i}(t)$ is the PDF for τ_i. Differentiating Eq. (3.42) with respect to s, we have

$$\frac{dP(\tau_i > s + t)}{ds} = -f_{\tau_i}(s)P(\tau_i > t).$$

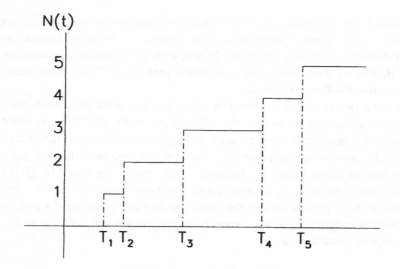

Figure 3.3.   A typical realization of a Poisson process $N(t)$.

Dividing both sides by $P(\tau_i > t)$ and setting $s = 0$, we have

$$\frac{dP(\tau_i > t)}{P(\tau_i > t)} = -f_{\tau_i}(0)ds.$$

Integrating s from 0 to t, we have

$$P(\tau_i > t) = e^{-f_{\tau_i}(0)t}.$$

Hence, the PDF is given by

$$f_{\tau_i}(t) = f_{\tau_i}(0)e^{-f_{\tau_i}(0)t}, \quad t \geq 0.$$

This is the exponential sojourn time that we have claimed.
Poisson process: Suppose we are observing customers arriving at a store, light bulbs burning out, or discharging of a neuron. In such situations, we are most interested in how many events occur in a given time interval and the distribution of the time interval between successive events. The Poisson process is the simplest model used to describe such phenomena. It forms one of the most important classes of stochastic processes and finds applications in areas of science and engineering as diverse as physics, biology, and teletraffic modeling. Let us now give the formal definition of the Poisson process (see also Fig. 3.3).

Definition: A Poisson process having rate λ is a sequence of events such that

1. $N(0) = 0$.

2. The process has independent increments. That is, for $t_1 < t_2 < \cdots < t_n$, the random variables $N(t_2 - t_1)$, $N(t_3 - t_2)$, $\cdots$, $N(t_n - t_{n-1})$ are all independent.

3. The number of events in any interval of length t is Poisson distributed with mean λt. That is, $\forall s, t \geq 0$,

$$P[N(t + s) - N(s) = n] = \frac{e^{-\lambda t}(\lambda t)^n}{n!}, \quad n = 0, 1, \cdots .$$

Let us now examine the distribution for the time interval between two successive events. This time interval is called interarrival time in teletraffic, interspike interval in neuroscience, etc. Denote this time interval by U. It is obvious that the event $U > t$ is equivalent to the event $N(t) = 0$. Therefore,

$$P(U > t) = P(N(t) = 0) = e^{-\lambda t}.$$

We thus see that this time interval has an exponential distribution. In fact, a Poisson process can also be defined via interevent intervals, as shown below.

Suppose that $\{U_j, j = 1, 2, \cdots, \}$ are iid exponential random variables with rate λ, that is,

$$P(U_j > t) = e^{-\lambda t}.$$

Let $T_0 = 0$ and $T_n = T_{n-1} + U_n$ for $n \geq 1$. We think of T_n as the time of the occurrence of some random event and U_n as the time between successive occurrences. We have found out that T_n follows an Erlang distribution (Eq. 3.14)). We define a counting process $N = \{N(t)\}$ as follows. For $n = 1, 2, \cdots ,$

$$N(t) < n \text{ if and only if } t < T_n.$$

In other words, $N(t) < n$ if the time of the nth occurrence is after t. Therefore,

$$P(N(t) \leq n) = P(N(t) < n + 1) = P(T_{n+1} > t) = \int_t^\infty \frac{\lambda(\lambda u)^n}{(n)!} e^{-\lambda u} du.$$

Repeatedly integrating by parts shows that for $n = 0, 1, 2, \cdots$ we have

$$P(N(t) \leq n) = \sum_{k=0}^n \frac{e^{-\lambda t}(\lambda t)^k}{k!}, \quad n = 0, 1, \cdots .$$

That is, $N(t)$ has a Poisson distribution with mean λt.

3.4 SPECIAL TOPIC: HOW TO FIND RELEVANT INFORMATION FOR A NEW FIELD QUICKLY

Nowadays phrases such as *interdisciplinary* or *multidisciplinary research* have become increasingly popular. When one is involved in such activities, often one finds it necessary to familiarize oneself with a completely new topic as quickly as possible. One can, of course, try to get started with a Google search. Sometimes information found that way may not be very specific. The authors' trick is to search the ISI Web of Knowledge. This little trick could be regarded as an effective application of the concept of power-law or scale-free networks, where the distribution of the number of links to a node in a network or graph follows a power law. In such a network, a few nodes have an extremely large number of links, while most others only have a few. The few nodes with a large number of links can be considered hubs. A citation network consists of nodes, which are published papers, and links, which are the citations of those papers in other papers. It has been found that such a network is a power-law network, with a few famous papers being cited thousands of times, while many unimportant papers have few or no citations. Of course, a paper with no citations may not be a bad paper. It could become a classic paper after many years. To get started in a new field as quickly as possible, it is wise to find some of the best papers in the field as well as a few important and recent papers on a topic that one is particularly interested in.

The url for the ISI Web of Knowledge is http://portal.isiknowledge.com/portal.cgi/ If one clicks on the first item, *Web of Science*, then one finds a page with multiple choices. One is *GENERAL SEARCH*. If one clicks on it, one again finds a page with multiple choices. Now, suppose we want to find out the characteristics of noise in nanotubes. Nanotubes are very appealing as bio-sensors. As a sensor, a noise level has to be very low. So this is a topic of considerable current interest. Now if one inputs

$$1/f \text{ noise in carbon nanotubes}$$

to the box *TOPIC*, then, by March 6, 2007, one finds that the earliest paper was written by Collins et al.:

> Collins PG, Fuhrer MS, Zettl A
> 1/f noise in carbon nanotubes
> APPLIED PHYSICS LETTERS 76 (7): 894-896 FEB 14 2000

It has been cited 40 times. Its 40 citations form a small network. It contains a paper with 65 citations:

> Sinnott SB, Andrews R
> Carbon nanotubes: Synthesis, properties, and applications
> CRITICAL REVIEWS IN SOLID STATE AND MATERIALS SCIENCES
> 26 (3): 145-249 2001

This paper can be considered the hub of the small network. With these two papers, one can be well on the way to research on the noise character of carbon nanotubes. Interestingly, the noise in nanotubes is a $1/f^\beta$ process, which we will study in depth in later chapters.

Research-oriented readers are strongly encouraged to use this little trick.

3.5 BIBLIOGRAPHIC NOTES

Readers new to probability theory may find the small book [257] by two masters, Khinchin and Gnedenko, very helpful. Feller's classic [136] is always a pleasure to read. For systematic textbooks, we refer to [205,335]. Readers wanting to know more about large-deviations theory are referred to [143,205]. Readers interested in lognormality in network traffic are referred to [30,54,63,311], while those interested in lognormality in switching times in ambiguous perceptions are referred to [496]. An entertaining as well as insightful paper to read about lognormal distribution and many other topics is [316]. Finally, readers interested in Zipf's law and power-law networks may want to browse an interesting website created by Dr. Wentian Li: http://www.nslij-genetics.org/wli/, and an excellent review article by Albert and Barabasi [7].

3.6 EXERCISES

1. Let z be $N(0, 1)$. Prove that $x = a + bz$ is $N(a, b^2)$ and that $w = e^{a+bz}$ has a log-normal distribution.

2. Prove that the geometrical distribution is the unique discrete memoryless distribution.

3. Let x and y be $N(0, \sigma)$ and independent. Define r and φ by the following equation:

$$x \cos(\omega t) + y \sin(\omega t) = r \cos(\omega t - \varphi), \quad |\varphi| < \pi.$$

Prove that the random variables r and φ are independent, φ is uniform in the interval $[-\pi, \pi]$, and r has a Rayleigh distribution.

4. Simulate 10^4 exponentially distributed random variables with parameter λ. Then estimate PDF and CDF. Plot PDF and CDF, both in linear scale and semilog scale. Estimate λ as the slope of the semilog plots. Is the estimated λ the same as that used in the simulation?

5. Simulate 10^4 Pareto-distributed random variables with parameters $\alpha = 0.5$, 1, 1.5, 2, 2.5, and 3. Estimate PDF and CDF using simple histograms. Then plot them out in log-log scale. Estimate α as the slope of the log-log plots. Are the estimated α the same as those used in the simulations?

6. This is the same as exercise 5, but with a different way of estimating PDF. The procedure in exercise 5 can be termed equal-linear-bin. An alternative is to use the equal-log-bin procedure: before forming the histogram, take the log of the random variables first; then plot out the estimated PDF or CDF using a semi-log plot. Determine if this procedure improves the accuracy in estimating α. If it does, explain why.

7. Simulate a number of AR(1) processes described by Eq. (3.35) for $a = 0.1$, $0.2, \cdots, 0.9$. Numerically compute the autocorrelation function for these processes and compare the result with the theory.

8. Construct a simple random walk process $Y = X_1 + X_2 + \cdots + X_n$, where $X_i \sim N(0, 1)$ are iid. Let $n = 1000$. Analytically find the mean and variance of Y. If you are asked to compute the mean and variance of Y numerically, how many realizations of Y would be needed? Compare your answer with the results of simulations.

9. Let X and Y be independent random variables with characteristic functions $\Phi_X(u)$ and $\Phi_Y(u)$, respectively. Let $Z = X + Y$. Prove that the characteristic function for Z is $\Phi_Z(u) = \Phi_X(u) \cdot \Phi_Y(u)$.

CHAPTER 4

FOURIER ANALYSIS AND WAVELET MULTIRESOLUTION ANALYSIS

In electrical engineering, frequency analysis of signals and systems is a central topic of signal processing. When used intelligently, Fourier analysis can be very revealing, even though the signals to be analyzed are very complicated. Unfortunately, even after several courses on signal processing, many electrical engineering graduate students still feel paralyzed when asked to analyze real data. We suspect that this problem may be even more acute among nonelectrical engineers. To help readers get started, especially those less experienced in data analysis, we devote Sec. 4.1 to a brief discussion of Fourier analysis, with emphasis on analysis of real data. We hope that after this treatment, readers will find complicated signals to be more friendly. Some of the examples discussed in Sec. 4.1 also illustrate fractal signals. In Sec. 4.2, we describe wavelet multiresolution analysis (MRA). In later chapters, wavelet MRA will be extensively used to estimate key parameters of random fractal processes. In order to focus on big pictures and to proceed quickly to later chapters, derivations of some key equations described below are omitted. They can be found in the Bibliographic Notes at the end of the chapter.

4.1 FOURIER ANALYSIS

4.1.1 Continuous-time (CT) signals

Suppose we are given a periodic signal $x(t)$ with period T_p. It can be represented as a linear combination of harmonically related complex exponentials of the form

$$x(t) = \sum_{k=-\infty}^{\infty} c_k e^{j2\pi k F_0 t}, \qquad (4.1)$$

where $F_0 = 1/T_p$ and $j^2 \equiv -1$. This is called Fourier series representation of periodic signals. One can think of the exponential signals

$$\{e^{j2\pi k F_0 t}, \quad k = 0, \pm 1, \pm 2, \cdots \}$$

as the basic building blocks. These building blocks are orthogonal, i.e.,

$$\int_{t_0}^{t_0+T_p} e^{j2\pi k F_0 t} e^{-j2\pi l F_0 t} dt = \begin{cases} 0, & k \neq l \\ T_p, & k = l, \end{cases}$$

where t_0 is arbitrary. Therefore, the coefficient c_k is given by

$$c_k = \frac{1}{T_p} \int_{T_p} x(t) e^{-j2\pi k F_0 t} dt, \qquad (4.2)$$

where integration is carried out over any interval of length T_p. In general, c_k is complex valued. It can be written as

$$c_k = |c_k| e^{j\theta_k}.$$

Graphically, one can plot $|c_k|$ and/or θ_k against the frequency kF_0. Such plots are called line spectra with equidistant lines. The line spacing is equal to the fundamental frequency, which in turn is the inverse of the fundamental period of the signal.

Now let us allow the period to increase without limit. The line spacing then tends toward zero. In the limit, when the period becomes infinite, the signal becomes aperiodic and its spectrum becomes continuous. Guided by this argument, and replacing summation by integration, one readily obtains the following Fourier transform of aperiodic signals:

$$x(t) = \int_{-\infty}^{\infty} X(F) e^{j2\pi F t} dF. \qquad (4.3)$$

The coefficient $X(F)$ is given by

$$X(F) = \int_{-\infty}^{\infty} x(t) e^{-j2\pi F t} dt. \qquad (4.4)$$

Comparing Eq. (4.4) with Eq. (4.2), we find that

$$T_p c_k = X(kF_0) = X\left(\frac{k}{T_p}\right).$$

Thus the Fourier coefficients are samples of $X(F)$ taken at multiples of F_0 and scaled by F_0 (multiplied by $1/T_P$). In other words, the spectrum of an aperiodic signal is the envelope of the line spectrum in the corresponding periodic signal obtained by repeating the aperiodic signal with period T_p.

Like c_k, $X(F)$ is complex valued. Its magnitude square,

$$S_{xx}(F) = |X(F)|^2,$$

is called the energy density spectrum. Parsavel's theorem states that

$$E_x = \int_{-\infty}^{\infty} |x(t)|^2 dt = \int_{-\infty}^{\infty} |X(F)|^2 dF.$$

4.1.2 Discrete-time (DT) signals

We start from a periodic sequence $x(n)$ with period N, that is, $x(n+N) = x(n)$ for all n. The Fourier series representation for $x(n)$ consists of N harmonically related exponential functions

$$e^{j2\pi kn/N}, \qquad k = 0, 1, \cdots, N-1$$

and is expressed as

$$x(n) = \sum_{k=0}^{N-1} c_k e^{j2\pi kn/N}. \tag{4.5}$$

The orthogonality of the basis functions for the CT case is now replaced by the following formula:

$$\sum_{k=0}^{N-1} e^{j2\pi kn/N} = \begin{cases} N, & k = 0, \pm N, \pm 2N, \cdots \\ 0 & \text{otherwise.} \end{cases}$$

With this equation, one can readily find

$$c_k = \frac{1}{N} \sum_{n=0}^{N-1} x(n) e^{-j2\pi kn/N}. \tag{4.6}$$

It is easy to see that $\{c_k\}$ is also periodic with period N. The sequence $|c_k|^2$, $k = 0, 1, \cdots, N-1$ is the distribution of power as a function of frequency and is called the power density spectrum (or power spectral density (PSD)). Parsavel's relation now becomes

$$P_x = \sum_{k=0}^{N-1} |c_k|^2 = \frac{1}{N} \sum_{n=0}^{N-1} |x(n)|^2. \tag{4.7}$$

Next, we discuss the Fourier transform of a finite-energy, aperiodic DT signal $x(n)$. It is defined as

$$X(\omega) = \sum_{n=-\infty}^{\infty} x(n)e^{-j\omega n}. \tag{4.8}$$

Physically, $X(\omega)$ represents the frequency content of the signal $x(n)$. One can readily prove that $X(\omega)$ is periodic with period 2π, that is,

$$X(\omega + 2\pi k) = X(\omega), \qquad k = \pm 1, \ \pm 2, \ \cdots.$$

This property indicates that the frequency range for any DT signal is limited to $(-\pi, \pi)$, or equivalently, $(0, 2\pi)$. In contrast, the frequency range for a CT signal is $(-\infty, \infty)$. When Eq. (4.8) is expressed in terms of frequency f, where $\omega = 2\pi f$, we have

$$X(f) = \sum_{n=-\infty}^{\infty} x(n)e^{-j2\pi f n}. \tag{4.9}$$

Since $X(\omega)$ is periodic with period 2π, $X(f)$ is periodic with period 1.

We now derive the inverse transform. Using the following orthogonality condition

$$\int_{-\pi}^{\pi} e^{i\omega(m-n)} d\omega = \begin{cases} 2\pi, & m = n \\ 0, & m \neq n \end{cases}$$

under suitable convergence conditions, we multiply both sides of Eq. (4.8) by $e^{j\omega n}$ and integrate; then we have

$$x(n) = \frac{1}{2\pi} \int_{2\pi} X(\omega)e^{j\omega n} d\omega. \tag{4.10}$$

The energy of a DT signal $x(n)$ can be defined as

$$E_X = \sum_{n=-\infty}^{\infty} |x(n)|^2.$$

It also can be expressed as integration of $X(\omega)$, yielding the following Parsavel's relation for DT aperiodic signals with finite energy:

$$E_X = \sum_{n=-\infty}^{\infty} |x(n)|^2 = \frac{1}{2\pi} \int_{2\pi} |X(\omega)|^2 d\omega < \infty.$$

As in CT aperiodic signals

$$S_{xx}(\omega) = |X(\omega)|^2$$

is called the energy density spectrum of $x(n)$. In Sec. 3.3.1, we defined the auto-correlation function of a stochastic process. Let $R_{xx}(m)$ denote the autocorrelation function of $x(n)$. We have the following famous theorem.

Wiener-Khintchine theorem: $R_{xx}(m)$ and S_{xx} form a Fourier transformation pair. Instead of proving this theorem, we illustrate it with an example.

Example 1: In Sec. 3.3.2, we discussed an AR(1) process, defined by

$$x_{n+1} = ax_n + \eta_n, \tag{4.11}$$

where, for simplicity, we took the mean $\bar{x}$ of the process to be zero (as will be discussed shortly, the mean contributes to a DC component in the Fourier transform). The autocorrelation function of the AR(1) process is given by

$$R_{xx}(m) = \frac{\sigma_\eta^2}{1 - a^2}\, a^{|m|}.$$

Taking the Fourier transform of $R_{xx}(m)$, one finds that

$$S_{xx}(\omega) = \sum_{m=-\infty}^{\infty} \frac{\sigma_\eta^2}{1 - a^2} a^{|m|} e^{-j\omega m} = \frac{\sigma_\eta^2}{1 + a^2 - 2a\cos\omega}, \quad -\pi \le \omega \le \pi.$$

An alternative way of obtaining S_{xx} is to take the Fourier transform of both sides of Eq. (4.11) to obtain

$$X(\omega) = ae^{-j\omega}X(\omega) + \sigma_\eta, \tag{4.12}$$

where $X(\omega)$ denotes the Fourier transform of the left-hand side of Eq. (4.11), and the coefficient $e^{-j\omega}$ on the right-hand side of Eq. (4.12) is due to delay by one unit of time. It is easy to see that

$$S_{xx}(\omega) = |X(\omega)|^2.$$

We thus verify the Wiener-Khintchine theorem.

4.1.3 Sampling theorem

Suppose we sample a CT signal $x(t)$ to obtain a DT signal $x(n)$ with a fixed time interval T. The inverse of T is called the sampling frequency, $F_s = 1/T$. Based on

$$t = nT = n/F_s,$$

what relation can be derived between the Fourier transform $X(f)$ of $x(n)$ and the Fourier transform $X_a(F)$ of $x(t)$? It is

$$X(f) = X\left(\frac{F}{F_s}\right) = F_s \sum_{k=-\infty}^{\infty} X_a(F - kF_s). \tag{4.13}$$

The right-hand side of Eq. (4.13) consists of a periodic repetition of the scaled spectrum $F_s X_a(F)$ with period F_s. This periodicity is necessary, since $X(f)$ is periodic with period 1. When the repeating copies of $F_s X_a(F)$ do not overlap, we say that the sampling frequency is high enough that no aliasing has occurred. This is ensured when $X_a(F)$ is band-limited to the interval $(-F_s/2, F_s/2)$. Formally, the sampling theorem can be stated that a band-limited CT signal, with highest frequency (bandwidth) B Hz, can be uniquely recovered from its samples provided that the sampling rate F_s is at least $2B$ samples per second.

4.1.4 Discrete Fourier transform

The discrete Fourier transform (DFT) and its inverse (IDFT) are defined by

$$\text{DFT}: \quad X(k) = \sum_{n=0}^{N-1} x(n)e^{-j2\pi kn/N} \quad k = 0, 1, 2, \cdots, N-1 \quad (4.14)$$

$$\text{IDFT}: \quad x(n) = \frac{1}{N}\sum_{k=0}^{N-1} X(k)e^{j2\pi kn/N} \quad n = 0, 1, 2, \cdots, N-1. \quad (4.15)$$

The derivations of DFT and IDFT involve periodic extension of the signal sequence $x(n)$, $n = 0, 1, \cdots, N-1$. Therefore, one can expect Eq. (4.14) to be similar to Eq. (4.6), while Eq. (4.15) is similar to Eq. (4.5). This is indeed the case. In fact, the difference is only by the factor $1/N$. This difference is by convention. It is not essential.

DFT has a number of interesting properties. We note one here. It indicates that for a real signal, $|X(k)|$ is symmetric about $N/2$. This is because

$$X(N-k) = X^*(k), \quad (4.16)$$

where the superscript * denotes a complex conjugate. Therefore, for a real signal, it is sufficient to plot out the magnitude and phase of $X(k)$ for $k = 0, 1, \cdots, N/2$, since $X(k)$ for $k = N/2 + 1, \cdots, N-1$ does not contain any new information. This property can also be considered an action of the sampling theorem.

4.1.5 Fourier analysis of real data

To truly understand Fourier analysis, it is essential to have hands-on experience with real data. We analyze four types of data here: parkinsonian tremor, quantum well device noise, DNA sequence, and EEG.

Example 1: A tremor is an involuntary, approximately rhythmic, and roughly sinusoidal movement of parts of the body. Broadly speaking, there are two classes of tremors: normal and pathologic. Pathologic tremors result from disorders of the central and peripheral nervous systems. The two most common ones are the essential tremor (ET) and the tremor in Parkinson's disease (PD). In PD, a typical 4–6 Hz resting tremor is observed, whereas in ET, the rhythmic activation of the muscles typically occurs at a higher frequency: 5–10 Hz. Figure 4.1(a) shows a signal of a parkinsonian tremor. The data were sampled at a rate of 1000 Hz. Figure 4.1(b) shows the PSD of the data.

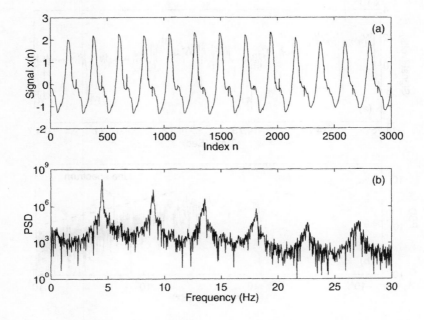

Figure 4.1. Parkinsonian tremor signal (a) and its PSD (b).

We now make a few comments.

1. In real data analysis, it is more useful to specify the unit for the frequency axis than to use the normalized frequency. This can be done as follows. Given a sequence of N points, sampled at a sampling rate of F_s, the total time interval monitored is N/F_s. The frequency f_k corresponding to the kth point $X(k)$ is

$$f_k = \frac{k}{N} F_s.$$

In particular, X_0 is called the DC component, contributed by the mean of the signal. In situations where the DC component is much stronger than other frequency components, it is desirable to remove the DC component before taking the Fourier transform (or equivalently, set $X_0 = 0$).

2. When $T_s = 1/F_s = 1$ unit and the signal is known to have a period T_i, one may expect to observe a sharp spectral peak at

$$i = \frac{N}{T_i}.$$

This feature will be made clearer when we discuss the period-3 feature in coding the DNA sequence in example 3.

3. In the digital signal processing class of one of the authors, J.B., many students plotted the spectrum for the entire frequency range $0 \le f \le 1000$ Hz. Then only

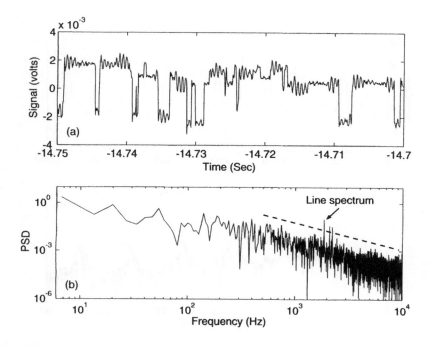

Figure 4.2.   (a) Noise recorded in a quantum well infrared photodetector; (b) the PSD of the noise.

nonzero spectrum could be seen along the two ends. After being reminded that at least the right half of the spectrum does not need to be plotted, many students plotted the spectrum for $0 \leq f \leq 500$ Hz. Still, not much could be seen. This example emphasizes that one needs to zoom in and out to find out which frequency band is significant.

4. In the so-called Bode plots, the frequency axis is plotted logarithmically. When Fig. 4.1(b) is plotted in log scale, the spacing between successive harmonics is no longer of equal distance. This hinders visual inspection. Therefore, the linear scale for the frequency axis is more appropriate for observing harmonics.

Example 2: We now discuss noise recorded in a quantum well infrared photodetector, with a sampling time of 50 μs. Figure 4.2(a) shows a segment of the noise trace. It is like an ON/OFF train. In the literature, such noise is called a random telegraph signal (RTS). Its PSD is shown in Fig. 4.2(b), where we observe that the PSD decays in a power-law manner. This is an example of so-called $1/f^\beta$ noise, a special type of random fractals. The dashed straight line shown in Fig. 4.2(b) has a slope approximately equal to -1. Therefore, $\beta \approx 1$ here. To more accurately estimate the β parameter, we must use methods to be discussed in Chapter 8.

Again, we make two comments.

1. Why is it important to study a spectrum of the $1/f^\beta$ shape? The reason is that a power-law decaying spectrum is a well-defined functional form, and therefore

warrants a physical explanation of how it is generated. The most popular model in the study of device noise (including this as well as the more general MOSFET noise) is the relaxation model. This model relates the spectrum to the distribution of the ON/OFF periods of the RTS. This will be explained in more detail in Chapter 11.

2. Figure 4.2(b) contains a few spectral lines indicated by an arrow. In J.B.'s digital signal processing class, some students thought that those lines correspond to the true signal, while the rest of the spectrum is due to noise. This is not correct. In fact, the spectral lines are not important at all. They are the noise from the measurement equipment.

Example 3: A DNA sequence is made up of four nucleotides: adenine (A), guanine (G), thymine (T), and cytosine (C). A and G are purines, while T and C are pyrimidines. A DNA sequence is highly complicated, containing many functional structures. On the coarsest level, a DNA sequence can be partitioned into alternating coding and noncoding segments. The coding region has a salient feature of period-3 (P3), resulting from the fact that three nucleotide bases encode an amino acid and that the usage of codons is highly biased. This P3 feature can be readily manifested by the Fourier transform. Specifically, one can map a DNA sequence of length N to a numerical sequence (for example, C and T to 1, A and G to -1), and then take the Fourier transform. One can then observe a strong peak at or around $N/3$. See Fig. 4.3(a). Usually, such a strong peak cannot be observed in noncoding regions, as shown in Fig. 4.3(b).

Another main feature of a DNA sequence is its randomness. It is not like simple white noise. One good way of analyzing its randomness is to form the so-called DNA walk,

$$y(n) = \sum_{k=0}^{n} u(k),$$

where $u(k)$ is the numerical sequence mapped from the DNA sequence. An example is shown in Fig. 4.4. The PSD of a DNA walk is also of the form $1/f^{\beta}$, similar to that shown in Fig. 4.2(b). It has been found that β is close to 2 for coding sequences but may differ from 2 quite significantly for noncoding sequences. Again, estimation of β using Fourier analysis may not be accurate enough. More accurate methods will be developed in Chapter 8. We should nevertheless emphasize that this does not mean that Fourier analysis is useless for analyzing fractal signals. In fact, qualitative understanding using Fourier analysis can serve as a guide in the further refined analysis.

Example 4: Brain waves (EEGs) result from electrical activity emanating from the brain. The signals can be highly nonstationary, as shown in Fig. 4.5(a). There are four categories of brain waves. Ranging from most to least active, they are beta, alpha, theta, and delta waves. The beta wave is associated with the brain's being aroused and actively engaged in mental activities. The alpha wave represents the nonarousal state of the brain. The theta wave is associated with drowsiness/idling,

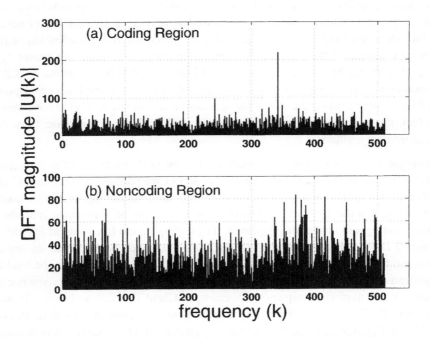

Figure 4.3.   Representative DFT magnitude for coding and noncoding regions in yeast chromosome I.

and the delta wave is associated with sleep/dreaming. The exact frequency ranges for the four waves may vary among researchers, however. For example, one Internet document states that the four waves correspond to frequency ranges $14 - 40$, $7.5 - 13$, $3.5 - 7.5$, and < 3 Hz (i.e., cycles per second), respectively. Another document states that they correspond to $15 - 40$, $9 - 14$, $5 - 8$, and $1.5 - 4$ Hz. Other research papers may give slightly different values.

One, especially a layman, could ask whether the four waves are characterized by distinct spectral peaks in the frequency domain. It turns out that they are not. One example is shown in Fig. 4.5(b) for a fairly stationary EEG segment. We do not observe any distinct spectral speaks. In Chapter 15, we shall discuss the physical meaning of these waves.

4.2   WAVELET MULTIRESOLUTION ANALYSIS

In situations where the time dependence of the frequency of a given signal is of interest, one could resort to windowed Fourier transformations (WFT). However, WFT is not always an effective solution, since sharp localizations in time and in frequency cannot be simultaneously achieved. The wavelet transform solves this problem by utilizing the scaled and shifted versions of a wavelet, $\psi(t)$, using the

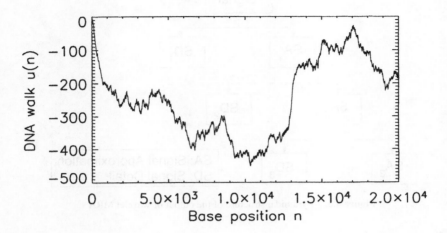

Figure 4.4. An example of a DNA walk constructed from the first 20,000 bases of the chromosome I of yeast.

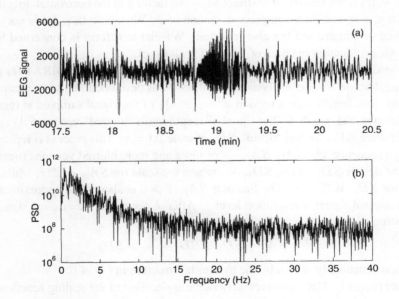

Figure 4.5. (a) A segment of EEG signal; (b) its PSD.

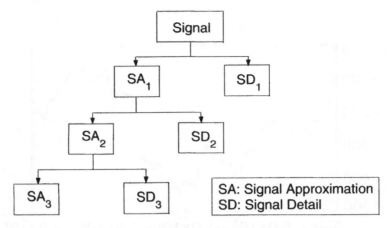

Figure 4.6.   Pyramidal structure of the output of wavelet MRA.

following equation:

$$\psi_{s,k}(t) \equiv \psi_s(t - k) = |s|^{-p}\psi\left(\frac{t - k}{s}\right),$$

where $p > 0$ is fixed, k is the translation, and s is the scale. The translation or shifting achieves time localization of the signal, while the use of scale s avoids commitment to any particular scale. To see the latter, let us take $k = 0$. Then, when $s > 1$, $\psi_s(t)$ is the version of ψ stretched by the factor s in the horizontal direction; when $0 < s < 1$, ψ_s is a compressed version of ψ. When $s < 0$, $\psi_s(t)$ is not only stretched or compressed but also reflected. Wavelet transform is concerned with expressing a function in terms of $\psi_{s,k}(t)$ for all (s, k).

Around 1986, a very powerful method, multiresolution analysis (MRA), for performing discrete wavelet analysis and synthesis was developed. This is a recursive method. One begins with a version $x^0 = \{x(n)\}$ of the signal sampled at regular time intervals $\Delta t = \tau > 0$. The signal x^0 is split into a "blurred" version SA_1 at the coarser scale $\Delta t = 2\tau$ and "detail" SD_1 at scale $\Delta t = \tau$. This process is repeated, giving a sequence x^0, SA_1, SA_2, $\cdots$ of more and more blurred versions together with the details SD_1, $SD_2, SD_3, \cdots$, where the scale for SA_m is $2^m\tau$, while the scale for SD_m is $2^{m-1}\tau$. The function SA_j, $j > 0$ is also called approximation of the original signal at resolution level j. After J iterations, the original signal is reconstructed as

$$x^0 = SA_J + SD_1 + SD_2 + \cdots + SD_J.$$

This is schematically shown by the pyramidal structure in Fig. 4.6.

More precisely, MRA involves an iterative application of the scaling function ϕ_0 and the wavelet function ψ_0. The scaling function is a low-pass filter. It satisfies the condition

$$\int_{-\infty}^{\infty} \phi_0(t)dt = 1.$$

The wavelet ψ_0 must have zero average and decay quickly at both ends. It is a high-pass filter. The scaled and shifted versions of ϕ_0 and ψ_0 are given by

$$\phi_{j,k}(t) = 2^{-j/2}\phi_0(2^{-j}t - k),$$

$$\psi_{j,k}(t) = 2^{-j/2}\psi_0(2^{-j}t - k),$$

where $j, k \in Z$ are the scaling (dilation) index and the shifting (translation) index, respectively. Different values of j correspond to analyzing different resolution levels of the signal. With the scaling function and the mother wavelet, MRA can be described by the following steps:

1. At the $j = 1$th resolution, for each $k = 0, 1, 2, \cdots$, compute the approximation coefficient $a_x(j, k)$ and the detailed coefficient $d_x(j, k)$ according to the following formula:

$$a_x(j, k) = \sum_n x(n)\phi_{j,k}(n) = \sum_n x(n)2^{-j/2}\phi_0(2^{-j}n - k), \quad (4.17)$$

$$d_x(j, k) = \sum_n x(n)\psi_{j,k}(n) = \sum_n x(n)2^{-j/2}\psi_0(2^{-j}n - k). \quad (4.18)$$

2. The signal approximation SA_j and the signal detail SD_j at the jth resolution level are computed as

$$SA_j = \sum_k a_x(j, k)\phi_{j,k}(n), \quad (4.19)$$

$$SD_j = \sum_k d_x(j, k)\psi_{j,k}(n). \quad (4.20)$$

Note that the resolution for SA_1 and SD_1 is 2τ and τ, respectively.

3. Repeat steps 1 and 2 for the $(j + 1)$th resolution level, using the signal approximation SA_j obtained in step 2 as the input signal. Note that the resolution for SD_{j+1} is the same as that for SA_j, which is $2^j\tau$, while the resolution for SA_{j+1} is $2^{j+1}\tau$, twice as coarse as that for SA_j.

Let the maximum scale resolution level chosen for analysis be J. The reconstruction formula is

$$x(n) = SA_J + \sum_{j=1}^J SD_j = \sum_k a_x(J, k)\phi_{J,k}(n) + \sum_{j=1}^J \sum_k d_x(j, k)\psi_{j,k}(n). \quad (4.21)$$

The first term represents the approximation at resolution level J, and the second term represents the details at resolution level J and lower.

To make the above procedure more concrete, let us take the Haar wavelet as an example. The scaling function and the mother wavelet of the Haar wavelet are defined as

$$\phi_0(t) = \begin{cases} 1, & 0 \le t < 1, \\ 0 & \text{elsewhere.} \end{cases}$$

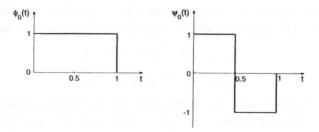

Figure 4.7.   The scaling function $\phi_0(t)$ and the mother wavelet $\psi_0(t)$ of the Haar wavelet.

$$\psi_0(t) = \begin{cases} 1, & 0 \leq t < 1/2, \\ -1, & 1/2 \leq t < 1, \\ 0 & \text{elsewhere.} \end{cases}$$

They are depicted in Fig. 4.7.

To understand how the Haar wavelet works, let us compute a few terms of $a_x(j, k)$ and $d_x(j, k)$ according to Eqs. (4.17) and (4.18). When $j = 1$, $k = 0$, we have

$$a_x(1, 0) = 2^{-1/2}[x(1) + x(2)], \quad \text{simple smoothing}$$

$$d_x(1, 0) = 2^{-1/2}[x(1) - x(2)], \quad \text{simple differencing.}$$

When $j = 1$, $k = 1$, we have

$$a_x(1, 1) = 2^{-1/2}[x(3) + x(4)], \quad \text{simple smoothing}$$

$$d_x(1, 1) = 2^{-1/2}[x(3) - x(4)], \quad \text{simple differencing.}$$

At the next resolution level $j = 2$, when $k = 0$, we have

$$a_x(2, 0) = 2^{-1}[x(1) + x(2) + x(3) + x(4)] = 2^{-1/2}[a_x(1, 0) + a_x(1, 1)],$$

$$d_x(2, 0) = 2^{-1}[x(1) + x(2) - x(3) - x(4)] = 2^{-1/2}[a_x(1, 0) - a_x(1, 1)].$$

Evidently, $\phi_0(t)$ is a low-pass filter, while $\psi_0(t)$ is a high-pass filter. For the Haar wavelet, for each resolution level j, the maximal k is $2^{-j}N$, where N is the length of the dataset. For wavelets with more complicated functional forms for $\phi_0(t)$ and $\psi_0(t)$, Eqs. (4.17) and (4.18) amount to smoothing and differencing with weights given by shifting and compressing/dilating $\phi_0(t)$ and $\psi_0(t)$, respectively.

We now further illustrate the Haar wavelet by analyzing an example signal $x(n)$, which consists of noisy blocks, as shown in Fig. 4.8(a). The signal approximations and details at resolution levels 1 through 3 are shown in Figs. 4.8(b,d,f) (left column) and Fig. 4.8(c,e,g) (right column), respectively. We have

$$x(n) = SA_1 + SD_1 = SA_2 + SD_2 + SD_1 = SA_3 + SD_3 + SD_2 + SD_1.$$

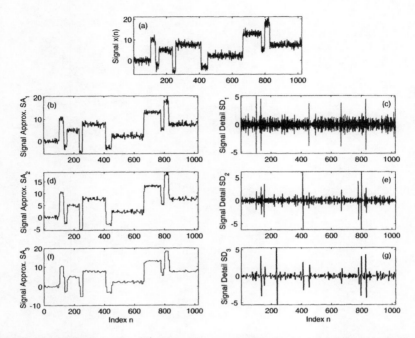

Figure 4.8.   (a) The input signal $x(n)$, (b,d,f) and (c,e,g) are the signal approximations and the signal details at resolution levels 1 through 3, respectively. $x(n) = SA_1 + SD_1 = SA_2 + SD_2 + SD_1 = SA_3 + SD_3 + SD_2 + SD_1$.

4.3 BIBLIOGRAPHIC NOTES

Fourier analysis is a central topic in numerous excellent textbooks on signal processing. We refer to [324, 355, 413] here. For wavelet MRA, we refer to [414]. For more information about tremor data, we refer to [150, 175, 435]; about quantum well noise data, we refer to [271]; about DNA sequence data, we refer to [154, 159]; and about EEG data, we refer to [61, 176].

4.4 EXERCISES

1. Prove the Parsavel's relation described by Eq. (4.7).

2. Prove the symmetric property of DFT given by Eq. (4.16).

3. Continuing with the example of the Haar wavelet, write expressions for the approximation and detailed coefficients at the level $j = 3$.

4. Using the data downloadable at the book's website (see Sec. A.4 in Appendix A), re-produce Figs. 4.1, 4.2, and 4.5. There are many other datasets at this website. You may want to take a look at some of them.

Figure 3.9 ...

5YG + 5GY + 5YD, 5GD.

3.6 BIBLIOGRAPHIC NOTES

Fourier analysis is a central topic in numerous excellent textbooks on signal processing. We refer to [16, 158, 215] here. For material on MRA, we refer to [161]. For more information about numerical analysis, we refer to [120, 135, 154]. About outliers, we will more generally refer to [271]. About DNA sequence data, we refer to [266, 269] and about LCD data, we refer to [63, 170].

3.7 EXERCISES

1. Draw the Random DAG/DG described by Eq. (3.7).

2. Prove the Independence structure of Fig. 3 given by Eq. (3.6).

3. Compared with the example of the Haar wavelet, write expressions for the approximation and detail coefficients at the three levels.

4. Draw the data distribution in the Fourier/wavelet top levels, with an approach as described for Eq. (3.2), p. 84, 85. There are more references throughout the website. The links in the book look at some of them.

CHAPTER 5

BASICS OF FRACTAL GEOMETRY

Fractal phenomena are situations in which a part of a system or object is exactly or statistically similar to another part or to the whole. Such phenomena may be associated with concrete geometrical objects such as clouds, mountains, trees, lightning, and so on, or associated with dynamical, spatial, or temporal variations such as turbulent wave motions in the ocean or variations of a stock market with time. Thus, it is no wonder that fractal geometry provides a solid and elegant framework for the study of many natural and man-made phenomena in various scientific and engineering domains.

Mathematically speaking, fractals are characterized by power-law relations over a wide range of scales. Such power-law relations are often called scaling laws. If a certain phenomenon can be fully characterized by a single power-law scaling relation, then it is called monofractal; otherwise, it is called multifractal. In the latter case, the number of scaling laws can be infinite. In this chapter, we describe several key concepts of fractal geometry.

5.1 THE NOTION OF DIMENSION

One of the key concepts in fractal geometry is the notion of dimension. While this concept was briefly touched on in Sec. 2.1, here we treat it in some depth.

(a) compass dimension

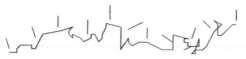

(b) box dimension

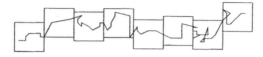

(c) grid dimension

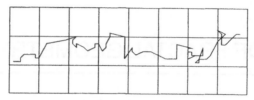

Figure 5.1.   Three ways to estimate the fractal dimension.

The major idea discussed in Sec. 2.1 is that the dimension of an object is obtained by covering it with small boxes. That leads us to Eq. (2.2) and an understanding of topological dimension for isolated points, a line, an area, and a volume (which is $0 - D$, $1 - D$, $2 - D$, and $3 - D$, respectively). We have also discussed the consequence of a jagged mountain trail with $1 < D < 2$. Let us now resume the discussion of that topic more systematically.

There are a number of ways of measuring the length of an irregular curve. Three methods are shown in Fig. 5.1. While in many situations they give very similar results, occasionally, they may yield different values. One example is fractional Brownian motion, which will be discussed in the next chapter.

To deepen our understanding, in the remainder of this section we shall discuss the concept of the Hausdorff-Besicovitch dimension. To facilitate this discussion, we remind readers that when covering an object by boxes, the size of the boxes does not have to be the same. The only requirement is that the size of the largest box is bounded (which serves as ϵ).

The Hausdorff-Besicovitch dimension is related to the α-covering measure. Since our purpose is to find the dimension of an object E, let us cover it with boxes of linear length ϵ. What is the measuring unit for such a procedure? If E is a line, the unit is simply ϵ; if E is an area, the unit is ϵ^2. In general, the unit is $\epsilon^{d(E)}$. Since the dimension of E is unknown, let us denote the measuring unit by $\mu = \epsilon^\alpha$ and try out a number of different α. Let us again consider a square $L \times L$

and partition it into grids of linear length ϵ. There are a total of $N = (L/\epsilon)^2$ boxes. Then we have the total measure

$$M = N\mu = L^2 \epsilon^{\alpha-2}.$$

Now if we choose $\alpha = 1$, then $M \to \infty$ as $\epsilon \to 0$. This means that the length of a square is infinite, which makes sense. If we try $\alpha = 3$, then $M \to 0$ as $\epsilon \to 0$. This means that the volume of a square is 0, which is also correct. By this argument, it is clear that a finite value for M can be obtained only when $\alpha = 2$, the true dimension of the square.

With the above argument, we can define the Hausdorff-Besicovitch dimension. First, we need the α-covering measure, which is simply the summation of the total measure in all the "boxes" denoted by V_i, $i = 1, 2, \cdots$, subject to the condition that the union of the boxes covers the object E, while the size of the largest V_i is not greater than ϵ:

$$m^\alpha(E) = \lim_{\epsilon \to 0} \inf\{\sum (diamV_i)^\alpha : \ \cup V_i \supset E, \ diamV_i \leq \epsilon\}. \tag{5.1}$$

The notation inf indicates that this covering is minimal: without this requirement, there would be infinitely many coverings. With Eq. (5.1), $dimE$ can then be defined by

$$dimE = \inf\{\alpha : \ m^\alpha(E) = 0\} = \sup\{\alpha : \ m^\alpha(E) = \infty\}. \tag{5.2}$$

The Hausdorff-Besicovitch dimension is the value of α such that the measure jumps from zero to infinity. For $\alpha = dimE$, this measure can be any finite value.

5.2   GEOMETRICAL FRACTALS

In this section, we discuss two examples of geometrical fractals, the Cantor set and its variants and the Koch curves.

5.2.1   Cantor sets

The standard Cantor set is one of the prototypical fractal objects. It is obtained by first splitting a line segment into thirds and deleting the middle third. This step is then repeated, deleting the middle third of each remaining segment iteratively. See Fig. 5.2(a). As we shall see in Chapter 13 when discussing chaotic maps, such a process can be related to the iteration of a nonlinear discrete map. The removed middle thirds can be related to the intervals that make the map diverge to infinity. Therefore, the remaining structures are invariant points of the map. At the limiting stage, $n \to \infty$, the Cantor set consists of infinitely many isolated points, with topological dimension 0. At any stage n, one needs $N(\epsilon) = 2^n$ boxes of length $\epsilon = (\frac{1}{3})^n$ to cover the set. Hence the fractal dimension for the Cantor set is

$$D = -\ln N(\epsilon)/\ln \epsilon = \ln 2/\ln 3.$$

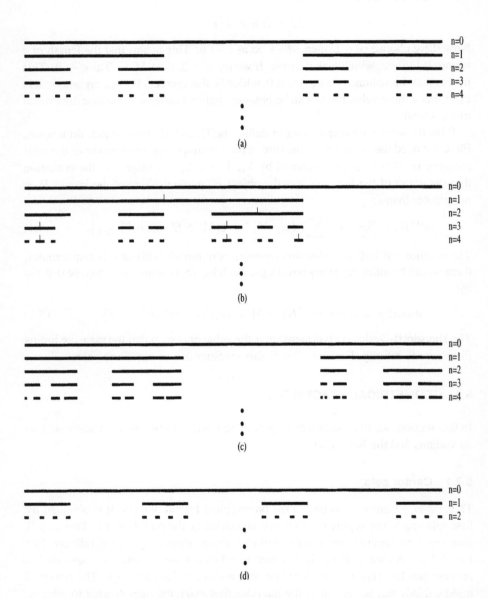

Figure 5.2. Examples of Cantor sets: (a) standard, (b) and (c) random, and (d) regular, with the same dimension as (a). See the text for details.

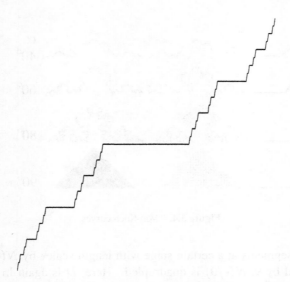

Figure 5.3. Devil's staircase constructed based on the sixth stage of the Cantor set.

Since $0 < D < 1$, the Cantor set is often called a dust. To better appreciate the self-similar feature of the Cantor set, one can construct a function, called a devil's staircase, which is simply the summation of the Cantor set — whenever there is a gap, the function remains constant. Figure 5.3 shows a devil's staircase based on the sixth stage of the Cantor set. The staircase is richest when the limiting Cantor set is used for the construction.

The fractal dimension of the Cantor set can also be computed by exploiting the self-similar feature. Denote the number of intervals needed to cover the Cantor set at a certain stage with scale ϵ by $N(\epsilon)$. When the scale is reduced by 3, $N(\epsilon/3)$ is doubled. Since $N(\epsilon/3)/N(\epsilon) = 3^D = 2$, one immediately gets $D = \ln 2 / \ln 3$.

A simple variation of the standard Cantor set is obtained by dividing each interval into three equal parts and deleting one of them at random (see Fig. 5.2(b)). Another random Cantor set is obtained by modifying the middle third of the Cantor set (Fig. 5.2(a)) so that a middle interval of random length (Fig 5.2(c)) is removed from each segment at each stage. When some regulations are imposed on the length distribution of these subintervals, the fractal dimension of these random Cantor sets can be readily computed. One way of imposing such regulations is to require that the ratio of the subinterval and its immediate parent interval follows some distribution that is stage-independent. Such a regulation is essentially a multifractal construction, as we shall see later.

The foregoing discussion suggests that two different geometrical fractals may have the same fractal dimension. To further appreciate this point, we have shown in Fig. 5.2(d) a different type of regular Cantor set. It is obtained by retaining four equally spaced segments whose length is 1/9th of the preceding segment. Denote

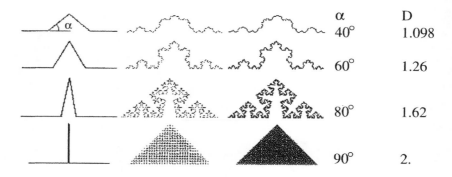

Figure 5.4. Von Koch curves

the number of segments at a certain stage with length scale ϵ by $N(\epsilon)$. When the scale is reduced by 9, $N(\epsilon/9)$ is quadrupled. Here, D is again $\ln 2/\ln 3$, since $N(\epsilon/9)/N(\epsilon) = 9^D = 4$.

5.2.2 Von Koch curves

The triadic Von Koch curve is constructed iteratively, just as the standard Cantor set is. One starts with an initiator, which is a straight line of unit length. One then superimposes an equilateral triangle on the middle third of the initiator. Finally, one removes the middle third. This completes one iteration. The resulting structure of four lines is called the generator. See the leftmost of the second row in Fig. 5.4. At the second iteration, each edge serves as an initiator and is replaced by the corresponding generator. Generators may be based on an equiangular triangle, with two edges being one third the length of the initiator and spanning an angle of α with the initiator. See the leftmost column of Fig. 5.4. The standard triadic Von Koch curve corresponds to $\alpha = 60^0$. Figure 5.4 shows the curves for four different α's of two fairly large stages. Such curves clearly are self-similar. What are their fractal dimensions?

Let us denote the length scale at iteration i by ϵ_i and the number of curves by $N(\epsilon_i)$. At iteration $i - 1$, the length is $\epsilon_{i-1} = 2\epsilon_i(1 + \cos \alpha)$, while $N(\epsilon_{i-1}) = N(\epsilon_i)/4$. Using Eq. (2.2), we immediately get

$$D = \frac{\ln 4}{\ln[2(1 + \cos \alpha)]}.$$

In particular, $D = \ln 4/\ln 3 = 1.2618\cdots$ for the triadic Von Koch curve, and $D = 2$ when $\alpha = 90°$. The latter implies that the resulting curve is plane-filling. A moment's thought should convince one that this is indeed the case, since the resulting curve densely covers the triangle shown in the rightmost column of the last row in Fig. 5.4.

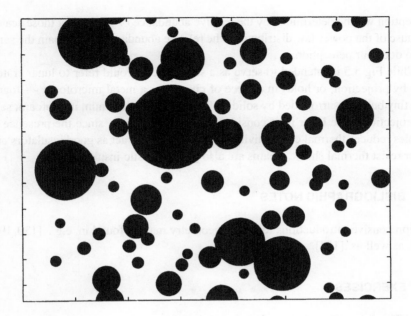

Figure 5.5. Random fractal of discs with a Pareto-distributed size: $P[X \geq x] = (1.8/x)^{1.8}$.

5.3 POWER LAW AND PERCEPTION OF SELF-SIMILARITY

The most salient property of fractal objects or phenomena is self-similarity. This is characterized by a power-law scaling relation in a wide range of temporal or spatial scales. Let us now examine how the power law is related to the perception of self-similarity.

Imagine a very large number of balls flying around in the sky. See Fig. 5.5. The size of the balls follows a power-law distribution,

$$p(r) \sim r^{-\alpha}.$$

Being human, we will instinctly focus on balls whose size is comfortable for our eyes — too small balls cannot be seen, while too large balls block our vision. Now let us assume that we are most comfortable with the scale r_0. Of course, our eyes are not sharp enough to tell the differences between scales r_0 and $r_0 + dr$, $|dr| \ll r_0$. Nevertheless, we are quite capable of identifying scales such as $2r_0$, $r_0/2$, etc. Which aspect of the flying balls may determine our perception? This is essentially given by the relevant abundance of the balls of sizes $2r_0$, r_0, and $r_0/2$:

$$p(2r_0)/p(r_0) = p(r_0)/p(r_0/2) = 2^{-\alpha}.$$

Note that the above ratio is independent of r_0. Now suppose we view the balls through a microscope, which magnifies all the balls by a scale of 100. Now our eyes will be focusing on scales such as $2r_0/100$, $r_0/100$, and $r_0/200$, and our perception will be determined by the relative abundance of the balls at those scales. Because of the power-law distribution, the relative abundance will remain the same — so does our perception.

While Fig 5.5 is intended to serve as a schematic, it could refer to lunar craters seen by projection, or holes in a piece of cheese, or a metal microfoam — foams with tiny bubbles surrounded by solid walls, such as an aluminum microfoam, seen by projection. The latter is of considerable current interest, since the presence of bubbles reduces thermal conductivity, allowing foams to act as good insulators and better resist thermal shock. Foams are also good acoustic insulators.

5.4 BIBLIOGRAPHIC NOTES

Comprehensive introductions to fractal geometry may be found in, e.g., [130, 198, 294], as well as [124].

5.5 EXERCISES

1. The dataset "crack.dat", downloadable at
   http://www.gao.ece.ufl.edu/GCTH_Wileybook/, describes a certain crack in a material. You could think of it as a crack on the ground or in a wall. Write a code (using whatever language you prefer) to compute $N(\epsilon)$ for different ϵ, plot $N(\epsilon)$ vs. ϵ on a log-log scale, and find the fractal dimension for this crack. How well is the scaling defined here?

2. Imagine that you are observing plants, breaking ocean waves, mountain ranges, ink stains, wrinkled or torn paper, etc. Discuss and write down qualitative observations about one or more of these objects. You might consider the following:
   How does the object fill space?
   Is its use of space dense or sparse?
   Are its edges smooth or jagged?
   What is similar throughout different parts of the object?
   What is random or different throughout different parts of the object?
   How does the object as a whole compare with its individual parts?
   What geometric shapes do you see in the object? Are the shapes similar to circles, lines, ovals, spheres?

3. Design a probability problem and associate each term on the left-hand side of the following equation with a probability:

$$\frac{1}{3} + 2 \cdot \left(\frac{1}{3}\right)^2 + 4 \cdot \left(\frac{1}{3}\right)^3 + \cdots = 1.$$

This is an excellent way of proving that the summation is 1.

4. Design another Cantor set with dimension $\ln 2/\ln 3$ (one possibility is to break each piece in stage i into eight pieces in stage $i + 1$).

5. Reproduce Fig. 5.5. (Hint: the centers of the circles could be chosen to be the regular grid points perturbed by Gaussian random variables; the radii of the circles follow a power law. Alternatively, you could use a Poisson process to go from the center of one circle to the center of another. Although seemingly simple, a similar and popular model has been developed for network traffic: messages arrive according to a Poisson process, and the size of the messages follows a power law.)

CHAPTER 6

SELF-SIMILAR STOCHASTIC PROCESSES

In Chapter 5, we discussed the standard Cantor set and briefly touched on its randomized variants. In real-world signal processing applications, random fractals are far more useful. In this chapter, we discuss one of the prototypical random fractal models, the fractional Brownian motion (fBm) model. Much will be said about its simulation. In Chapter 7, we shall discuss a different type of random fractal, the Levy motions.

6.1 GENERAL DEFINITION

A continuous-time stochastic process, $X = \{X(t), t \geq 0\}$, is said to be self-similar if

$$X(\lambda t) \overset{d}{=} \lambda^H X(t), \ \ t \geq 0 \tag{6.1}$$

for $\lambda > 0, 0 < H < 1$, where $\overset{d}{=}$ denotes equality in distribution. H is called the self-similarity parameter or the Hurst parameter.

Before proceeding, we note that, more rigorously speaking, processes defined by Eq. (6.1) should be called self-affine processes instead of self-similar processes, since X and t have to be scaled differently to make the function look similar. A simple physical explanation for this is that the units for X and t are different. This distinction is particularly important if one wishes to discuss the fractal dimensions of such processes, as we will see in Sec. 6.4. In general, however, we will continue to call such processes self-similar in order to follow the convention used in certain engineering disciplines, such as network traffic engineering.

The following three properties can be easily derived from Eq. (6.1):

$$E[X(t)] = \frac{E[X(\lambda t)]}{\lambda^H} \quad \text{Mean,} \tag{6.2}$$

$$Var[X(t)] = \frac{Var[X(\lambda t)]}{\lambda^{2H}} \quad \text{Variance,} \tag{6.3}$$

$$R_x(t,s) = \frac{R_x(\lambda t, \lambda s)}{\lambda^{2H}} \quad \text{Autocorrelation.} \tag{6.4}$$

Note that $Var[X(t)] = R_x(t,t)$; hence, Eq. (6.3) can be simply derived from Eq. (6.4); we have listed it as a separate equation for convenience of future reference. If we consider only second-order statistics, or if a process is Gaussian, Eqs. (6.2)–(6.4) can be used instead of Eq. (6.1) to define a self-similar process.

A very useful way of describing a self-similar process is by its spectral representation. Strictly speaking, the Fourier transform of $X(t)$ is undefined due to the nonstationary nature of $X(t)$. One can, however, consider $X(t)$ in a finite interval, say, $0 < t < T$:

$$X(t,T) = \begin{cases} X(t) & 0 < t < T \\ 0 & \text{otherwise} \end{cases} \tag{6.5}$$

and take the Fourier transform of $X(t,T)$:

$$F(f,T) = \int_0^T X(t)e^{-2\pi jft}dt \tag{6.6}$$

$|F(f,T)|^2 df$ is the contribution to the total energy of $X(t,T)$ from those components with frequencies between f and $f + df$. The (average) power spectral density (PSD) of $X(t,T)$ is then

$$S(f,T) = \frac{1}{T}\left|F(f,T)\right|^2,$$

and the spectral density of X is obtained in the limit as $T \to \infty$

$$S(f) = \lim_{T \to \infty} S(f,T).$$

Noting that the PSD for $\lambda^{-H}X(\lambda t)$ is

$$\lambda^{-2H} \lim_{T \to \infty} \frac{1}{\lambda T}\left|\int_0^{\lambda T} X(\lambda t)e^{-2\pi jft}dt\right|^2 = \lambda^{-2H-1}S(f/\lambda),$$

and that the PSD for $\lambda^{-H}X(\lambda t)$ and $X(t)$ are the same, we have

$$S(f) = S(f/\lambda)\lambda^{-2H-1}.$$

The solution to the above equation is

$$S(f) \sim f^{-\beta} \tag{6.7}$$

with

$$\beta = 2H + 1. \tag{6.8}$$

Processes with PSDs as described by Eq. (6.7) are called $\frac{1}{f}$ processes. Typically, the power-law relationships that define these processes extend over several decades of frequency. Such processes have been found in numerous areas of science and engineering. See the partial list of relevant literature in the Bibliographic Notes at the end of the chapter. The ubiquity of $\frac{1}{f}$ processes is one of the most stimulating puzzles in physics, astrophysics, geophysics, engineering, biology, and the social sciences, and many efforts have been made to discover the mechanisms for generating such processes. However, none of the mechanisms proposed so far has been considered universal. We shall return to this issue in Chapter 11.

6.2 BROWNIAN MOTION (BM)

As an example, let us consider the $\frac{1}{f}$ process known as Brownian motion (Bm). It is defined as a (nonstationary) stochastic process $B(t)$ that satisfies the following criteria:

1. All Brownian paths start at the origin: $B(0) = 0$.

2. For $0 < t_1 < t_2 < t_3 < t_4$, the random variables $B(t_2) - B(t_1)$ and $B(t_4) - B(t_3)$ are independent.

3. For all $(s, t) \geq 0$, the variable $B(t + s) - B(t)$ is a Gaussian variable with mean 0 and variance s.

4. $B(t)$ is a continuous function of t.

We may infer from the above definition that the probability distribution function of B is given by

$$P\{[B(t + s) - B(t)] \leq x\} = \frac{1}{\sqrt{2\pi s}} \int_{-\infty}^{x} e^{-\frac{u^2}{2s}} du. \tag{6.9}$$

This function also satisfies the scaling property:

$$P\{[B(t + s) - B(t)] \leq x\} = P\{B(\lambda t + \lambda s) - B(\lambda t) \leq \lambda^{1/2}x\}. \tag{6.10}$$

In other words, $B(t)$ and $\lambda^{-1/2}B(\lambda t)$ have the same distribution. *Thus, we see* from Eq. (6.1) that Bm is a self-similar process with Hurst parameter 1/2.

Suppose we have measured $B(t_1)$, $B(t_2)$, $t_2 - t_1 > 0$. What can we say about $B(s)$, $t_1 < s < t_2$? The answer is given by the Levy interpolation formula:

$$B(s) = B(t_1) + \frac{(s - t_1)}{(t_2 - t_1)}[B(t_2) - B(t_1)] + \left[\frac{(t_2 - s)(s - t_1)}{t_2 - t_1}\right]^{1/2}W, \quad (6.11)$$

where W is a zero-mean and unit-variance Gaussian random variable. The first two terms are simply a linear interpolation. The third term gives the correct variances for $B(s) - B(t_1)$ and $B(t_2) - B(s)$, which are $s - t_1$ and $t_2 - s$, respectively.

A popular way of generating Bm is by using the following random midpoint displacement method, which is essentially an application of the Levy interpolation formula. Suppose we are given $B(0) = 0$ and $B(1)$ as a sample of a Gaussian random variable with zero mean and variance σ^2. We can obtain $B(1/2)$ from the following formula:

$$B(1/2) - B(0) = \frac{1}{2}[B(1) - B(0)] + D_1,$$

where D_1 is a random variable with mean 0 and variance $2^{-2}\sigma^2$. D_1 is simply the third term of Eq. (6.11), and the coefficient 2^{-2} equals $(t_2 - s)(s - t_1)/(t_2 - t_1)$, with $t_1 = 0$, $t_2 = 1$, $s = 1/2$. Similarly, we have

$$B(1/4) - B(0) = \frac{1}{2}[B(1/2) - B(0)] + D_2,$$

where D_2 is a random variable with zero mean and variance $2^{-3}\sigma^2$. We can apply the same idea to obtain $B(3/4)$. In general, the variance for D_n is $2^{-(n+1)}\sigma^2$.

We note that a Brownian path is not a differentiable function of time. Heuristically, this can be readily understood: consider the variable $B(t + s) - B(t)$ with variance s. Its standard deviation, which is a measure of its order of magnitude, is $\sim \sqrt{s}$. Thus, the derivative of B at t behaves like the limit $\sqrt{s}/s = s^{-1/2}$ as $s \to 0$.

Although $B(t)$ is almost surely not differentiable in t, symbolically one still often writes

$$B(t) = \int_0^t w(\tau)d\tau, \quad (6.12)$$

where $w(t)$ is stationary white Gaussian noise, and extends the above equation to $t < 0$ through the convention

$$\int_0^{-t} \equiv -\int_t^0.$$

It should be understood that integrals with respect to the differential element $w(t)dt$ should be interpreted more precisely as integrals with respect to the differential element $dB(t)$ in the Riemann-Stieltjes integral sense. This has profound consequences when numerically integrating Eq. (6.12).

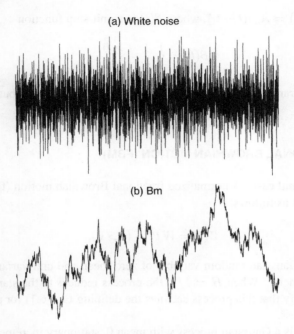

Figure 6.1.   White noise (a) and Bm (b). Axes are arbitrary.

To see how this works, let us partition $[0, t]$ into n equally spaced intervals, $\Delta t = t/n$. Since $w(t)$ are Gaussian random variables with zero mean and unit variance, in order for $B(t)$ to have variance t, we should have

$$B(t) \approx \sum_{i=1}^{n} w(i\Delta t) \cdot (\Delta t)^{1/2}. \qquad (6.13)$$

That is, the coefficient is $(\Delta t)^{1/2}$ instead of Δt, as one might have guessed. Typically, $\Delta t \ll 1$; hence, $(\Delta t)^{1/2} \gg \Delta t$. Thus, if one incorrectly uses Δt instead of $(\Delta t)^{1/2}$, one is severely underestimating the noise term.

When the time is genuinely discrete, or when the units of time can be arbitrary, one can take Δt to be 1 unit, and Eq. (6.13) becomes

$$B(t) = \sum_{i=1}^{n} w(i). \qquad (6.14)$$

Equation (6.14) provides perhaps the simplest method of generating a sample of Bm. An example is shown in Fig. 6.1.

Equation (6.14) is also known as a random walk process. A more sophisticated random walk (or jump) process involves summing up an infinite number of jump functions: $J_i(t) = A_i\beta(t - t_i)$, where $\beta(t)$ is a unit-step function

$$\beta(t) = \begin{cases} 1 & t \geq 0 \\ 0 & t < 0 \end{cases} \tag{6.15}$$

and A_i, t_i are random variables with Gaussian and Poisson distributions, respectively.

6.3 FRACTIONAL BROWNIAN MOTION (FBM)

One-dimensional case: A normalized fractional Brownian motion (fBm) process, $Z(t)$, is defined as follows:

$$Z(t) \overset{d}{=} W t^H \quad (t > 0), \tag{6.16}$$

where W is a Gaussian random variable of zero mean and unit variance and H is the Hurst parameter. When $H = 1/2$, the process reduces to the standard Bm. It is trivial to verify that this process satisfies the defining Eq. (6.1) for a self-similar stochastic process.

FBm $B_H(t)$ is a Gaussian process with mean 0, stationary increments, variance

$$E[(B_H(t))^2] = t^{2H}, \tag{6.17}$$

and covariance

$$E[B_H(s)B_H(t)] = \frac{1}{2}\left\{ s^{2H} + t^{2H} - |s - t|^{2H} \right\}, \tag{6.18}$$

where H is the Hurst parameter. Due to its Gaussian nature, according to Eqs. (6.2)–(6.4), the above three properties completely determine its self-similar character. Figure 6.2 shows several fBm processes with different H.

Roughly, the distribution for an fBm process is

$$B_H(t) = \frac{1}{\Gamma(H + 1/2)} \int_{-\infty}^{t} (t - \tau)^{H-1/2} dB(\tau). \tag{6.19}$$

It is easy to verify that fBm satisfies the following scaling property:

$$P(B_H(t + s) - B_H(t) \leq x) = P\{B_H(\lambda t + \lambda s) - B_H(\lambda t) \leq \lambda^H x\}. \tag{6.20}$$

In other words, this process is invariant for transformations conserving the similarity variable x/t^H.

The integral defined by Eq. (6.19) diverges. The more precise definition is

$$B_H(t) - B_H(0) = \frac{1}{\Gamma(H + 1/2)} \int_{-\infty}^{t} K(t - \tau) dB(\tau), \tag{6.21}$$

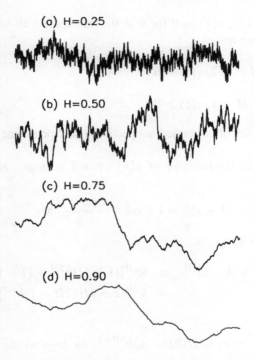

(a) H=0.25

(b) H=0.50

(c) H=0.75

(d) H=0.90

Figure 6.2.   Several fBm processes with different H.

where the kernel $K(t - \tau)$ is given by

$$K(t - \tau) = \begin{cases} (t - \tau)^{H-1/2}, & 0 \le \tau \le t \\ (t - \tau)^{H-1/2} - (-\tau)^{H-1/2}, & \tau < 0. \end{cases} \quad (6.22)$$

The increment process of fBm, $X_i = B_H((i+1)\Delta t) - B_H(i\Delta t)$, $i \ge 1$, where Δt can be considered a sampling time, is called the fractional Gaussian noise (fGn) process. It is a zero mean stationary Gaussian time series. Noting that

$$E(X_i X_{i+k}) = E\{[B_H((i+1)\Delta t) - B_H(i\Delta t)][B_H((i+1+k)\Delta t) - B_H((i+k)\Delta t)]\},$$

by Eq. (6.18), one obtains the autocovariance function $\gamma(k)$ for the fGn process:

$$\gamma(k) = E(X_i X_{i+k})/E(X_i^2) = \frac{1}{2}\left\{(k+1)^{2H} - 2k^{2H} + |k-1|^{2H}\right\}, \ k \ge 0. \quad (6.23)$$

Notice that the expression is independent of Δt. Therefore, without loss of generality, one could take $\Delta t = 1$. In particular, we have

$$\gamma(1) = \frac{1}{2}\left(2^{2H} - 2\right).$$

Let us first note a few interesting properties of $\gamma(k)$:

1. When $H = 1/2$, $\gamma(k) = 0$ for $k \neq 0$. This is the well-known property of white Gaussian noise.

2. When $0 < H < 1/2$, $\gamma(1) < 0$.

3. When $1/2 < H < 1$, $\gamma(1) > 0$.

Properties 2 and 3 are often termed antipersistent and persistent correlations, respectively.

Next, we consider the behavior of $\gamma(k)$ when k is large. Noting that when $|x| \ll 1$,

$$(1 + x)^\alpha \approx 1 + \alpha x + \frac{\alpha(\alpha - 1)}{2} x^2,$$

we have, when $k \gg 1$,

$$
\begin{aligned}
(k + 1)^{2H} + |k - 1|^{2H} &= k^{2H}[(1 + 1/k)^{2H} + (1 - 1/k)^{2H}] \\
&= k^{2H}[2 + 2H(2H - 1)k^{-2}].
\end{aligned}
$$

Hence,

$$\gamma(k) \sim H(2H - 1)k^{2H-2} \quad \text{as } k \to \infty \tag{6.24}$$

when $H \neq 1/2$. When $H = 1/2$, $\gamma(k) = 0$ for $k \geq 1$, the X_i's are simply white noise.

High-dimensional case: Similar to the one-dimensional case, a fBm in a n-dimensional Euclidean space, $B_H(x_1, x_2, \cdots, x_n)$, is a scalar function of $(x_1, x_2, \cdots, x_n)$. It can be defined by the following property:

$$
\begin{aligned}
&\left\langle [B_H(x'_1, x'_2, \cdots, x'_n) - B_H(x_1, x_2, \cdots, x_n)]^2 \right\rangle \\
&\propto \left[(x'_1 - x_1)^2 + (x'_2 - x_2)^2 + \cdots + (x'_n - x_n)^2 \right]^H.
\end{aligned}
\tag{6.25}
$$

The process B_H is Gaussian:

$$
\begin{aligned}
&Pr\{B_H(x'_1, x'_2, \cdots, x'_n) - B_H(x_1, x_2, \cdots, x_n) < \bar{\zeta}\} \\
&= \frac{1}{\sqrt{2\pi}\sigma r^H} \int_{-\infty}^{\bar{\zeta}} \exp\left(\frac{\zeta^2}{2\sigma^2 r^{2H}} \right) d\zeta,
\end{aligned}
\tag{6.26}
$$

where

$$r = \sqrt{(x'_1 - x_1)^2 + (x'_2 - x_2)^2 + \cdots + (x'_n - x_n)^2},$$

H is the Hurst parameter, and σ^2 measures the variance of the process in unit "distance."

6.4 DIMENSIONS OF BM AND FBM PROCESSES

Due to their self-affine nature, Bm and fBm have several different dimensions. We first discuss their box-counting dimension.

Suppose we only consider a Bm or fBm process in the unit interval [0,1]. We partition the unit interval into N bins, the length of each bin being $\epsilon = \Delta t = 1/N$. Since $B_H(\Delta t) \propto \Delta t^H > \Delta t$, when $0 < H < 1$, for each bin, we will need $B_H(\Delta t)/\Delta t = \Delta t^{H-1}$ square boxes to cover the function. Overall, we need $B_H(\Delta t)/\Delta t \cdot N = \Delta t^{H-2} = \epsilon^{-(2-H)}$ boxes to cover the function. By Eq. (2.2), we find that the box-counting dimension for fBm is $D = 2 - H$. For Bm, $H = 0.5$, $D = 1.5$. The dimension $D = 2 - H$ is sometimes called the graph dimension of the random function. The smaller the Hurst parameter is, the larger the dimension of the process. This can be easily appreciated from Fig. 6.2.

The above argument can be easily extended to the n-dimensional case. To see how, let us replace Δt by Δr. We have $\epsilon = \Delta r = 1/N$, and $B_H(\Delta r) \propto \Delta r^H > \Delta r$. Now we need $B_H(\Delta r)/\Delta r \cdot N^n = \Delta r^{H-1-n}$ hypercubes of linear scale Δr to cover B_H. Therefore,

$$D = n + 1 - H.$$

Next, we consider the dimension obtained by measuring the length of the curve defined by Bm or fBm using a sequence of rulers. Consider a total time span of T_{total}, and partition it into equal time steps Δt. Then the length of the curve within each interval Δt is

$$\Delta l = \sqrt{\Delta t^2 + \Delta B_H^2} \propto \Delta t \sqrt{1 + \Delta t^{2H-2}}.$$

This value varies for different time intervals, but we assume that, for each Δt, it is bounded. When $\Delta t \ll 1$, $\Delta t^{2H-2} \gg 1$; hence, $\Delta l \propto \Delta t^H$. When $\Delta t \gg 1$, $\Delta l \propto \Delta t$. The number $N(\Delta l)$ of rulers of size Δl, which is $1/\Delta t$, is expected to vary as Δl^{-D}. Hence, for $\Delta t \gg 1$, $D = 1$, while for $\Delta t \ll 1$, $D = 1/H$. The latter is called the latent dimension.

We should point out that the above argument is also valid for an n-dimensional fBm process. Therefore, so far as the length of the fBm process is concerned, the latent dimension is always $D = 1/H$, regardless of the dimension of the space where fBm is defined.

To summarize, the same $B_H(t)$ can have an apparently self-similar dimension D of either 1, $1/H$, or $2 - H$, depending on the measurement technique and the choice of length scale. By now, the distinction between self-similarity and self-affinity should be clearer; self-affinity means that $B_H(t)$ and t can be scaled differently because they have different units.

One can obtain a number of interesting self-similar processes from a self-affine process. One simple process involves the so-called zerosets, defined by the time instants when $B_H(t) = 0$. A generalization of zerosets is the levelsets, obtained by the crossings of $B_H(t)$ with an arbitrary constant line, $B_H(t) = c$. Levelsets have a fractal dimension of $1 - H$, which is simply one dimension lower than the box-counting dimension of $B_H(t)$. The gaps between successive points in a levelset have a power-law distribution, $P(r) \sim r^{-H}$, where r denotes the size of the gaps.

Another type of simple self-similar process can be obtained by using a number of independent fBm processes to form a vector process,

$$\mathbf{V}(t) = [B_H^{(1)}(t), B_H^{(2)}(t), ..., B_H^{(m)}(t)].$$

Now the units for all the components are the same. This is called an fBm trail or trajectory. Its fractal dimension is $1/H$, corresponding to the $\Delta t \ll 1$ case for the compass dimension (Now $\Delta t \gg 1$ case is no longer relevant.) $1/H$ is also its box-counting dimension if one realizes that the boxes used to cover the trace have to be square boxes with linear length $\Delta l (\propto \Delta t^H)$ instead of Δt.

The box-counting dimension for an fBm trail can be easily computed by the correlation dimension, originally developed by Grassberger and Procaccia for the characterization of chaotic systems. The algorithm works as follows. First, one samples $\mathbf{V}(t)$ to obtain $\mathbf{V}(1), \mathbf{V}(2), \cdots, \mathbf{V}(N)$. Then one computes the correlation integral,

$$C(N, r) = \frac{2}{N(N-1)} \sum_{i,j=1}^{N} \theta(r - \|\mathbf{V}(i) - \mathbf{V}(j)\|), \tag{6.27}$$

where θ is the Heaviside step function, N is the total number of points in the time series, and r is a prescribed small distance. The correlation dimension μ of the time series is then given by

$$C(N, r) \sim r^\mu, \quad r \to 0. \tag{6.28}$$

While in general the correlation dimension cannot be greater than the box-counting dimension, here it is the same as the box-counting dimension, $1/H$, since fBm is a monofractal.

In the chaos research community, for quite some time, it was thought that an estimated positive Lyapunov exponent or entropy, or a finite nonintegral correlation dimension would suffice to indicate that the time series under study is deterministically chaotic. $1/f$ processes, especially fBm processes, have been key counterexamples invalidating that assumption. Very recently, it has been clarified that the dimension estimated by the Grassberger and Procaccia algorithm corresponds to the box-counting dimension of the sojourn points in a neighborhood of an arbitrary point. In light of that result, we see that while the danger of misinterpreting a $1/f$ process to be deterministically chaotic can be readily avoided, one can in fact use the Grassberger and Procaccia algorithm to conveniently estimate the key parameter for the $1/f$ process.

The possibility of misinterpreting $1/f$ processes as being deterministic chaos never occurs if one works within the framework of power-law sensitivity to initial conditions that will be developed in Chapter 14. In that framework, we monitor the evolution of two nearby trajectories. If these trajectories diverge exponentially fast, then the time series is chaotic; if the divergence increases in a power-law manner, then the trajectories belong to $1/f$ processes. These ideas can be easily expressed precisely. Let $||\mathbf{V}(s_1) - \mathbf{V}(s_2)||$ be the initial separation between two trajectories. This separation is assumed to be not larger than some small prescribed distance r. After time t, the distance between the two trajectories will be $||\mathbf{V}(s_1 + t) - \mathbf{V}(s_2 + t)||$. For truly chaotic systems,

$$||\mathbf{V}(s_1 + t) - \mathbf{V}(s_2 + t)|| \propto ||\mathbf{V}(s_1) - \mathbf{V}(s_2)||e^{\lambda t},$$

where $\lambda > 0$ is called the largest Lyapunov exponent. On the other hand, for $1/f$ processes,

$$||\mathbf{V}(s_1 + t) - \mathbf{V}(s_2 + t)|| \propto ||\mathbf{V}(s_1) - \mathbf{V}(s_2)||t^H.$$

The latter equation thus provides another simple means of estimating the Hurst parameter.

In practice, one often only has a scalar time series, possibly belonging to the family of $1/f$ processes. How may one get a vector time series so that one can use chaos theory to study the scalar time series? There is a very powerful tool called the time delay embedding technique, also developed for the study of chaotic systems, which is very easy to use. It will be discussed in detail in Chapter 13.

6.5 WAVELET REPRESENTATION OF FBM PROCESSES

In Chapter 4, Sec. 2, we introduced wavelet MRA. Using the scaling function and the wavelet function, fBm can be expressed as

$$B_H(t) = \sum_{k=-\infty}^{\infty} \phi_H(t - k)S_k^{(H)} + \sum_{j=0}^{\infty} \sum_{k=-\infty}^{\infty} 2^{-jH}\psi_H(2^j t - k)\epsilon_{j,k} - b_0, \quad (6.29)$$

where ϕ_H and ψ_H are defined through their Fourier transform

$$\hat{\phi}_H(x) = \left(\frac{1 - e^{-ix}}{ix}\right)^{H+1/2} \hat{\phi}(x),$$

$$\hat{\psi}_H(x) = (ix)^{-(H+1/2)}\hat{\psi}(x),$$

where $\hat{\phi}(x)$ and $\hat{\psi}(x)$ are the Fourier transforms of a chosen scaling function $\phi(x)$ and a wavelet $\psi(x)$, respectively. $S_k^{(H)}$, $k \in Z$, is a partial sum process of a $FARIMA(0, H - 1/2, 0)$ sequence, $\epsilon_{j,k}, j \geq 0, k \in Z$, are independent Gaussian $N(0, 1)$ random variables, and b_0 is a random constant such that $B_H(0) = 0$.

The term $\psi_H(2^j t - k)$ in Eq. (6.29) seems to contradict the term $\psi_{j,k}(t) = 2^{-j/2}\psi_0(2^{-j}t - k)$ in Eq. (4.21), given the same j. To understand this, we note that in practical implementation of wavelet decomposition, one usually chooses a maximum scale resolution level J for analysis. If one redefines $j^* = J - j$, then Eqs. (6.29) and (4.21) become consistent. The transformation $j^* = J - j$ amounts to viewing the wavelet multiresolution decomposition by two different ways: bottom-up or top-down.

The summands in the second term of Eq. (6.29) can be expressed as $d(j, k)\tilde{\psi}_{j,k}$, where

$$d(j, k) = 2^{-j(H+1/2)}\epsilon_{j,k} = 2^{j^*(H+1/2)}2^{-J(H+1/2)}\epsilon_{j,k} \qquad (6.30)$$

and

$$\tilde{\psi}_{j,k} = 2^{j/2}\psi_H(2^j t - k). \qquad (6.31)$$

For a specific scale j^*, let us denote the variance for the coefficients $d(j, k)$ by Γ_{j^*}. Then by Eq. (6.30), we have

$$\log_2 \Gamma_{j^*} = (2H + 1)j^* + c_0, \qquad (6.32)$$

where c_0 is some constant.

6.6 SYNTHESIS OF FBM PROCESSES

In the past several decades, a number of algorithms have been developed for synthesizing the fBm processes. We will describe some of them here, based on the consideration that they are either computationally very fast or will deepen our understanding of the fBm processes.

Random midpoint displacement (RMD) method. This method is similar to that for a Bm process. Suppose we are given $B_H(0) = 0$ and $B_H(1)$ as a sample of a Gaussian random variable with mean 0 and variance σ^2. To obtain $B_H(1/2)$, we also write

$$B_H(1/2) - B_H(0) = \frac{1}{2}[B_H(1) - B_H(0)] + D_1.$$

In order to have

$$Var(B_H(1/2) - B_H(0)) = Var(B_H(1) - B_H(1/2)) = (1/2)^{2H}\sigma^2,$$

D_1 must be a Gaussian random variable with zero mean and variance

$$(1/2)^{2H}\sigma^2(1 - 2^{2H-2}).$$

Applying this idea recursively, we find that D_n must have the variances

$$\Delta_n^2 = \sigma^2 \left(\frac{1}{2}\right)^{2Hn}(1 - 2^{2H-2}). \qquad (6.33)$$

It turns out that the method does not yield the true fBm when $H \neq 1/2$. One can easily verify that $Var(B_H(3/4) - B_H(1/4)) \neq (1/2)^{2H}\sigma^2$. Hence, the process does not have stationary increments. This defect causes the graphs of $B_H(t)$ to show some visible traces of the first few stages in the recursion, especially when H is close to 1. Visually, such features are not very pleasant.

Successive random addition (SRA) method. To overcome the shortcomings of the RMD method, Voss developed the SRA method. It has become quite popular due to its speed, efficiency, and flexibility in generating various random fractal surfaces and processes. The idea is to add a random variable to every point instead of only to the middle points at any stage. More concretely, the method works as follows. Start at $t_i = 0$, $1/2$, 1. First, let $B_H(t_i) = 0$ be the initial condition. Now add a Gaussian random variable of zero mean and unit variance, $\sigma_1^2 = 1$, to all three points. We refer to these preparations as stage 1. At the next stage, we consider $t_i = 0$, $1/4$, $1/2$, $3/4$, 1. $B_H(1/4)$ and $B_H(3/4)$ are first approximated by linear interpolations from $B_H(t_i)$, $t_i = 0$, $1/2$, 1. Then a Gaussian random variable of zero mean and variance $\sigma_2^2 = (1/2)^{2H}\sigma_1^2$ is added to all five points. This is stage 2. At stage 3, the midpoints of these five points are again obtained by interpolation. Then Gaussian random variables of zero mean and variance $\sigma_3^2 = (1/2)^{2H}\sigma_2^2$ are added to all nine points. The method is applied recursively. At stage n, we have $\sigma_n^2 = (1/2)^{2H}\sigma_{n-1}^2 = (1/2)^{2H(n-1)}\sigma_1^2$.

It should be noted however, that the random processes or fields simulated by SRA are not truly statistically homogeneous. To overcome the difficulty, Elliott et al. and Majda et al. have recently introduced hierarchical Monte Carlo methods, which are capable of generating self-similarity over 12 decades of scales with about 100 realizations.

Fast Fourier transform (FFT) filtering. This is a simple and efficient method. By Eqs. (6.7) and (6.8), we have, for an fBm process, $S(f) \propto f^{-(2H+1)}$. One thus can proceed as follows: Start with a "white noise" sequence $W(t)$, whose spectral density is a constant, and filter the sequence with a transfer function $T(f) \propto f^{-(H+1/2)}$; then the output, $B_H(t)$, has the desired spectral density, $S_B(f) \propto |T(f)|^2 S_W(f) \propto f^{-(2H+1)}$. In computer simulations, we want to obtain a discrete sequence, B_n, defined at discrete times $t_n = n\Delta t$, $n = 0, 1, \cdots, N-1$. We can write:

$$B_n = \sum_{m=0}^{N-1} v_m e^{2\pi i f_m t_n}, \tag{6.34}$$

where

$$f_m = \frac{m}{N\Delta t} \text{ for } m = 0 \text{ to } N-1,$$

and

$$< |v_m|^2 > \propto f^{-(2H+1)} \propto m^{-(2H+1)}. \tag{6.35}$$

There are several ways to obtain v_m: (1) multiply the Fourier coefficients of a white noise sequence by $f^{-(H+1/2)}$; (2) directly choose complex random variables with mean square amplitude given by Eq. (6.35) and random phases; (3) simply set

$$|v_m|^2 = m^{-(2H+1)}$$

and randomize phases. The third method is perhaps the most popular.

Since the FFT-based algorithm is periodic in time, $B_n = B_{n+N}$, at most half of the sequence is usable. Typically, one generates a long sequence and only retains a portion with length $N/4$ to $N/2$. Another drawback of the above algorithms is that the autocorrelation function for short time lags will not exactly match that for the fBm.

Weierstrass-Mandelbrot function–based method. The Fourier series of Eq. (6.34) involves a linear progression of frequencies. There is an interesting function, called the Weierstrass-Mandelbrot function,

$$f(t) = \sum_{k=1}^{\infty} r^{(2-D)k} \sin(2\pi r^{-k}t), \tag{6.36}$$

where D is the box-counting dimension of the curve, which involves a geometric progression of frequencies. This function is continuous but nowhere differentiable. Since $H + D = 2$, we find that the amplitude for the frequency $f = r^{-k}$ is f^{-H}. The square amplitude is thus $\propto f^{-2H}$. In a bandwidth of $\Delta f \propto f$, we thus have the spectral density $S(f) \propto amplitude^2/\Delta f \propto f^{-(2H+1)}$. In order to build an fBm generator based on the Weierstrass-Mandelbrot function, one only needs to randomize the function in a certain fashion. A general form was given by Mandelbrot:

$$V_{WM}(t) = \sum_{k=-\infty}^{\infty} A_n r^{kH} \sin(2\pi r^{-k}t + \phi_k), \tag{6.37}$$

where A_n is a Gaussian random variable with the same variance for all n and ϕ_k is a random phase uniformly distributed on $[0, 2\pi]$. In numerical simulations, after one specifies the low and high cutoff frequencies, the series becomes finite. To speed up computation, one can simply set $\phi_k = 0$.

Inverse Fourier transform method. In applications where the exact autocovariance function of fBm or fGn (Eq. (6.18) or (6.23)) is required, one may use the FFT-based approach proposed by Yin. This method involves first computing an fGn process, and then obtaining an fBm process by summing the fGn. The fGn process is obtained by noting that its autocovariance function and spectral density form a Fourier transform pair. Since the former is known, its spectral density can be exactly computed. Generating a sequence from a known spectral density is then an easy matter.

Wavelet-based method. Starting from Eq. (6.29), Abry and Sellan proposed a fast algorithm to synthesize an fBm process. It works as follows. First, one generates a Gaussian $FARIMA(0, s, 0)$ sequence of finite length as an approximation of fBm at some coarse scale, where $s = H + 1/2$. Then one applies, recursively, a fast wavelet transform, which involves two filters, one called a fractional low-pass filter, and the other a fractional high-pass filter, to obtain a much longer $FARIMA(0, s, 0)$ sequence. Finally, the $FARIMA(0, s, 0)$ sequence is properly normalized and taken for an approximation of fBm at some finer scale. Abry and Sellan have made their implementation available on the World Wide Web.

6.7 APPLICATIONS

In recent years, the fBm and related models have found numerous applications. In this section, we discuss network traffic modeling by fBm in some depth and then briefly describe modeling of rough surfaces by two-dimensional fBm or related processes.

6.7.1 Network traffic modeling

Network traffic modeling is an important yet challenging problem. Traditionally, traffic is modeled as a Poisson or Markovian process. A significant discovery in the early 1990s was that network traffic often has very long temporal correlations. In this section, we briefly describe how network traffic is specified; quantitatively show how miserably the Poisson traffic model fails to describe the measured traffic trace; and finally show that the fBm model is much more successful than the Poisson process model for network traffic.

6.7.1.1 Description of traffic patterns Network traffic is often measured by collecting interarrival times, $\{T_j\}$, where T_j denotes the jth interarrival time between two successive packet arrivals, and the packet-length sequence, $\{B_j\}$, where B_j represents the length of the jth packet. This is a point-process description. See Fig. 6.3(a). Sometimes one may only be concerned with the number of packets. Then one could set $B_j = b$, where b is the average size of the packets. This is shown in Fig. 6.3(b). The Poisson traffic model is a point-process approach.

Sometimes network traffic is specified in terms of aggregated traffic flows measured at a network node. This is the counting process description. Corresponding to Figs. 6.3(a,b), we have Figs. 6.3(c,d), respectively, designating the total number of bytes and packets in each time window Δt. The fBm model is a counting-process approach.

It is found that as long as the time window used for aggregating network traffic is not longer than the average message delay time, the point and counting processes are equivalent in terms of network performance.

6.7.1.2 Fractal behavior of network traffic Until about the late 1980s, researchers in the field of traffic engineering were quite happy with Poisson or Markovian traffic modeling. Since then, however, a number of traffic traces have been collected, first by the Bellcore (now Telcordia) people. Researchers were surprised to find out that real traffic differs greatly from Poisson traffic. The aggregation of LAN traffic data at different time scales is shown in the left panel of Fig. 6.4. The corresponding Poisson traffic, also aggregated at the same time scales, is shown in

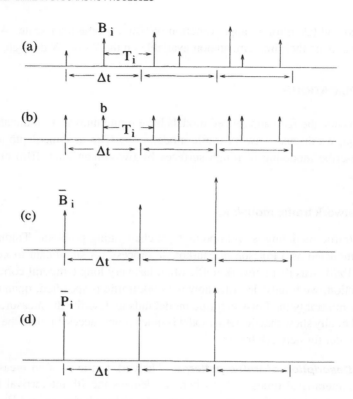

Figure 6.3.   Schematics for the pattern of (a) Traffic B, (b) Traffic P, (c) Traffic $\overline{B}$, and (d) Traffic $\overline{P}$.

Fig. 6.4 in the right panel. We notice that the real traffic varies over the time scales from 0.01 to 100 s. However, the Poisson traffic varies only on time scales smaller than 1 s. The "little smoothing" behavior of real traffic can be characterized by the Hurst parameter, as we shall explain in Chapter 8.

6.7.1.3   Failure of the Poisson model   In a Poisson model, interarrival times between two successive messages follow one exponential distribution; the size of the messages follows another exponential distribution. In queuing theory, this model is denoted as $M/M/1$. If a measured traffic trace is indeed Poissonian, then the two parameters, the mean of the message size and the interarrival times, can be easily estimated from the measured data.

To show quantitatively how bad the Poisson traffic model can be, let us study a single server queuing system with first-in-first-out (FIFO) service discipline and an infinite buffer, driven by both measured and Poisson traffic, and focus on three of the most important performance measures — throughput (which is also called utilization level), delay, and packet loss probability. Figure 6.5 shows a typical result of queue size tail distribution. To understand the result, the x axis can be

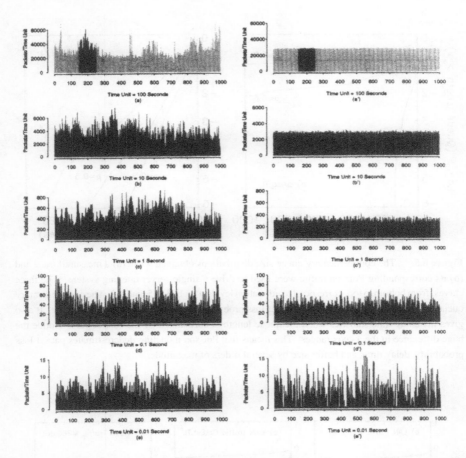

Figure 6.4.   Left panel, (a)-(e), measured LAN traffic trace pAug.TL aggregated on different time scales; right panel, (a')-(e'), the corresponding (compound) Poisson traffic aggregated on the same time scales. Courtesy of Leland et al. [281].

considered the buffer size. Let us fix it at 600 Kbytes. Then for a given utilization level ρ, say, 0.5, the tail probability measures packet loss probability, which is above 1% for real traffic but much smaller than 10^{-4}% for the Poisson model. This means that Poisson modeling underestimates packet loss probability, delay time, and buffer size by several orders of magnitude.

6.7.1.4 *Performance of the fBm model*   FBm is one of the most intensively studied models exhibiting the long-range dependence property in traffic engineering. After Norros introduced fBm as a traffic model, Erramilli et al. checked the complementary queue length distribution formula of Norros and found excellent agreement with simulations for a single server queuing system driven by some measured traffic trace operating at utilization $\rho = 0.5$. Gao and Rubin extended Erramilli

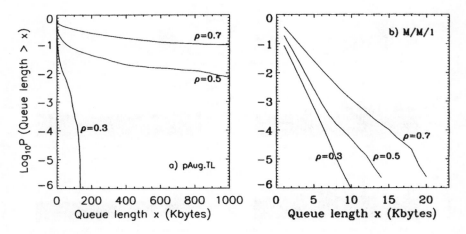

Figure 6.5.   The complementary queue size distributions obtained when (a) a measured trace and (b) its corresponding Poisson traffic were used to drive a single-server queuing system. The three curves, from bottom to top, correspond to utilization levels $\rho = 0.3, 0.5$, and 0.7. The complementary queue size distribution estimates the packet loss probability when the buffer is of finite size. It also gives the delay time statistics when the queue length is normalized by the service rate. Notice the huge difference in the x-axis ranges. This means that Poisson modeling underestimates packet loss probability, delay time, and buffer size by several orders of magnitude.

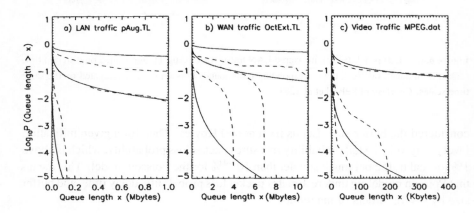

Figure 6.6.   Comparison of complementary queue length distributions of single-server FIFO queuing systems driven by measured data (a) local area network (LAN) traffic, (b) wide area network (WAN) traffic, and (c) video traffic (dashed lines) and corresponding fBm traffic processes (solid lines). Three curves, from top to bottom, correspond to $\rho = 0.7, 0.5$, and 0.3, respectively. Note that for video traffic MPEG.data, the solid curve for $\rho = 0.3$ is too close to the y-axis to be seen.

et al.'s study by considering several utilization levels simultaneously. Figure 6.6 shows a typical result for three different types of traffic processes. When it is compared with Fig. 6.5, it is clear that fBm is indeed a much improved model. The key reason is that fBm has captured the long-range correlation properties of the traffic processes. We shall discuss in detail how to estimate long-range correlation properties in real data in Chapter 8.

6.7.2   Modeling of rough surfaces

One of the most vivid examples of high-dimensional fractals is the beautiful, irregular mountainous massifs. It is thus no wonder that two-dimensional fBm has been used to produce beautiful images. Recently, it has been found that scattering from natural rough surfaces modeled by two-dimensional fBm fits measurement data better than when the surface is modeled by nonfractal classic models. In related work, rough sea surface has been modeled by a Weierstrass-Mandelbrot function augmented with parameters describing wave motion on the sea surface. As we have explained, the fractal dimension of a rough surface and the Hurst parameter are related by a simple equation, $D = 3 - H$. It is thus clear that an important problem in such studies is to accurately estimate the Hurst parameter. In later chapters, we shall discuss in depth many methods for estimating the Hurst parameter; therefore, we shall not go into any of the details of rough surface modeling here.

6.8   BIBLIOGRAPHIC NOTES

For more in-depth treatment of Bm, we refer to [76]. Early studies of $1/f$ processes can be found, for example, in Press [353], Bak [22], and Wornell [483]. Some of the more recently discovered $1/f$ processes are found in traffic engineering [44, 89, 281, 337], DNA sequences [284, 340, 463], genomes of DNA copy number changes associated with tumors [225], human cognition [188], ambiguous visual perception [151, 496], coordination [71], posture [81], dynamic images [50, 51], the distribution of prime numbers [479], and carbon nanotube devices [82, 406], among many others. Mechanisms for proposed $1/f$ processes range from the superposition of many independent relaxation processes [199, 315] to self-organized criticality [23, 24]. A mechanism for fractal traffic is proposed by Willinger et al. [476], based on the superposition of many independent heavy-tailed ON/OFF sources, somewhat similar to the relaxation processes of [199, 315]. This approach will be briefly reviewed in Chapter 11, where we shall also show how the observed traffic can be generated by a conventional communication network driven by Poisson traffic. Readers interested in the study of $1/f$ processes using state space and chaos theory are referred to [145, 149, 155, 158, 178–180, 328, 334, 357, 421, 431]. Relevant references for the representation and synthesis of fBm processes using wavelets include [3, 307, 347, 394]. References [295, 387, 462] discuss the RMD method, [130, 461] discuss the SRA method, [118, 291] discuss multi-wavelet expansion (MWE) Monte

Carlo methods, [387, 462, 488] discuss Fourier transform–based approaches for synthesizing fBm processes, and [48, 294] introduce the Weierstrass-Mandelbrot function. Finally, for traffic modeling, we refer to [119, 171–173, 321, 322], and for rough surface modeling and scattering on rough surfaces, we refer to [139, 207, 302].

6.9 EXERCISES

1. Let

$$Y = \sum_{i=1}^{n} X_i,$$

   where X_i, $i = 1, 2, \cdots$ are independent uniformly distributed random variables on the unit interval $[0,1]$. For each fixed $n = 2, 3, 4, \cdots$, check to see if Y follows the normal distribution. What is the "minimal" n such that Y is almost normally distributed?

2. Imagine you are tossing a coin, and set $X = 1$ when you get a head and $X = -1$ when you get a tail. Assume that head and tail occur with an equal probability of 0.5. Now assume that at time $t = 0, 1, 2, \cdots$ you toss the coin. Form the sum $Y = \sum_{i=1}^{n} X_i$. Simulate a path of $Y(t)$. This is the standard Bm, often used to model how a complete drunk walks (in one-dimensional space).

3. Rework problem 2 with $p = 0.6$ for a head and $q = 1 - p = 0.4$ for a tail.

4. Develop a two-dimensional version of problems 2 and 3.

5. Prove Eq. (6.7) by starting from the defining property of Eq. (6.4).

6. Implement one or a few of the methods described in the text for synthesizing the fBm model. This will be handy in future.

CHAPTER 7

STABLE LAWS AND LEVY MOTIONS

The central limit theorem states that the sum of many iid random variables of finite variance is a Gaussian random variable. What happens if the condition of finite variance is dropped? In particular, one can ask: When will the distribution for the sum of the random variables and those being summed have the same functional form? This is exactly the problem asked by Paul Levy in the 1920s, and the answer is given by Levy stable laws. Consequently, the central limit theorem has been generalized. This subject has found numerous applications in fields as diverse as astrophysics, physics, engineering, finance, economics, and ecology, among many others. In fact, the applications are so broad that a book much bigger than ours will be needed to cover all of them. For this reason, no specific applications will be discussed here. Instead, we shall focus on the basic theory and provide some (casual) references about applications in the Bibliographic Notes at the end of the chapter.

To stimulate readers' curiosity about the material in this chapter, let us consider a random event, which can be described by a random variable Y and is a function of a number of unknown factors $X_1, X_2, \cdots, X_n$, $Y = f(X_1, X_2, \cdots, X_n)$. If $f(0, \cdots, 0) = 0$, then, in the first approximation using Taylor series expansion, we have

$$Y = \sum_{i=1}^{n} c_i X_i,$$

where $c_i = \left.\frac{\partial f}{\partial X_i}\right|_{X_1=X_2=\cdots=X_n=0}$. If the unknown factors play equivalent roles, then $c_1 = c_2 = \cdots = c_n = c$, and Y is simply the summation of the iid random variables $X_1, \cdots, X_n$. In some sense, normal distributions and the central limit theorem describe daily mundane life. Many lucky people live such a life happily. However, occasionally one has to take an unplanned journey, during which many unexpected and exciting (or terrible) things happen. Such a journey could be related to hate, love, patriotism, and so on, as illustrated by numerous classic poems, novels, and movies. Could stable laws and Levy motions describe some aspects of such unusual journeys?

7.1 STABLE DISTRIBUTIONS

Let us first introduce the equivalence relation: Two random variables X and Y are equivalent if their distributions are the same. This is denoted as

$$X \overset{d}{=} Y.$$

Example 1: If X is a $[0, 1]$ random variable, then

$$1 - X \overset{d}{=} X.$$

From example 1, it is clear that X and Y are not required to be pairwise independent.

We can also introduce the similarity relation: Two random variables X and Y are similar

$$Y \overset{s}{=} X$$

if there exist constants a and $b > 0$ such that

$$Y \overset{d}{=} a + bX.$$

Example 2: If $X \sim N(0, 1)$, $Y \sim N(\mu, \sigma^2)$, then $Y \overset{d}{=} \mu + \sigma X$; hence, $Y \overset{s}{=} X$.

We can now state the definition of stable random variables. There are a few equivalent forms. While some can be derived from others, for simplicity we simply state them here.

Definition 1. A random variable Y is stable if

$$\sum_{i=1}^{n} Y_i \overset{d}{=} a_n + b_n Y, \tag{7.1}$$

where a_n is real, $b_n > 0$, and $Y_1, Y_2, \cdots$ are independent random variables, each having the same distribution as Y.

Definition 2. A stable random variable is called strictly stable if Eq. (7.1) holds with $a_n = 0$:

$$\sum_{i=1}^{n} Y_i \overset{d}{=} b_n Y, \tag{7.2}$$

where $b_n > 0$ and $Y_1, Y_2, \cdots$ are independent random variables, each having the same distribution as Y.

Observing Eq. (7.1) and rewriting

$$\sum_{i=1}^{n} Y_i = \sum_{i=1}^{m} Y_i + \sum_{i=1+m}^{n} Y_i,$$

one readily sees that $\sum_{i=1}^{m} Y_i$ and $\sum_{i=1+m}^{n} Y_i$ both have to be similar to Y. We thus arrive at the following definition.

Definition 3. A random variable Y is stable if and only if for any arbitrary positive constants b' and b'' there exist constants a and $b > 0$ such that

$$b'Y_1 + b''Y_2 \overset{d}{=} a + bY, \tag{7.3}$$

where Y_1 and Y_2 are independent and $Y_1 \overset{d}{=} Y_2 \overset{d}{=} Y$.

Example 3: Let us examine the stability of Gaussian random variables. Assume that $Y \sim N(\mu, \sigma^2)$; then $a + bY \sim N(b\mu + a, (b\sigma)^2)$. Similarly, $b'Y_1 \sim N(b'\mu, (b'\sigma)^2))$ and $b''Y_1 \sim N(b''\mu, (b''\sigma)^2))$. By the addition rule for independent Gaussian random variables, we have $b\mu + a = (b' + b'')\mu, (b\sigma)^2 = (b'\sigma)^2 + (b''\sigma)^2$.

Example 4: Cauchy distribution. In Chapter 3, we derived the Cauchy distribution centered at $x = 0$:

$$f(x) = \frac{l}{\pi(l^2 + x^2)}.$$

The more general form is centered at $x = \delta$ and is given by

$$f(x) = \frac{l}{\pi[l^2 + (x - \delta)^2]}, \quad -\infty < x < \infty.$$

It is often denoted by Cauchy(l, δ). It is a stable distribution, as can be readily verified by using convolution (see exercise 2).

Example 5: Levy distribution. Let $X \sim N(0, 1)$. Consider the random variable $Y = X^{-2}$. The distribution for Y is

$$P[Y < x] = P[X^{-2} < x] = 2P[X > x^{-1/2}] = \sqrt{\frac{2}{\pi}} \int_{1/\sqrt{x}}^{\infty} e^{-u^2/2} du.$$

Differentiating the above expression with respect to x, we have

$$p_Y(y) = \frac{1}{\sqrt{2\pi}} \frac{1}{y^{3/2}} e^{-\frac{1}{2y}}, \quad y > 0. \tag{7.4}$$

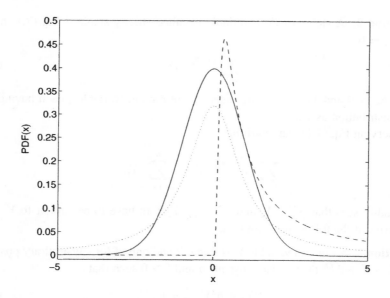

Figure 7.1.   The normal (solid), Cauchy (dotted), and Levy (dashed) distributions.

This is the Levy distribution. The most general form, denoted by Levy(γ, δ), can be parameterized as

$$p_Y(y) = \sqrt{\frac{\gamma}{2\pi}} \frac{1}{(y-\delta)^{3/2}} e^{-\frac{\gamma}{2(y-\delta)}}, \quad y > \delta,$$

which corresponds to $X \sim N(\mu, \sigma^2)$. Its tail is

$$p_Y(y) \sim y^{-3/2}, \qquad y \to \infty.$$

The Levy distribution is also stable (see exercise 3).

The normal, Cauchy, and Levy distributions are the only stable distributions with closed form formulas for the densities. To better appreciate their functional dependence, Fig. 7.1 plots these three distributions. While the lack of a closed form formula for general stable distributions may be considered a drawback, fortunately characteristic functions for the stable distributions have a simple form. Although the general form for the characteristic functions can be derived based on the above definitions, for simplicity we present them as another definition. This definition enables one to readily compute stable distributions using the FFT method.

Definition 4. A random variable X is stable if and only if $X \overset{s}{=} Z$, where Z is a random variable with the characteristic function

$$\Phi_Z(u) = E[e^{juZ}] = \begin{cases} \exp\left(-|u|^\alpha[1 - j\beta \tan(\pi\alpha/2)\text{sign}(u)]\right), & \alpha \neq 1 \\ \exp\left(-|u|^\alpha[1 + j\beta\frac{2}{\pi}\log|u|\text{sign}(u)]\right), & \alpha = 1, \end{cases}$$

$$(7.5)$$

where $0 < \alpha \le 2$, $-1 \le \beta \le 1$, and sign is the sign function defined by

$$\text{sign}(u) = \begin{cases} -1 & u < 0 \\ 0 & u = 0 \\ 1 & u > 0. \end{cases}$$

The parameter α is called the stability index. As we shall see later, it governs how fast the tail probability decays, and $\alpha > 2$ is not allowed. The parameter β characterizes the skewness of the distribution. When $\beta = 0$, the distribution for Z is symmetric around zero. Since $X \overset{d}{=} aZ + b$, where $a > 0$, the distribution for X has two additional parameters a and b. They are called scale and location parameters. It is clear that the characteristic function for X is simply

$$\Phi_X(u) = e^{jbu}\Phi_Z(au).$$

More explicitly, we have

$$\Phi_X(u) = E[e^{juX}] = \begin{cases} \exp(jbu - |au|^{\alpha}[1 - j\beta\tan(\pi\alpha/2)\text{sign}(u)]), & \alpha \ne 1 \\ \exp(jbu - |au|^{\alpha}[1 + j\beta\frac{2}{\pi}\log|au|\text{sign}(u)]), & \alpha = 1. \end{cases}$$

$$(7.6)$$

In the above expression, we have used $\text{sign}(u) = \text{sign}(au)$ because $a > 0$.

To appreciate better the characteristic function representation, we note that the characteristic functions for the normal $N(\mu, \sigma^2)$, Cauchy(l, δ), and Levy(γ, δ) distributions are obtained by recognizing that

$$(\alpha = 2, \beta = 0, a = \sigma/\sqrt{2}, b = \mu),$$

$$(\alpha = 1, \beta = 0, a = l, b = \delta),$$

and

$$(\alpha = 1, \beta = 1, a = \gamma, b = \delta),$$

respectively.

7.2 SUMMATION OF STRICTLY STABLE RANDOM VARIABLES

Let us now examine more closely Eq. (7.1). When $Y_i \sim N(0, \sigma)$, $b_n = n^{1/2}$. We wish to find b_n for arbitrary stable distributions.

To find the answer, let us use the characteristic function representation. Denote the characteristic function for Y by $\Phi_Y(u)$. Then the characteristic function for $\sum_{i=1}^{n} Y_i$ is $[\Phi_Y(u)]^n$, while that for $b_n Y$ is $\Phi_Y(b_n u)$. Using Eq. (7.5), we then obtain $n = b_n^{\alpha}$. Therefore,

$$\sum_{i=1}^{n} Y_i \overset{d}{=} n^{1/\alpha}Y. \tag{7.7}$$

Equation (7.7) states that $(Y_1 + \cdots + Y_n)/n$ has the same distribution as $Y_i n^{-1+1/\alpha}$, $i = 1, \cdots, n$. Therefore, $n^{-1+1/\alpha}$ acts as the normalization factor, just as $1/\sqrt{n}$ does so for the normal distributions. If we denote the distribution for $\sum_{i=1}^{n} Y_i$ as $p_n(y)$, i.e., the n-step distribution, then Eq. (7.7) means that

$$p_n(y) = p_1\left(\frac{y}{n^{1/\alpha}}\right)\frac{1}{n^{1/\alpha}}.$$

This is the basic scaling property governing the self-similarity of the processes defined by summation of stable laws (they are called Levy flight and will be discussed soon).

While Eq. (7.7) appears to be very simple, it actually shows that $\alpha > 2$ is not a valid parameter. To see this, we note that the variance for the left-hand side is $n\text{Var}Y$, while that for the right-hand side is $n^{2/\alpha}\text{Var}Y$. We thus have

$$n\text{Var}Y = n^{2/\alpha}\text{Var}Y.$$

We identify three cases:

1. When $\text{Var}Y$ is finite but not zero, α has to be 2. This is the normal case.

2. When $0 < \alpha < 2$, $\text{Var}Y = \infty$.

3. When $\alpha > 2$, $\text{Var}Y = 0$. This means that the density function, as a function of y, must be negative for some y and thus is not a valid probability density function. Therefore, $\alpha > 2$ is not allowed.

There is another way of deriving $\text{Var}Y = 0$ for $\alpha > 2$. Starting with the characteristic function $\Phi_Y(u) = \exp(-|au|^\alpha)$ and noting that

$$\text{Var}Y = -\frac{d^2\Phi_Y(u)}{du^2}\bigg|_{u=0},$$

one indeed sees that $\text{Var}Y = 0$. The details are left as exercise 4.

7.3 TAIL PROBABILITIES AND EXTREME EVENTS

So far as tail probabilities are concerned, one can simply examine the symmetric Levy laws with the characteristic function $\exp(-|au|^\alpha)$. Let us denote the corresponding density function by $l_\alpha(a, x)$. It can be obtained by the inverse Fourier transform and is given by

$$
\begin{aligned}
l_\alpha(a, x) &= \frac{1}{2\pi}\int_{-\infty}^{\infty} \exp(-jux)\exp(-|au|^\alpha)du \\
&= \frac{1}{\pi}\int_{0}^{\infty} \cos(ux)\exp[-(au)^\alpha]du. \quad (7.8)
\end{aligned}
$$

We can expand $\exp[-(au)^\alpha]$ to obtain

$$
\begin{aligned}
l_\alpha(a, x) &= \frac{1}{\pi} \int_0^\infty \cos(ux) \sum_{m=0}^\infty \frac{[-(au)^\alpha]^m}{m!} du \\
&= \frac{1}{\pi} \sum_{m=0}^\infty \frac{(-1)^m a^{\alpha m}}{m!} \int_0^\infty \cos(ux) u^{\alpha m} du.
\end{aligned}
$$

Utilizing the identity

$$
\int_0^\infty \cos(t) t^{v-1} dt = \cos\left(\frac{\pi v}{2}\right) \Gamma(v),
$$

we obtain

$$
l_\alpha(a, x) = \frac{1}{\pi x} \sum_{m=1}^\infty \frac{a^{\alpha m} \sin(\alpha m \pi/2) \Gamma(\alpha m + 1)(-1)^{m-1}}{m! x^{\alpha m}}.
$$

The leading order term is

$$
l_\alpha(a, x) \sim \frac{a^\alpha \sin(\alpha \pi/2) \Gamma(\alpha) \alpha}{\pi x^{1+\alpha}}, \quad \text{as} \quad x \to \infty,
$$

which is a power law. If one is also concerned with the tail probability for the negative x, one can replace x by $|x|$.

We now consider extreme events. For comparison purposes, we consider both power-law tails and exponential distributions.

1. Extreme events with the power-law tail.
Suppose

$$
p(x) = \frac{A}{x^{1+\alpha}} \quad \text{as} \quad x \to \infty \quad \text{(PDF)}
$$

$$
P(x) = 1 - \frac{A}{\alpha x^\alpha} \quad \text{as} \quad x \to \infty \quad \text{(CDF)}.
$$

Consider iid steps Δx_i. Let

$$
\Delta x_{(n)} = \max(\Delta x_1, \Delta x_2, \cdots, \Delta x_n).
$$

Then

$$
\text{Prob}\{\Delta x_{(n)}\} < x\} = P(x)^n = \left(1 - \frac{A}{\alpha x^\alpha}\right)^n \quad \text{as} \quad x \to \infty.
$$

Normalizing $\Delta x_{(n)}$ by the factor $\left(\frac{A}{\alpha} n\right)^{1/\alpha}$, we have

$$
Z_n = \frac{\Delta x_{(n)}}{\left(\frac{A}{\alpha} n\right)^{1/\alpha}}.
$$

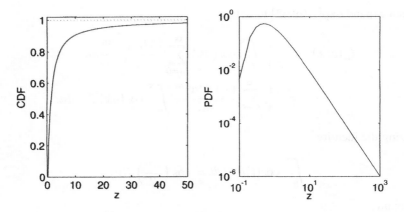

Figure 7.2.   Frechet distribution with $\alpha = 1$.

Then

$$\frac{A}{\alpha x^\alpha} = z^{-\alpha}/n.$$

Using the identity

$$\lim_{n \to \infty} (1 - v/n)^n \to e^{-v},$$

we obtain

$$\text{Prob}(Z_n < z) = e^{-z^{-\alpha}}.$$

This is the Frechet distribution. See Fig. 7.2. For large z, the density decreases as a power law, $z^{-(\alpha+1)}$. This is evident from the right-hand plot of Fig. 7.2. Since the mean and standard deviation of $\Delta x_{(n)}$ are the mean and standard deviation of Z multiplied by the scaling factor $\left(\frac{A}{\alpha}n\right)^{1/\alpha}$, the mean and standard deviation of $\Delta x_{(n)}$ thus have the same scaling law,

$$< \Delta x_{(n)} > \sim n^{1/\alpha},$$

$$\sigma_{\Delta x_{(n)}} \sim n^{1/\alpha}.$$

Finally, let us consider the mth moment of the iid steps:

$$\frac{1}{n}\sum_{i=1}^{n} \Delta x_i^m.$$

The largest term, which is on the order of $\frac{n^{m/n}}{n}$, dominates the sum and is infinite when $m > \alpha$.

2. Extreme events for exponential distributions.
Now the CDF is

$$P(\Delta x < x) = 1 - e^{-x/x_0},$$

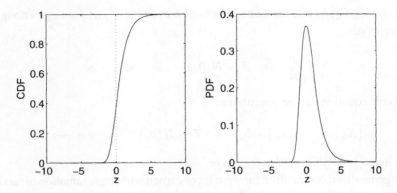

Figure 7.3.   The Fischer-Tippet distribution.

where x_0 is the mean of the random variable X. Then

$$\text{Prob}\{\Delta x_{(n)} < x\} = (1 - e^{-x/x_0})^n.$$

Rewriting e^{-x/x_0} as

$$\exp\left(-\frac{x - x_0 \ln n}{x_0}\right) \Big/ n,$$

letting

$$Z_n = \frac{\Delta x_{(n)} - x_0 \ln n}{x_0},$$

and taking large n limit, we get

$$P(Z_n \le z) = e^{-e^{-z}}.$$

This is the Fischer-Tippet distribution. See Fig. 7.3. Its density function is $e^{-z}e^{-e^{-z}}$, and the mean is the Euler constant $\gamma = -\int_0^\infty e^{-x} \ln x\, dx$. Therefore, to leading order,

$$< \Delta x_{(n)} > \sim x_0(\ln n + \gamma) + \cdots.$$

Similarly, to leading order, the variance of $\Delta x_{(n)}$ is

$$\sigma^2_{\Delta x_{(n)}} \sim x_0^2 + \pi^2/6 + \cdots.$$

7.4   GENERALIZED CENTRAL LIMIT THEOREM

The classical central limit theorem says that the normalized sum of independent identical terms with a finite variance converges to a normal distribution. To be more precise, let $X_1, X_2, \cdots$ be iid random variables with mean μ and variance σ^2. Then the classical central limit theorem states that the sample mean $\overline{X} =$

$(X_1 + \cdots + X_n)/n$ has the same distribution as $N(\mu, \sigma^2/n)$ for large enough n. In other words,

$$\frac{\overline{X} - \mu}{\sigma/\sqrt{n}} \overset{d}{\to} Z \sim N(0, 1) \quad \text{as} \quad n \longrightarrow \infty.$$

The above equation can be rewritten as

$$a_n(X_1 + \cdots + X_n) - b_n \overset{d}{\to} Z \sim N(0, 1) \quad \text{as} \quad n \longrightarrow \infty,$$

where $a_n = 1/(\sigma\sqrt{n})$ and $b_n = \sqrt{n}\mu/\sigma$.

The generalized central limit theorem is concerned with the summation of random variables with infinite variance. Note that we can always group the summation $X_1 + X_2 + \cdots + X_n$ into two terms, where the first is the summation of the first k terms, while the second is the summation of the remaining $l = n - k$ terms. If the limit for the summation exists, then each term has to converge to a stable random variable and has a similar distribution. More concretely, the generalized central limit theorem can be stated as follows.

Generalized Central Limit Theorem. Let $X_1, \cdots, X_n$ be iid random variables. There exist constants $a_n > 0$, $b_n \in R$ and a nondegenerate random variable Z with

$$a_n(X_1 + \cdots + X_n) - b_n \overset{d}{\to} Z$$

if and only if Z is α-stable for some $0 < \alpha \le 2$.

The generalized central limit theorem makes it clear that an α-stable distribution can be considered an attractor: even though the iid random variables $X_1, \cdots, X_n$ may not be α-stable, when the sum is suitably normalized and shifted, it converges to an α-stable distribution. Because of this, the random variable is said to be in the domain of attraction of the α-stable distribution. All the random variables with finite variance are in the domain of attraction of the normal distribution.

7.5 LEVY MOTIONS

A stochastic process $\{L_\alpha(t), t \ge 0\}$ is called (standard) symmetric α-stable Levy motion if

1. $L_\alpha(t)$ is almost surely 0 at the origin $t = 0$;

2. $L_\alpha(t)$ has independent increments; and

3. $L_\alpha(t) - L_\alpha(s)$ follows an α-stable distribution with characteristic function $e^{-(t-s)|u|^\alpha}$, where $0 \le s < t < \infty$.

Observe that α-stable Levy motion is simply Bm when $\alpha = 2$. More precisely, $L_2(t) = \sqrt{2}B(t)$. The symmetric α-stable Levy motion is $1/\alpha$ self-similar. That is, for $c > 0$, the processes $\{L_\alpha(ct), t \geq 0\}$ and $\{c^{1/\alpha}L_\alpha(t), t \geq 0\}$ have the same finite-dimensional distributions. By this argument as well as Eq. (7.7), it is clear that the length of the motion in a time span of Δt, $\Delta L(\Delta t)$ is given by the following scaling:

$$\Delta L(\Delta t) \propto \Delta t^{1/\alpha}.$$

This contrasts with the scaling for fractional Brownian motion:

$$\Delta B_H(\Delta t) \propto \Delta t^H.$$

We thus note that $1/\alpha$ plays the role of H. Following the arguments of Sec. 6.4, we see that there are also two dimensions for Levy motions. One is the graph dimension, $2 - 1/\alpha$; the other is the self-similarity dimension, α.

One can also define asymmetric Levy motions by utilizing arbitrary α-stable distributions. When each step takes the same time regardless of length, the process is usually called a Levy flight. Figure 7.4 shows two examples of Levy flights with $\alpha = 1.5$ and 1. The difference between Bm's and Levy motions is that Bm's appear everywhere similarly, while Levy motions are comprised of dense clusters connected by occasional long jumps: the smaller α is, the more frequently the long jumps appear.

Let us now consider where Levy flight-like esoteric processes could occur. One situation could be this: a mosquito headed toward a giant spider web and got stuck; it struggled for a while, and luckily, with the help of a gust of wind, escaped. When the mosquito was struggling, the steps it could take were tiny; but the step leading to its fortunate escape had to be huge. As another example, let us consider (American) football games. For most of the time during a game, the offense and defense are fairly balanced, and the offense team may only be able to advance a few yards. But during an attack that leads to a touchdown, the hero getting the touchdown often "flies" tens of yards — he has somewhat escaped the defense. While these two simple examples are only analogies, remembering them could be helpful when reading research papers searching for Levy statistics in various types of problems, such as animal foraging. At this point, we should also mention that Levy flight-like patterns have been used as a type of screen saver for Linux operating systems.

A Levy motion is called a Levy walk when the time taken for each step is proportional to its length. Levy flights and walks have very different scaling behaviors: the former are characterized by memoryless jumps governed by a heavytail, while the latter, now having a fixed step size, have attained serial correlations within each huge jump.

7.6 SIMULATION OF STABLE RANDOM VARIABLES

In Chapter 3 (see Eq. (3.28)), we discussed simulation of normal random variables. For the Cauchy distribution, the inverse function method discussed in Chapter 3 can

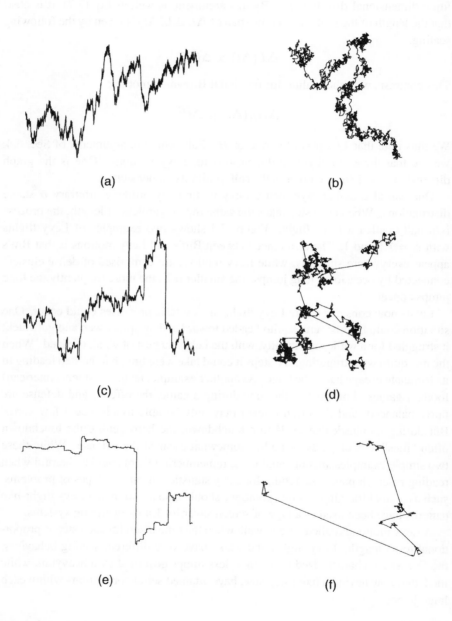

Figure 7.4. One- and two-dimensional Bm's (a,b) vs. Levy flights with $\alpha = 1.5$ and 1 for (c,d) and (e,f), respectively.

be readily applied. More precisely,

$$X = l \tan[\pi(U - 1/2)] + \delta \tag{7.9}$$

is Cauchy(l, δ).

Simulation of the Levy distribution using the definition is also straightforward. That is,

$$X = \gamma \frac{1}{Z^2} + \delta \tag{7.10}$$

is Levy(γ, δ) if $Z \sim N(0,1)$.

The first breakthrough in the simulation of stable distributions was made by Kanter [247], who provided a direct method for simulating a stable distribution with $\alpha < 1$ and $\beta = 1$. The approach was later generalized to the general case by Chambers et al. [68]. We first describe the algorithm for constructing a standard stable random variable $X \sim S(\alpha, \beta; 1)$.

Theorem: Let Θ and W be independent, with Θ uniformly distributed on $(-\pi/2, \pi/2)$ and W exponentially distributed with mean 1. For $0 < \alpha \le 2$, $-1 \le \beta \le 1$, and $\theta_0 = \arctan[\beta \tan(\pi\alpha/2)]/\alpha$,

$$Z = \begin{cases} \frac{\sin \alpha(\theta_0 + \Theta)}{(\cos \alpha\theta_0 \cos \Theta)^{1/\alpha}} \left\{ \frac{\cos[\alpha\theta_0 + (\alpha-1)\Theta]}{W} \right\}^{(1-\alpha)/\alpha} & \alpha \ne 1 \\ \frac{2}{\pi} \left[\left(\frac{\pi}{2} + \beta\Theta \right) \tan \Theta - \beta \ln \left(\frac{\frac{\pi}{2} W \cos \Theta}{\frac{\pi}{2} + \beta\Theta} \right) \right] & \alpha = 1 \end{cases} \tag{7.11}$$

has a $S(\alpha, \beta; 1)$ distribution.

For a symmetric stable distribution with $\beta = 0$, the above expression can be greatly simplified. To simulate an arbitrary stable distribution $S(\alpha, \beta, \gamma, \delta)$, we can simply take the following linear transform,

$$Y = \begin{cases} \gamma Z + \delta & \alpha \ne 1 \\ \gamma Z + \frac{\pi}{2}\beta\gamma \ln \gamma + \delta & \alpha = 1, \end{cases} \tag{7.12}$$

where Z is given by Eq. (7.11).

7.7 BIBLIOGRAPHIC NOTES

There is a huge literature on the subject discussed here. It includes a few books, such as [381, 455], among others. An interesting general paper is [239]. An interesting website with a good discussion of stable laws is
http://www.quantlet.com/mdstat/scripts/csa/html/node235.html.
In fact, this is part of a handbook about statistics in finance. See also John Nolan's website http://academic2.american.edu/~jpnolan/stable/stable.html, where one can find his draft first chapter for a book being written, and a file with many references

emphasizing applications in finance. The 2001 class notes of Professor Martin Z. Bazant of MIT also have some materials relevant to this chapter. Readers keen on a physics perspective are referred to two very entertaining general review articles by a few pioneers [260, 403], as well as two topical reviews focusing on fractional dynamics [305, 306]. See also [472, 473] on the application of Levy statistics to random water waves. For simulation of stable random variables, refer to [68, 247].

Since the publication of Mandelbrot's pioneering work on applying Levy statistics to economics [296], Levy statistics have found numerous applications in many fields. In fact, the book by Uchaikin and Zolotarev [455] has nine chapters on applications. Here are some of the applications:

- Fluid mechanics: In 1993, Henry Swinney considered a rotating container of fluid shaped like a washer. When the container rotated faster, vortices appeared in the fluid. It was shown that the tracer particles followed Levy flights between the vortices with $\alpha = 1.3$ [407].

- Device physics: Levy statistics have been found in Josephson junctions [184], at liquid-solid interfaces [411], and, more recently, in electrical conduction properties due to chaotic electron trajectories [308].

- Foraging movements of animals: This is an active area of research. The animals studied include albatrosses, microzooplankton, seabirds, spider monkeys, and others. See [35, 144, 362, 458, 459]. One of the most important topics here is how to optimize the success of random searches. An interesting finding is that when the food or target density is low, a Levy walk is an optimal solution.

- Art: In the so-called automatic painting developed by the Surrealist art movement, artists paint with such speed that any conscious involvement is thought to be eliminated. One of the masterpieces created by this technique is Jackson Pollock's *Autumn Rhythm*, produced by dripping paint onto large horizontal canvases. Taylor et al. [429] found that Pollock's paintings are fractal and that his motions can be described as Levy flights. This might offer a clue to the mysterious behavior of unconsciousness.

7.8 EXERCISES

1. Using the convolution formula, show that the normal distribution is stable.

2. Using the convolution formula, show that the Cauchy distribution is stable.

3. Using the convolution formula, show that the Levy distribution is stable.

4. Taking the second derivative of the structure function $\Phi_Y(u) = \exp(-|au|^\alpha)$, show that $\text{Var} Y = 0$ when $\alpha > 2$.

5. In Sec.7.3, we considered tail distribution. Starting from Eq. (7.8) and expanding $\cos(ux)$, show that for small x,

$$l_\alpha(a, x) = \frac{1}{\pi \alpha a} \sum_{m=0}^{\infty} \frac{(-1)^m}{(2m)!} \Gamma\left(\frac{2M + 1}{\alpha}\right) \left(\frac{x}{a}\right)^{2m}.$$

Further, show that the leading order term is

$$l_\alpha(a, x) \sim \frac{\Gamma(1/\alpha)}{\pi \alpha a}.$$

Thus, the center gets sharper and higher as $\alpha \to 0$.

6. Reproduce Fig. 7.4.

CHAPTER 8

LONG MEMORY PROCESSES AND STRUCTURE-FUNCTION–BASED MULTIFRACTAL ANALYSIS

In Chapter 6, we introduced fractional Brownian motion (fBm) model and briefly talked about persistent and antipersistent correlations in a time series. Persistent and antipersistent correlations may be collectively called longmemory. In this chapter, we describe the general theory of longmemory in a time series and discuss various methods of quantifying longmemory. We then introduce structure-function–based multifractal analysis, which provides a more comprehensive characterization of a complex time series than the concept of longmemory. To appreciate the power as well as the limitations of the concepts and methodologies, we shall discuss a number of different models as well as applications. Special attention is paid to two important notions, fractal scaling break and consistency of different methods of quantifying memory, as well as multifractal properties of a time series data. We shall also examine the meaning of dimension reduction in a fractal time series using principal component analysis.

8.1 LONG MEMORY: BASIC DEFINITIONS

Let $X = \{X_t : t = 0, 1, 2, \ldots\}$ be a covariance stationary stochastic process with mean μ, variance σ^2, and autocorrelation function $r(k), k \geq 0$. Assume $r(k)$ to be of the form

$$r(k) \sim k^{2H-2} \ as \ k \to \infty, \tag{8.1}$$

where $0 < H < 1$ is the Hurst parameter and measures the persistence of the correlation:

- When $0 < H < 1/2$, the process is said to have antipersistent correlation.

- When $1/2 < H < 1$, the process has persistent correlation; the larger the H value, the more persistent the correlation is. In this case, we have

$$\sum_k r(k) = \infty.$$

Because of this property, the time series is said to be long-range-dependent (LRD).

- When $H = 1/2$, the time series is said to be either memoryless or short-range-dependent (SRD).

Do processes with persistent and antipersistent correlations exist? The answer is yes. For example, the hydrologist Hurst found that the water level of Niles has an $H \approx 0.74$. For real network traffic, one often finds $0.7 \leq H \leq 0.9$. The best-known example of antipersistence is perhaps Kolmogorov's energy spectrum of turbulence with $H = 1/3$ (this will be discussed in some depth in Sec. 9.4). In Chapter 6, we discussed a simple model, fractional Gaussian noise (fGn), and showed in Eq. (6.24) that the autocorrelation function, $r(k)$, for an fGn process satisfies the following equation:

$$\lim_{k \to \infty} \frac{r(k)}{k^{2H-2}} = H(2H - 1). \tag{8.2}$$

Hence, an fGn process is an example of a process with longmemory, and as noted in the following, it is also an *exact* longmemory process in a strict mathematical sense.

Next, we construct a new covariance stationary time series

$$X^{(m)} = \{X_t^{(m)} : t = 1, 2, 3, \dots \}, \ m = 1, 2, 3, \dots,$$

obtained by averaging the original series X over nonoverlapping blocks of size m,

$$X_t^{(m)} = (X_{tm-m+1} + \cdots + X_{tm})/m, \ t \geq 1. \tag{8.3}$$

There are several useful relationships between the autocorrelation functions of the original LRD process and its averaged version. Using the stationarity properties of the processes, a general formula for the autocorrelation function, $r^{(m)}(k)$, of $X^{(m)}$, can be stated as

$$r^{(m)}(k) = \frac{(k+1)^2 V_{(k+1)m} - 2k^2 V_{km} + (k-1)^2 V_{(k-1)m}}{2V_m}, \tag{8.4}$$

where $V_m = var(X^{(m)})$. Using this relationship, it is straightforward to verify that the variance of $X^{(m)}$ satisfies

$$var(X^{(m)}) = \sigma^2 m^{2H-2} \tag{8.5}$$

if and only if the autocorrelation function of the LRD process satisfies

$$r(k) = \frac{1}{2}\left[(k+1)^{2H} - 2k^{2H} + (k-1)^{2H}\right]. \tag{8.6}$$

Moreover, one can verify that if X satisfies Eq. (8.5), then the autocorrelation function, $r^{(m)}(k)$, of the process $X^{(m)}$ satisfies

$$r^{(m)}(k) = r(k), \quad k \geq 0. \tag{8.7}$$

A process X that satisfies Eq. (8.5) (or equivalently, Eqs. (8.1), (8.5), and (8.7)) is often referred to as an exactly second-order self-similar process. Note that such a process always satisfies Eq. (8.2). On the other hand, instead of satisfying Eq. (8.5), if one has

$$\lim_{k \to \infty} \frac{r(k)}{k^{2H-2}} = c_1,$$

where $0 < c_1$ is an arbitrary constant, then one can show that

$$\lim_{k \to \infty} \frac{var(X^{(m)})}{m^{2H-2}} = c_2 \tag{8.8}$$

for some constant $c_2 > 0$. Such a process is often referred to as an asymptotically second-order self-similar process.

Equation (8.5) (or, more generally, Eq. (8.8)) is often called the variance-time relation. It provides a simple and precise way of quantifying the "little smoothing" behavior depicted in Fig. 6.4. For ease of exposition, let us assume that $H = 0.75$ for real traffic, and at the smallest time scale, real traffic and Poisson traffic have the same variance σ_0^2. For Poisson traffic, since $H = 0.5$, $var(X_i^{(m)})$ drops to $10^{-2}\sigma_0^2$ when $m = 100$. When $H = 0.75$, however, we need $m = 10,000$ to have the same effect. That is exactly what Fig. 6.4 has shown us.

To further illustrate the significance of the Hurst parameter, let us digress to consider a wellknown physical phenomenon: scattering of light. As is well-known, a harmonic oscillator may be described by $A_1 \cos(\omega t + \phi_1)$. Then the intensity of light is $\sim A_1^2$. When there are two oscillators described as $A_1 \cos(\omega_1 t + \phi_1)$ and $A_2 \cos(\omega t + \phi_2)$, the intensity of light is $\sim A_1^2 + A_2^2 + 2A_1 A_2 \cos(\phi_1 - \phi_2)$. When the phase difference is fixed, we have interference; otherwise, there is no interference. This corresponds to averaging out of the cross term. When there are m oscillators, under the condition $A_i = A = \text{const}$, if there is no interference, then the intensity of light is $\sim mA^2 = m^{2H}A^2$, $H = 1/2$; when interference is maximal, the intensity of light is $\sim (mA)^2 = m^{2H}A^2$, $H = 1$. This is how a laser works.

8.2 ESTIMATION OF THE HURST PARAMETER

Using a proof similar to that developed in Sec. 6.1, one can prove that Eq. (8.1) implies power-law power spectral density (PSD) for the process x, $1/f^{2H-1}$, $f \to 0$. Therefore, processes with longmemory are a type of $1/f$ processes. Note the difference between this expression and Eq. (6.7). The latter is for the PSD of a corresponding random walk process, e.g., for fBm that is a random walk process generated by integrating the increment process fGn.

Due to the ubiquity of longmemory processes, estimation of the Hurst parameter has been a much-studied topic in various fields. In this section, we describe five simple estimators. Discussion of other methods for estimating the Hurst parameter will be postponed until the next section after we introduce random walk formulation and structure-function–based multifractal analysis.

1. **Variance-time plot**. With this method, one checks if Eq. (8.5) holds. If it does, then in a log-log plot of $var(X^{(m)})$ vs. the aggregate block size m, the curve should be linear for large m with slope larger than -1. By contrast, an SRD process has $var(X^{(m)}) \sim m^{-1}$, so that in a log-log plot of $var(X^{(m)})$ vs. block size m, the slope is -1. This is perhaps the most widely used method in traffic engineering.

2. **R/S statistic**: For a given set of observations $\{X_k, k = 1, 2, \cdots, n\}$ with sample mean $\overline{X}(n)$ and sample variance $S(n)^2$, the R/S statistic is given by

$$\frac{R(n)}{S(n)} = \frac{1}{S(n)} \left[\max(0, W_1, W_2, \cdots, W_n) - \min(0, W_1, W_2, \cdots, W_n) \right],$$

where

$$W_k = \sum_{i=1}^{k} [X_i - \overline{X}(n)], \tag{8.9}$$

and the factor $S(n)$ is introduced for the normalization purpose. As will be explained in Sec. 8.3, Eq. (8.9) defines a random walk. Therefore, $R(n)/S(n)$ essentially characterizes the normalized extent or range of the process W_k. One expects that the square of this extent scales with n as n^{2H}, similar to the scaling between the variance of a random walk and n. Indeed, we have

$$E\left[\frac{R(n)}{S(n)}\right] \sim n^H, \quad \text{as} \quad n \to \infty.$$

With small datasets, this method works better than the variance–time plot, since now one is required to compute the mean instead of the variance. Perhaps for this reason, the method is more popular in physiology. There, typically the datasets are small.

3. **Autocorrelation function–based estimator**: $r(k) \sim k^{-\beta}$ as $k \to \infty$. This estimator is, however, seldom used.

4. **Spectral density–based estimator**: As stated at the beginning of this section, the PSD for x is $S(f) \sim f^{1-2H}$ as $f \to 0$. This expression is very important in examining the consistency of different H estimators.

5. **Wavelet-based estimator**: In Sec. 4.2, we introduced wavelet MRA. Furthermore, in Sec. 6.5, we applied MRA to represent an fBm process. Eq. (6.32) is an effective method of estimating the Hurst parameter of an fBm process. If we start from a stationary time series, then we can expect that Eq. (6.32) will be modified to give a slope of $2H - 1$ instead of $2H + 1$, since the time series under study is like the fGn rather than the fBm. This is indeed the case, as shown by the following equations.

In Sec. 4.2, we find that wavelet MRA decomposes a signal $x(t)$ into

$$x(t) = \sum_k a_x(J,k)\phi_{J,k}(t) + \sum_{j=1}^{J} \sum_k d_x(j,k)\psi_{j,k}(t).$$

Let

$$\Gamma_j = \frac{1}{n_j} \sum_{k=1}^{n_j} |d_x(j,k)|^2 , \qquad (8.10)$$

where n_j is the number of coefficients at level j; then the Hurst parameter is given by

$$\log_2 \Gamma_j = (2H - 1)j + c_0, \qquad (8.11)$$

where c_0 is some constant.

We note that either (1), (3), or (4) can be used to define the long-range dependence in a time series. Thus, methods (1), (3), and (4) actually constitute one independent estimator. Some researchers recommend using periodogram-based spectral estimation combined with Whittle's approximate maximum likelihood estimator to estimate the H parameter. With this approach, however, it may be difficult to determine a suitable region to define the power-law scaling.

8.3 RANDOM WALK REPRESENTATION AND STRUCTURE-FUNCTION–BASED MULTIFRACTAL ANALYSIS

In this section, we explain, first, how to construct a random walk process from a time series and, second, how to carry out multifractal analysis of the constructed random walk process based on the structure-function technique. The multifractal formulation will help us gain deeper understanding of the conventional methods for estimating the Hurst parameter and develop new means of estimating H.

8.3.1 Random walk representation

Let us denote the time series we want to study by $\{X_i, i = 1, 2, \cdots, N\}$. It can be any time series. For example, it can be a numerical sequence representing a

DNA sequence, or the interspike interval of neuronal firings, or switching times in ambiguous visual perceptions. For network traffic, it may be the interarrival time series, the packet length sequence, or the counting processes. For illustration purposes, in this section we use the counting process of a network traffic trace.

First, we subtract the mean from the time series. Denote the new time series as $\{x_i, i = 1, 2, \cdots, N\}$, where

$$x_i = X_i - \frac{1}{N} \sum_{j=1}^{N} X_j,$$

and consider it as a process similar to the fGn process. Then we form the partial summation of $\{x_i, i = 1, 2, \cdots\}$ to get the random walk process $\{y_k, k = 1, 2, \cdots\}$, where

$$y_k = \sum_{i=1}^{k} x_i. \tag{8.12}$$

Note that Eq. (8.12) is equivalent to Eq. (8.9). Figure 8.1 shows an example of the random walk process constructed from the counting process $\{\overline{B}_i, i = 1, 2, 3, \dots\}$ of the network traffic trace we analyzed in Sec. 6.7. Even though the length of the random walk process shown in the figure is 2^{18} points long, it crosses a specific level, say $y = 0$, very rarely. This is a consequence of the fact that the H parameter for the process is much larger than 1/2. For this reason, levelset analysis of the process would require unrealistically long time series.

8.3.2 Structure-function–based multifractal analysis

After the random walk process is constructed, we can compute $Z^{(q)}(m)$ defined by

$$Z^{(q)}(m) = \langle |y(i+m) - y(i)|^q \rangle \sim m^{\zeta(q)}, \tag{8.13}$$

where the average is taken over all possible pairs of $(y(i+m), y(i))$, and examine whether the power-law scaling laws for different values of real q exist or not. Negative and positive q values emphasize small and large absolute increments of $y(n)$, respectively. Thus, the approach allows us to focus on different aspects of the data by using different q. In particular, when $q = 2$, the method is called fluctuation analysis (FA). When power-law scaling for some q exists, we say that the process under study is a fractal process. Furthermore, if $H(q)$, defined by

$$H(q) = \zeta(q)/q, \tag{8.14}$$

is not a constant function of q, we say that the process is multifractal. Note that when Eqs. (8.13) and (8.14) are combined, we can write

$$F^{(q)}(m) = \langle |y(i+m) - y(i)|^q \rangle^{1/q} \sim m^{H(q)}. \tag{8.15}$$

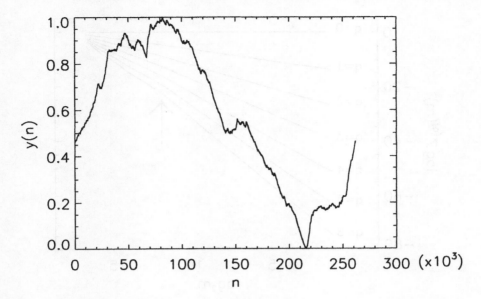

Figure 8.1. The random walk process constructed from the counting process of the network traffic trace, $\{\overline{B}_i, i = 1, 2, 3, \cdots\}$, analyzed in Sec. 6.7.

To make the idea concrete, Fig. 8.2 shows the result of multifractal analysis of the random walk process shown in Fig. 8.1. We observe that overall the curves are fairly linear; thus, the dataset can be classified as a fractal. However, if we look more carefully, we find that there is a knee point near $m = 2^{12}$. This means that there are two scaling regimes, $m \in (1, 2^{12})$ and $m \in (2^{12}, 2^{17})$. The $H(q)$ curves estimated for these two scaling regimes are shown in Fig. 8.3. We observe that for the scaling region $m \in (1, 2^{12})$, $H(q)$ slightly decreases with q, whereas for the scaling region $m \in (2^{12}, 2^{17})$, $H(q)$ decreases with q quite significantly. Thus, we conclude that for this specific random walk process, for scaling region $m \in (1, 2^{12})$, the process is a weak multifractal (i.e., more like a monofractal), while for the scaling region $m \in (2^{12}, 2^{17})$, the process is a multifractal. Since the scaling regions $m \in (1, 2^{12})$ and $m \in (2^{12}, 2^{17})$ correspond to short and long time scales, respectively, we can say that for this traffic trace data, it is almost a monofractal for short time scales and a multifractal for long time scales. This conclusion, however, may not be universal. In other words, some other traffic trace data may be multifractals for short time scales and monofractals for long time scales. It is also possible that some traffic data may be different types of mono– and/or multifractals on different time scales. We term this behavior nonuniversal fractal scaling for network traffic.

8.3.3 Understanding the Hurst parameter through multifractal analysis

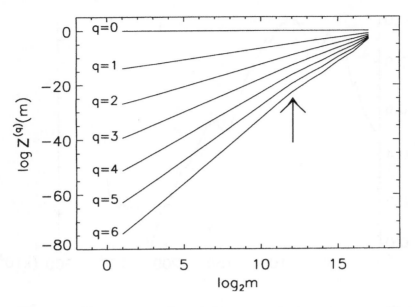

Figure 8.2. $Z^{(q)}(m)$ vs. m (in log-log scale) for the random walk process of Fig. 8.1.

In this subsection, we consider the meaning of the $H(q)$ spectrum from the multi-fractal analysis to develop new means of estimating the Hurst parameter. First, we note that $H(2)$ is simply the Hurst parameter H. To see this, we note that when $q = 2$, Eq. (8.13) reduces to the variance-time relation described by Eq. (8.5) if one notices that $\langle |y(i + m) - y(i)|^2 \rangle = m^2 var(X^{(m)})$.

Closely related to the variance-time relation is Fano factor analysis, which is quite popular in neuroscience. In the context of analysis of the interspike interval of neuronal firings, the Fano factor is defined as

$$F(T) = \frac{Var[N_i(T)]}{Mean[N_i(T)]},\tag{8.16}$$

where $N_i(T)$ is the number of spikes in the ith window of duration T. For a Poisson process, $F(T)$ is 1, independent of T. For a fractal process, $Var[N_i(T)] \propto T^{2H}$ and $Mean[N_i(T)] \propto T$. Therefore, $F(T) \sim T^{2H-1}$. In other words, the Fano factor can be viewed as the relation between $[\langle |y(i + m) - y(i)|^2 \rangle /m]$ and m instead of the relation between $[\langle |y(i + m) - y(i)|^2 \rangle]$ and m .

We now discuss methods that employ $H(1)$ to estimate H. Two such approaches are reviewed by Taqqu et al. [424], namely, the Absolute Values of the Aggregated Series Approach and Higuchi's method. In the former, one determines if the following scaling law holds:

$$\frac{1}{[N/m]} \sum_{k=1}^{[N/m]} \left| X^{(m)}(k) \right| \sim m^{H-1},$$

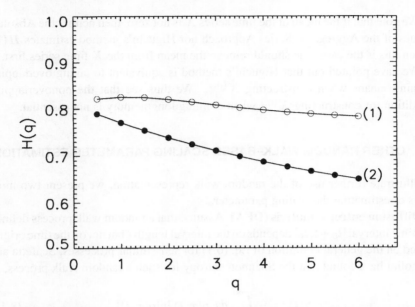

Figure 8.3. The variation of $H(q)$ vs. q computed from Fig. 8.2. Curves (1) and (2) represent $H(q)$ curves for two different scaling regimes, $m \in (1, 2^{12})$ and $m \in (2^{12}, 2^{17})$.

where N is the length of the time series, $X^{(m)}$ is the nonoverlapping running mean of X of block size m, as defined by Eq. (8.3), and [ ] denotes the greatest integer function. Higuchi's method, on the other hand, examines if the following scaling law is true:

$$L(m) = \frac{N-1}{m^3} \sum_{i=1}^{m} \left[(N-i)/m\right]^{-1} \sum_{k=1}^{[(N-i)/m]} \left| y(i+km) - y(i+(k-1)m) \right|$$
$$\sim m^{H-2},$$

where N again is the length of the time series, m is essentially a block size, and $y(i)$ is the random walk process constructed from the X instead of the x time series. Note that the two methods are quite similar. In fact, the first summation of Higuchi's method divided by m is equivalent to the Absolute Values of the Aggregated Series Approach. The second summation, $\sum_{i=1}^{m}$, is another moving average, equivalent to taking overlapping running means of the original time series X. By now, it should be clear that both methods estimate the $H(1)$ parameter instead of the $H(2)$ parameter when the time series X has mean zero. The reason $H(1)$ can be used to estimate $H(2)$ is that typically they are quite close, even if the time series X is a multifractal. When the time series X is very much like a monofractal, or is only weakly multifractal, then we see that any $H(q)$, $q \neq 2$, can be used to estimate $H(2)$. In this case, the structure-function–based technique provides infinitely many ways of estimating the Hurst parameter.

We note that if the mean of the time series X is not zero, then neither the Absolute Values of the Aggregated Series Approach nor Higuchi's method estimates $H(1)$. When this is the case, one should remove the mean from the X time series first.

We have pointed out that Higuchi's method is equivalent to taking overlapping running means when constructing $X^{(m)}$. We thus see that the nonoverlapping condition for constructing $X^{(m)}$ when defining longmemory is not essential.

8.4 OTHER RANDOM WALK–BASED SCALING PARAMETER ESTIMATION

To illustrate further use of the random walk representation, we present two more ways of estimating the scaling parameter.

1. **Diffusion entropy analysis (DEA)**: Assume that a random walk process defined in a time interval $[t_0 + t, t_0]$ depends on the interval length t but not on the time origin. Based on the general definition of Eq. (6.1) for self-similar processes, Scafetta and Grigolini have found that the Shannon entropy for such a random walk process,

$$S(t) = - \int_{-\infty}^{\infty} dx \, p(x, t) \ln[p(x, t)], \qquad (8.17)$$

where $p(x, t)$ is the PDF for the random walk, increases with t logarithmically,

$$S(t) \sim \delta \ln t, \qquad (8.18)$$

where δ equals the Hurst parameter for fBm and Levy walk processes and equals the α parameter for Levy flight processes. To estimate $S(t)$, one can partition a random walk using a maximal overlapping window and then estimate $p(x, t)$ based on all the segments of the random walk process. While computationally the method is simple, it has a drawback: for a given time series of N points, the scaling behavior can be resolved at most up to $t = N/100$. This is because estimation of $p(x, t)$ for a random walk requires many (say, 100) sample realizations of the random walk.

2. **Phase space–based methods**: At the end of Sec. 6.4, we briefly mentioned that the Hurst parameter can also be estimated by monitoring the divergence of two nearby trajectories in a reconstructed phase space. The ideas will be easier to understand after we explain related concepts in Chapters 13 to 15. Therefore, we shall postpone this discussion until Chapters 14 and 15.

8.5 OTHER FORMULATIONS OF MULTIFRACTAL ANALYSIS

There are other formulations of multifractal, such as those based on detrended fluctuation analysis (DFA) and wavelet analysis. Let us first discuss the latter.

In Sec. 8.2, we discussed a wavelet-based method for estimating the Hurst parameter. The simple generalization of FA to the structure-function–based multifractal formulation discussed in Sec. 8.3 suggests that we generalize Eq. (8.10) to consider

$$\gamma_j(q) = \left[\frac{1}{n_j} \sum_{k=1}^{n_j} |d_x(j,k)|^q \right]^{1/q} \tag{8.19}$$

and examine whether the following scaling relations hold or not:

$$\gamma_j(q) \sim 2^{j[H(q)-1/2]}. \tag{8.20}$$

If they do, then we have

$$\log_2 \gamma_j(q) = j[H(q) - 1/2] + c_q, \tag{8.21}$$

where each c_q is some constant. This generalization is straightforward.

Next, we consider the DFA-based multifractal formulation. We first describe DFA. It is developed to first remove linear or other trends and then perform scaling analysis. Linear or other trends often exist in experimental data obtained under conditions that may be nonstationary.

DFA works as follows: First, divide a given random walk of length N into $\lfloor N/l \rfloor$ nonoverlapping segments (where the notation $\lfloor x \rfloor$ denotes the largest integer that is not greater than x); then define the local trend in each segment to be the ordinate of a linear least-squares fit for the random walk in that segment; finally, compute the "detrended walk," denoted by $y_l(n)$, as the difference between the original walk $y(n)$ and the local trend. Then one examines

$$F_d(l) = \left\langle \sum_{i=1}^{l} y_l(i)^2 \right\rangle^{1/2} \sim l^H, \tag{8.22}$$

where the angle brackets denote the ensemble average of all the segments and $F_d(l)$ is the average variance over all segments.

The extension of DFA to a multifractal formulation is also straightforward. Depending on how the deviation from a straight line in each window is characterized, one can have (at least) two forms. One is given by

$$F_d^{(q)}(l) = \left\langle \sum_{i=1}^{l} |y_l(i)|^q \right\rangle^{1/q} \sim l^{H(q)}, \tag{8.23}$$

where q is real, taking on both negative and positive values. Another is given by

$$F_d^{(q)}(l) = \left\langle \left[\sum_{i=1}^{l} |y_l(i)|^2 \right]^{q/2} \right\rangle^{1/q} \sim l^{H(q)}. \tag{8.24}$$

The first formulation amounts to using the l_1 norm, while the second uses the l_2 or Euclidean norm.

8.6   THE NOTION OF FINITE SCALING AND CONSISTENCY OF H ESTIMATORS

To facilitate our discussion on the correlation structure of ON/OFF intermittency and Levy motions in this section, as well as numerous applications in Sec. 8.8, we consider two important issues. One is finite scaling behavior. The other is consistency of the estimated H using different estimators. Let us illustrate both issues with simple examples.

Imagine two people playing a simple game: Hanna tosses a fair coin, and Albert guesses whether the outcome is a head or tail and compares his guess with the actual outcome. Let head and tail be denoted by 1 and -1, respectively. Occasionally, they observe a long sequence of heads or tails. Being an intelligent adult, Albert reasons that whenever a sequence of heads (or tails) appears, it would be better to guess the next one to be a tail (or head), since the probability for a head (or tail) to appear after a sequence (say, $n = 4$) of heads (or tails) has occurred is very low. Intuitively, one expects that a sequence of 1 and -1 guessed by Albert yields $H < 1/2$, i.e., antipersistence, for a small scale but $H = 1/2$ for not too small a scale. This is indeed so, as can be readily verified by carrying out such an experiment. To save time, however, let us use a computer and perform a simple simulation.

First, we generate a sequence of 1 and -1 equivalent to tossing a fair coin. Next, we sequentially search the sequence to check whether a sequence of 1 (or -1) has occurred with length greater than a parameter n. Whenever we find such a patch of 1 (or -1), we replace the nth 1 (or -1) of that sequence by -1 (or 1), and then resume our search, starting from the position that has just been modified, until the end of the sequence. The resulting sequence would be equivalent to Albert's sequence of 1 and -1. Figure 8.4 shows $\log_2 F(m)$ vs. $\log_2 m$ for $n = 3$ (denoted by the circle) and 4 (denoted by the asterisk), respectively. Clearly we observe that $H < 1/2$ for small m but is equal to 1/2 when m is not too small. This is an excellent example showing the meaning of both antipersistence and finite scaling (or multiple fractal scaling).

Let us now consider time series generated from an AR(1) model, $x_{n+1} - \overline{x} = a(x_n - \overline{x}) + \eta_n$, discussed in Chapter 3. For this model, the autocorrelation decays exponentially: $C(m) = \sigma^2 a^{|m|}$ (Eq. (3.36)). When the time lag m is large, the correlation is essentially zero; we can expect H to be 1/2. However, when the coefficient a is only slightly smaller than 1, $C(m)$ will be close to 1 for a considerable range of m. In this case, we have almost perfect correlation. One thus might expect $H \approx 1$ for not too large a time scale. This seems to be verified when one applies the variance-time relation to analyze the generated time series or, equivalently, applies FA to the random walk process constructed from the data. The latter is shown in Fig. 8.5(a). However, there is a problem here: If we employ DFA, then we obtain $H = 1.5$ for not too large a time scale, as shown in Fig. 8.5(b). What is going on?

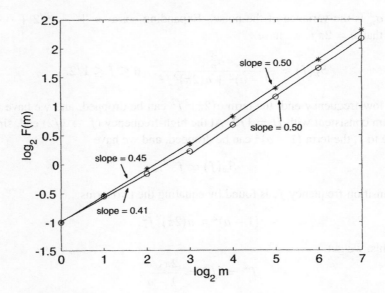

Figure 8.4. Antipersistence in guessing the outcome of a coin toss.

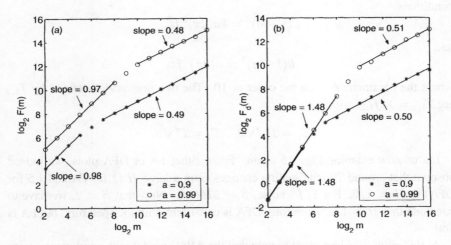

Figure 8.5. H parameter for the AR(1) model.

Which result should be trusted? To find the answer, let us examine which one is consistent with the PSD of the AR(1) process.

As we discussed in Sec. 4.12, the PSD of an AR(1) process is

$$S_x(f) = \frac{\sigma_n^2}{1 + a^2 - 2a\cos\omega}, \quad 0 \le \omega \le \pi, \tag{8.25}$$

where σ_n^2 is the variance of the noise. Expanding $\cos \omega = 1 - \omega^2/2 + \cdots$ and noting that $\omega = 2\pi f$, we have

$$S_x(f) \approx \frac{\sigma_n^2}{(1 - a)^2 + a(2\pi)^2 f^2}, \quad 0 \le f \le 1/2.$$

At the low-frequency end, the term $a(2\pi)^2 f^2$ can be dropped, and we have a flat spectrum consistent with $H = 1/2$. At the high-frequency ($f \to 1/2$) end, since a is close to 1, the term $(1 - a)^2$ can be dropped, and we have

$$S_x(f) \propto f^{-2}.$$

The transition frequency f_* is found by equating the two terms:

$$(1 - a)^2 \approx a(2\pi)^2 f_*^2.$$

From this, we get

$$T^* = 1/f_* = \frac{2\pi\sqrt{a}}{1 - a}.$$

To find more precisely the frequency ranges where the PSD is flat or decays as f^{-2}, we may require $(1 - a)^2 \gg a(2\pi)^2 f^2$ when $f \le f_1$ and $(1 - a)^2 \ll a(2\pi)^2 f^2$ when $f_2 \le f \le 1/2$. Quantitatively, $f_{1,2}$ may be defined by the following conditions:

$$(1 - a)^2 = \theta a(2\pi)^2 f_1^2$$

and

$$\theta(1 - a)^2 = a(2\pi)^2 f_2^2,$$

where the parameter θ is on the order of 10. The two time scales defined by $f_{1,2}$ are $T_{1,2} = 1/f_{1,2}$, with

$$T_1 = T^*/\sqrt{\theta}, \quad T_2 = T^*\sqrt{\theta}.$$

Let us now examine Fig. 8.5 again. From either FA or DFA plots, we indeed observe that around T^*, the scaling changes from a large H (1 for FA and 1.5 for DFA) to $H = 1/2$. For $1/f^\beta$ noise, $\beta = 2H - 1$. Now that $\beta = 2$, we have to conclude that $H = 1.5$. Therefore, DFA is consistent with the spectrum, but FA is not!

At this point, it is important to note that the AR(1) model with coefficient a very close to 1 has been proposed as a (pseudo) model for LRD traffic with $H = 1$ and a convenient model for exact $1/f$ noise. The former mis-interpretation is due to misuse of FA (or the variance-time relation) on the data; the latter may occur because the magnitude response of the Fourier transform of the process, $|X(\omega)|$, scales with f as f^{-1} when $f \to 1/2$. When $|X(\omega)|$ is mistaken for PSD, one would claim that the model generates the exact $1/f$ spectrum.

Now let us understand why FA gives $H = 1$ for the AR(1) model. We approach the problem from a broader perspective by asking this question: Given a measured time series $x_1, x_2, \cdots$, shall we treat it as a noise process like fGn or as a random walk process like fBm? If the data show a noise process, then we have to form a partial summation according to Eq. (8.12) if we wish to apply the structure-function–based multifractal analysis. On the other hand, if the data show a random walk process, then no partial summation is needed when applying the structure-function–based multifractal analysis. However, one has to obtain the noise process by differencing the original data when applying the variance-time relation, Fano factor analysis, or the R/S statistic.

To answer the question posed above, let us be more quantitative. Denote an ideal fractal process by $x_1, x_2, \cdots$. Let the process obtained by removing the mean of x and integrating it be denoted by the y process. If the PSD for the x time series is $1/f^{\beta_x}$, then the PSD for y is $1/f^{\beta_y}$, where $\beta_y = \beta_x + 2$. As we pointed out earlier, $\beta_x = 2H_x - 1$ and $\beta_y = 2H_x + 1$ (where the subscript x is used to emphasize that this H is associated with the x process). Here it appears that we are all right, since the two equations appear to be equivalent (but there is a hidden complexity, as we shall point out soon).

Now let us start from the y process and integrate it to obtain the z process. The PSD for z is $1/f^{\beta_z}$, with $\beta_z = \beta_y + 2$. We expect to get $\beta_y = 2H_y - 1$ and $\beta_z = 2H_y + 1$. Now what is the relation between H_x and H_y? It is simple and is given by

$$H_y = H_x + 1. \tag{8.26}$$

It seems that everything is still all right. By this argument, one expects H to increase by 1 each time the process is integrated. So where is the complexity?

The trouble lies in the simple fact that our theory demands $0 < H < 1$. As we shall show shortly, FA and other equivalent methods for estimating H only return an H that belongs to the unit interval, no matter what types of data are analyzed. For example, if we estimate H_y by the variance-time relation, or Fano factor analysis, or the R/S statistic, or FA, the value of H_y will be at most 1! However, if $0 < H_x < 1$, then by Eq. (8.26), we should have $1 < H_y < 2$. This is the difficulty!

To understand why H estimated by FA or other equivalent methods saturates at 1, let us assume that $y(n) \sim n^\gamma$, $\gamma > 1$. Then $\langle |y(n+m) - y(n)|^2 \rangle = \langle [(n+m)^\gamma - n^\gamma]^2 \rangle$ is dominated by terms with large n. When this is the case, $(n+m)^\gamma = [n(1+m/n)]^\gamma \approx n^\gamma[1+\gamma m/n]$. One then sees that $\langle |y(n+m) - y(n)|^2 \rangle \sim m^2$, i.e., $H(2) = 1$. Similarly, one can prove that the smallest H given by FA is 0.

By now, it should be clear why FA gives $H = 1$ for the high-frequency end of an AR(1) process, while the PSD and DFA give $H = 1.5$. It turns out that DFA saturates at $H = 2$ from above and $H = 0$ from below. Wavelet analysis is the most versatile; H can be both negative and larger than 2, and multiple integrations or differentiations of a measured time series are all fine. However, for practical purposes, DFA can be considered adequate.

For ease of practical applications, we list a few rules of thumb:

Rule of thumb 1: When a time series is treated as a noise process and the estimated H parameter is close to 1, question your result; redo the analysis by treating the data as a random walk process.

Rule of thumb 2: When a time series is treated as a random walk process and the estimated H parameter is close to 0, do not trust your result; redo the analysis by integrating the data.

Rule of thumb 3: To be safe, perform DFA or the wavelet-based analysis on your data, along with FA (or other equivalent methods), and check the consistency of the results based on different methods.

8.7 CORRELATION STRUCTURE OF ON/OFF INTERMITTENCY AND LEVY MOTIONS

ON/OFF intermittency is a ubiquitous and important phenomenon. For example, a queuing system or a network can alternate between idle and busy periods; a fluid flow can switch from a turbulent motion to a regular (called laminar) one. In this section, we study the correlation structure of an ON/OFF train. It turns out that the correlation structures of ON/OFF models and Levy walks are closely related. Therefore, we shall also study the latter in this section. At the end of this section, the distinction between FA and DFA will be further illustrated.

8.7.1 Correlation structure of ON/OFF intermittency

Let us denote an ON period by 1 and an OFF period by 0. We study three types of ON/OFF trains where ON and OFF periods are independent and both have the same (1) exponential distribution, (2) Pareto distribution (Eq. (3.24)), and (3) Pareto distribution with truncation. For Pareto distributions, we choose two α: 1.6 and 0.6. Truncation is achieved by simply requiring $x \leq L$, where L is a parameter. When $1 \leq \alpha \leq 2$, it can be proven that

$$H = (3 - \alpha)/2. \tag{8.27}$$

One of our purposes is to check whether Eq. (8.27) can be numerically verified. To do this, we apply FA and DFA to the integrated data of an ON/OFF train. The ON/OFF train is sampled in such a way that in a total of about 1000 ON/OFF periods, on average a few tens of points of an ON or OFF period are sampled. The results for FA and DFA are shown in Figs. 8.6(a,c) and (b,d), respectively. We observe that for these three cases, for a small time scale (determined by the average length of an ON or OFF period), H (as the slopes of the lines) is close to 1 by FA and 1.5 by DFA. By simple analytical analysis or numerical simulation, one can readily find that for high frequency, the PSD for an ON/OFF train scales with the frequency as f^{-2}, just like the high-frequency end of an AR(1) model. Therefore, for time scales not longer than the average ON or OFF period, DFA is consistent

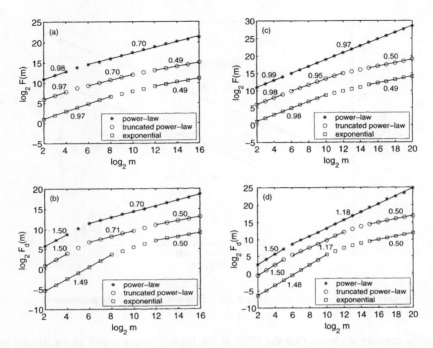

Figure 8.6. H parameter for the ON/OFF model. The α parameter is 1.6 for (a,b) and 0.6 for (c,d).

with PSD but FA is not. For larger scales, for case (1), we observe that H from both FA and DFA is 0.5 (with regard to Eq. (8.27), this amounts to taking $\alpha = 2$), while for cases (2) and (3), we observe that Eq. (8.27) is correct with FA when $1 \leq \alpha \leq 2$ and correct with DFA for the entire range of admissible α: $0 \leq \alpha \leq 2$. When $0 \leq \alpha < 1$ due to saturation, FA always gives $H = 1$. When the power-law distribution is truncated, H eventually becomes 1/2, both by FA and DFA.

8.7.2 Correlation structure of Levy motions

In Chapter 7, we studied stable laws and Levy motions. A stable law is a distribution with heavy tails. There are two types of Levy motions. One is Levy flights, which are random processes consisting of many independent steps, each step characterized by a stable law. The other is Levy walks, where the time consumed on each step is proportional to its length. As we noted in Chapter 7, a Levy walk can be obtained by resampling a Levy flight. Intuitively, we expect Levy flights to be memoryless, simply characterized by $H = 1/2$, irrespective of the value of the exponent α characterizing the stable laws. This is indeed the case, as is shown in Fig. 8.7. The correlation structure of a Levy walk, however, is more complicated. We observe from Fig. 8.8 that when the scale is small, corresponding to "walking" along a single step of a Levy flight, H is close to 1 by FA and close to 1.5 by DFA. Analysis by

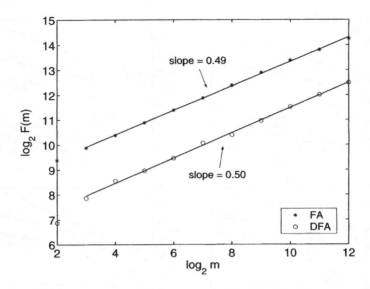

Figure 8.7. H parameter for Levy flights. H is independent of the parameter α.

Fourier transform shows that the PSD at the high-frequency end again decays as f^{-2}. Therefore, on smaller scales, DFA is consistent with PSD but FA is not. On larger scales, corresponding to constantly "switching" from one step of a Levy flight to another, H is given by Eq. (8.27) for FA when $1 \leq \alpha \leq 2$ and for DFA when $0 \leq \alpha \leq 2$. Again due to saturation, FA always yields $H = 1$ when $0 \leq \alpha < 1$. While these observations are similar to those found for the ON/OFF trains discussed above, we note a difference between Figs. 8.6 and 8.8. That is, for a Levy walk, the transition from a larger H at a small scale to a smaller H at a large scale is more gradual. This difference is due to the difference between a stable law and a Pareto distribution.

To further illustrate the similarity between Levy walks and ON/OFF trains, we have shown in Fig. 8.9 the data obtained by differencing a Levy walk. They resemble an ON/OFF train. Indeed, a moment's thought should convince one that when walking within a step of a Levy flight, one will be on one of the ON (or OFF) levels. However, depending on the stepsize of the Levy walk, around the transitions from one step of the original Levy flight to another, the pattern may deviate slightly from the ON/OFF train.

8.8 DIMENSION REDUCTION OF FRACTAL PROCESSES USING PRINCIPAL COMPONENT ANALYSIS

Principal component analysis (PCA) is a method for reducing the dimension of the original data by projecting the raw data onto a few dominant eigenvectors with large

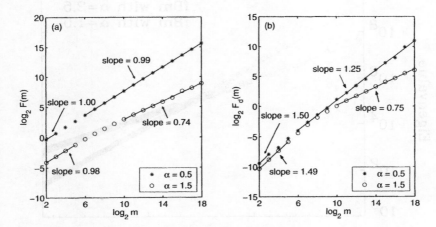

Figure 8.8. H parameter for Levy walks.

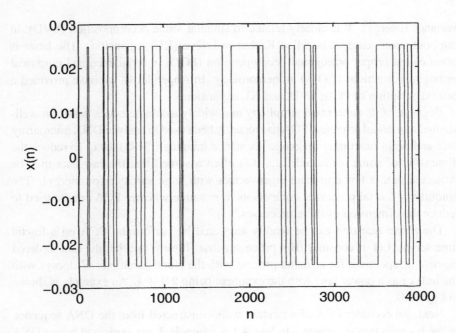

Figure 8.9. Difference data for a Levy walk.

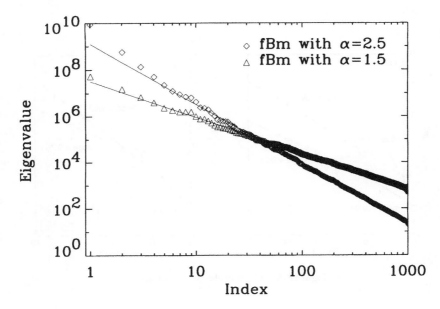

Figure 8.10. Eigenvalue spectrum for the fBm processes with $H = 0.25$ and 0.75.

variance (energy). It is closely related to singular value decomposition (SVD). In the continuous case, it is called Karhunen-Loève (KL) expansion. The latter is often called proper orthogonal decomposition (POD) in turbulence and empirical orthogonal functions (EOFs) in meteorology. In Appendix B, we have provided a brief description of PCA, SVD, and KL expansion.

Because of its conceptual simplicity and widely available codes based on well-studied numerical schemes, PCA has recently been used to analyze DNA microarray data and brain functional magnetic resonance imaging (fMRI) data. To reduce the dimension of some measured data, it is often assumed that the raw data may be projected onto a few dominant eigenvectors with large variance (or energy). The ubiquity of fractal processes compels us to examine whether PCA can be used to reduce the dimension of these processes.

The above question can be readily answered by performing PCA on a fractal time series. Let us consider fBm processes first. It turns out that the rank-ordered eigenvalue spectrum for a fractal process with the Hurst parameter H decays with the index i as a power law, with the exponent being $2H + 1$. An example is shown in Fig. 8.10.

Next, we consider PCA of a random walk constructed from the DNA sequence of the bacteria lambda phage. In Sec. 4.1.5, example 3, we explained how a DNA walk can be constructed. For the DNA walk of the bacteria lambda phage shown in Fig. 8.11(a), the rank-ordered eigenvalue spectrum is shown in Fig. 8.11(b). Again we observe a power-law decaying eigenvalue spectrum with exponent $\beta = 2.034$.

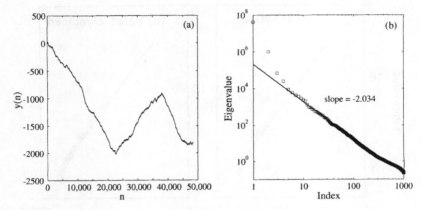

Figure 8.11.   The random walk for the DNA sequence of the bacteria lambda phage (a) and its eigenvalue spectrum (b).

The Hurst parameter is then $H = (\beta - 1)/2 = 0.517$, very close to the value estimated using DFA, which is 0.51.

It is interesting to note that the bacteria lambda phage DNA has long patches. FA of its DNA walk yields $H \approx 0.56$, leading to the interpretation that this DNA sequence has longmemory. The original motivation for developing DFA was to remove this patchy effect. DFA achieves this goal. PCA can serve the same purpose (except that it is much slower than DFA).

The analysis of the above two examples is easy to interpret: fractal time series cannot be represented by a few dominant eigenvectors and eigenvalues. Let us now perform a more complicated analysis — apply PCA to network data.

A network can be represented by an adjacency matrix A, where the elements a_{ij} are zeros when there is no interaction between the nodes i and j. For an undirected network, a_{ij} can be assigned a value of 1 when node i interacts with node j. This results in a symmetric matrix A and can be readily analyzed by eigen-decomposition. For a directed network, a_{ij} can be assigned a value of either 1 or -1, depending on whether i "controls" or is being "controlled" by j. The matrix A is then asymmetric. Note that the elements of the matrix may be assigned numbers other than ± 1 or 0 to explicitly reflect coupling strength. This is the case for microarray data. To analyze such matrices with eigen-decomposition, one can form either AA^T or $A^T A$, where the superscript T denotes transpose, and perform PCA. Alternatively, one can simply perform SVD on A.

Recently, it has been found that a power-law eigenvalue spectrum can be generated by a power-law network such as the global Internet. When PCA is applied to molecular interaction network data, such as microarray data (which monitor the expression of thousands of genes simultaneously under various conditions) and protein-protein interaction data, a powerlaw-like eigenvalue spectrum can also be observed. A few examples are shown in Fig. 8.12, where the data of (a) were used

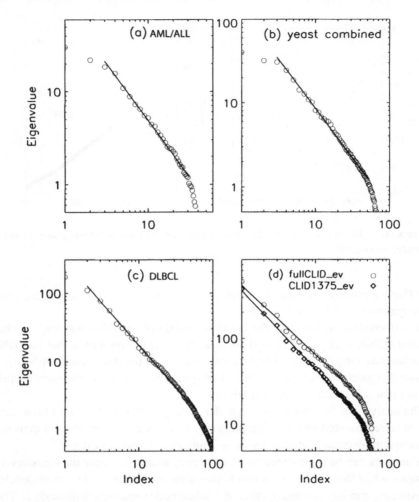

Figure 8.12. Eigenvalue spectrum, in log-log scale, for four microarray datasets.

to classify leukemia into acute myeloid leukemia (AML) and acute lymphoblastic leukemia (ALL), the data of (b) were used to study sporulation in budding yeast, the data of (c) were used for the classification of diffuse large B-cell lymphoma (DLBCL), and the data of (d) were used to study systematic variation in gene expression patterns in 60 human cancer cell lines. There are two curves in (d): one is for the full dataset, the other for a subset of the dataset (which was considered to have higher quality than the rest of the data). The power-law distributed eigenvalue spectra observed here may be a signature of dynamic correlations among different pathways expressed by microarray data in a gene transcriptional network.

We should emphasize here that although a power-law eigenvalue spectrum implies that the data cannot be described by a few dominant eigenvectors and eigenvalues, when one's purpose is to classify data into a few groups, one can try to perform PCA and retain only a few eigenvalues and eigenvectors, so long as one's goal can be effectively achieved. This is different than data modeling. Of course, in this regard, it is possible that eigenvalues and eigenvectors other than the few dominant ones may serve one's purpose better.

At this point, we should also mention that the eigenvalue spectrum for a chaotic time series decays exponentially. However, the number of eigenvalues following the exponential decay is typically larger than the dimension of the chaotic attractor. In this case, for the purpose of modeling the chaotic data, one has to retain all the exponentially decaying eigenvalues and their corresponding eigenvectors. However, for the purpose of pattern classification, one can use a smaller number of eigenvalues and eigenvectors, as noted earlier.

8.9 BROAD APPLICATIONS

The materials discussed in this chapter have found numerous applications in fields as diverse as physics, geophysics, physiology, bioinformatics, neuroscience, finance, and traffic engineering, among many others. Analysis and modeling of traffic data has already been discussed in this chapter and in Chapter 6. Below we examine three more examples: target detection within sea clutter radar returns, deciphering the causal relation between neural inputs and movements by analyzing neuronal firings, and gene finding from DNA sequences. These examples will further illustrate the importance of the two notions discussed in Sec. 8.6: fractal scaling break and consistency of the H estimators. This way, our appreciation of the power as well as the limitations of the concepts and methodologies will be greater.

8.9.1 Detection of low observable targets within sea clutter

Sea clutter is the backscattered returns from a patch of the sea surface illuminated by a radar pulse. Robust detection of targets from sea clutter radar returns is an important problem in remote sensing and radar signal processing applications. This is a difficult problem because of the turbulent wave motions on the sea surface as well as multipath propagation of radar pulses. In the past several decades, great efforts have been made to understand the nature of sea clutter as well as to detect targets within sea clutter. However, novel, simple, and reliable methods for target detection are yet to be developed. In this subsection, we show that the $H(q)$ spectrum together with the finite fractal scaling range offer a very simple and effective method to detect low observable targets within sea clutter.

First, we note that the details of the sea clutter data can be found in Sec. A.2 of Appendix A. The following analysis is based on 392 sea clutter time series, each containing 2^{17} complex numbers, sampled with a frequency of 1000 Hz. Visually, all these time series are similar to the one shown in Figs. 1.4(a–d). Similar signals

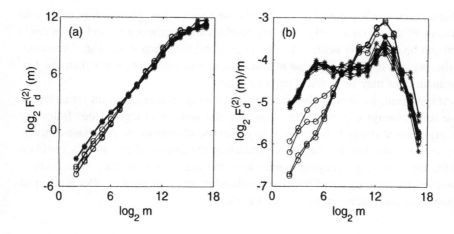

Figure 8.13. Target detection within sea clutter using DFA. Open circles designate data with a target, while asterisks designate data without a target.

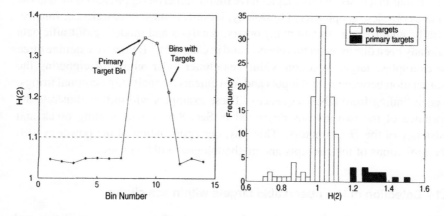

Figure 8.14. Left: $H(2)$ parameter for the 14 range bins of a measurement; Right: Histogram (equivalent to PDF) for the $H(2)$ parameter for all the measurements.

have been observed in many different fields. Therefore, the analysis below may also be applicable to those fields.

Let us denote the sea clutter amplitude data by $u_1, u_2, \cdots$, the integrated data by $v_1, v_2, \cdots$, and the differenced data by $w_1, w_2, \cdots$. First, we apply DFA to $v_1, v_2, \cdots$. A typical result for a measurement (which contains 14 range bins) is shown in Fig. 8.13(a). From it, one would conclude that the data have excellent fractal scaling behavior. However, this is an illusion due to the large y-axis range in the figure. If one reduces the y-axis range by plotting $\log_2[F_d^{(2)}(m)/m]$ vs. $\log_2 m$ (which can be viewed as detrended Fano factor analysis; see Sec. 8.3.3), then one finds that the curves for sea clutter data without a target change abruptly around

$m_1 = 2^4$ and $m_2 = 2^{12}$. Since the sampling frequency is 1000 Hz, they correspond to time scales of about 0.01 and 4 s. It turns out that if one fits a straight line to the $\log_2[F_d^{(2)}(m)/m]$ vs. $\log_2 m$ curves in this m range, then the H parameter can completely separate data with and without a target, as shown in Fig. 8.14. The last statement simply says that the H-based method achieves very high accuracy in detecting targets within sea clutter. We note that $H(q)$, $q \neq 2$ is similarly effective in detecting targets. Below we shall focus on $H(2)$, and simply abbreviate it as H.

Let us now make a few comments: (1) The time scales of 0.01 s and a few seconds have specific physical meanings: below 0.01 s, the data are fairly smooth and hence cannot be fractal; above a few seconds, the wave pattern on the sea surface may change; hence, the data may have a different behavior (possibly another type of fractal). With the available length of the data (which is about 2 min), the latter cannot be resolved, however. (2) If one tries to estimate H from other intervals of time (which would be the case when one tries to apply, say, maximum likelihood estimation), then H fails to detect targets within sea clutter. (3) The fractal scaling in the identified time scale range is not well defined, especially for data without a target. This implies that sea clutter data are too complicated for fractal scaling to characterize. (4) If one applies DFA to the u_i process, the original sea clutter amplitude data, then the estimated H_u is about $H_v - 1$, and the H-based method for target detection still works. (5) When FA is applied to the u_i process, the obtained H are similar to those obtained by DFA. Hence, FA is consistent with DFA. However, FA fails to work when it is applied to the integrated data, the v_i process, since all the estimated H_v cannot be larger than 1. (6) The wavelet H estimator is the most versatile. The H values obtained by applying the method to the u_i and v_i processes as well as the w_i process can all be used to detect the target, as shown in Fig. 8.15. In fact, H is increased by 1, progressing from w_i to u_i and from u_i to v_i. Neither FA nor DFA gives useful results when applied to the w_i process because of saturation of H at 0. (7) H for some datasets with targets is close to $1/3$, the H corresponding to the famous Kolmogorov energy spectrum of turbulence. This may be due to the development of wave-turbulence interactions around the target under favorable weather and sea conditions.

At this point, we ask a question: Why should the H from wavelet analysis of the u_i process (Fig. 8.15(b)) be compared to the H from DFA of the v_i process (Fig. 8.14)? We leave this as exercise 5.

8.9.2   Deciphering the causal relation between neural inputs and movements by analyzing neuronal firings

Biological systems have many properties such as complexity, robustness, reliability, and degeneracy. Such properties arise from the interactions of the components of a system as well as from the features of those components. Even without deep understanding of the dependence of those properties on the structure and features

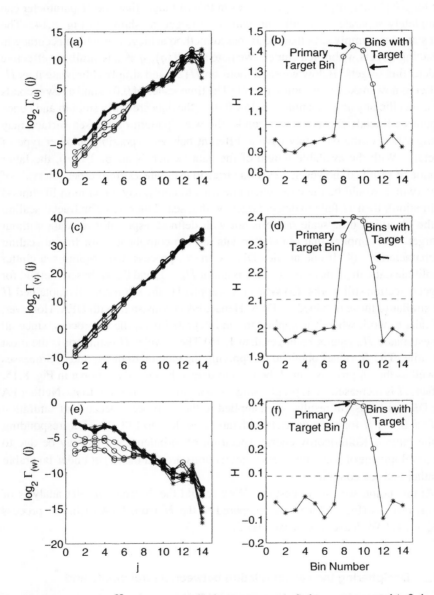

Figure 8.15.   Wavelet H estimator on (a,b) the u_i process, (c,d) the v_i process, and (e,f) the w_i process.

of biological systems, those properties have been providing guidelines for the design of many artificial systems. One fascinating artificial system is the brain-machine interface (BMI), whose purpose is to provide a method for people with damaged sensory and motor functions to use their brain to control artificial devices and restore their lost ability via the devices.

In recent years, many BMI researchers have demonstrated the feasibility of using adaptive input-output models to map the fundamental timing relations between neural inputs and hand movement trajectory. To achieve the mapping, model parameters are chosen in such a way that the difference between model output and hand movements is minimized using a statistical criterion such as meansquare error. The adaptive models used so far usually contain a very large number of parameters and require extensive training. This limits the optimal correlation between model output and hand trajectory to around 70–80% and prevents researchers from gaining deep understanding of the causal relation between neural inputs and hand movements. Furthermore, adaptive models assume that neuronal firings in the cortex are stationary, while in fact they are not (as will be shown shortly). To fundamentally advance the state of the art of BMI research, it has become increasingly important to develop new theoretical frameworks and methods to better understand neural information processing through characterization of the spatial-temporal dynamics of the firing patterns of a population of neurons. In this subsection, we show that DFA can help reveal causal relations between neuronal firings and hand trajectory. For details of the data, see Sec. A.3 of Appendix A.

8.9.2.1 Varying degree of correlation between neuronal firings and hand trajectory
To understand how neuronal firings control hand trajectory, it is instructive to examine the correlation between them. For this purpose, three consecutive hand movements are shown in Fig. 8.16(a), while neuronal firings of five neurons associated with those hand movements are shown in Figs. 8.16(b–f). A number of interesting features can be observed from Figs. 8.16(b–f):

1. The firing rate varies considerably among the neurons. For example, neuron 1, plotted in Fig. 8.16(b), fired far more than most other neurons.

2. The firing of neuron 1 in Fig. 8.16(b) does not have much correlation with hand trajectory. In fact, out of 104 neurons whose firings were recorded, more than half had little correlation with hand trajectory.

3. While the firing patterns of neurons 2–4, plotted in Figs. 8.16(c–e), appear to have strong correlations with the hand movement trajectory, the degree of correlation varies considerably with time. For example, neurons 2, 3, and 4 did not fire much during the monkey's first, second, and third periods of hand movement, respectively. It should be mentioned that although neuron 5, shown in Fig. 8.16(f), fired often during all three periods, it also had "quiet" periods even though the monkey was actively grabbing food and bringing it to its mouth.

These observations suggest that (1) different neurons have different degrees of importance in determining the causal relation between neural inputs and hand movements and that (2) even for the same neuron, the degree of importance varies considerably with time.

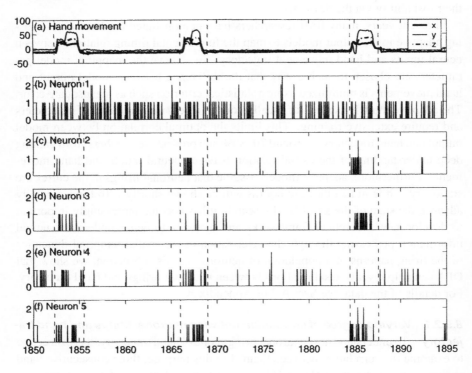

Figure 8.16. (a): X, Y, Z components of the monkey's hand movements. Dashed lines indicate time intervals when the monkey stretched its hand to grab food and subsequently placed the food in its mouth. (b–f): firings of five neurons associated with the hand movements plotted in (a).

8.9.2.2 *Heterogeneity of neuronal firings revealed by distributional analysis*
Conventionally, neuronal interspike interval data are modeled by exponential and gamma distributions (see Eqs. (3.12) and (3.15)). Besides these two distributions, many other distributions have been observed from the monkey's neuronal firing data, such as log-normal (Eq. (3.20)) and power-law distributions (Eq. (3.23)). Four examples are shown in Figs. 8.17(a–d) for exponential, gamma, log-normal, and power-law distributions, respectively, for four different neurons. This simple distributional analysis clearly indicates that the interspike interval data of different neurons may follow very different distributions and, therefore, that the firing patterns of the neurons can be considered very heterogeneous. Existence of multiple distributions implies existence of different stochastic processes underlying neuronal firings in the cortex.

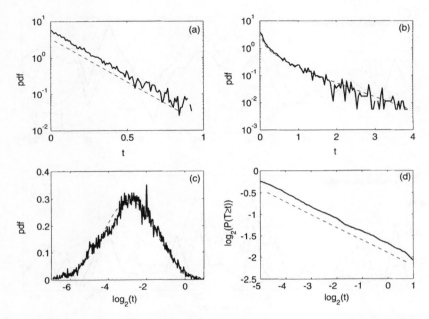

Figure 8.17. Four types of neuronal interspike interval distributions: (a) exponential, (b) gamma, (c) log-normal, and (d) power-law. Plotted in (a–c) and (d) are PDFs and the complementary cumulative distribution function (CCDF), respectively.

8.9.2.3 *Nonstationary neuronal firings revealed by linear and nonlinear correlation analysis*

The nonstationarity of observed neuronal firing patterns can be quantified by characterizing the correlations between neuronal firings and hand trajectory. Simple linear correlations can be assessed by cross-correlation analysis between the firing of a specific neuron and the hand trajectory. More general correlations, including nonlinear correlations, can be characterized by mutual information. To compute the dependence of cross-correlation and mutual information with time, one can partition the data into many small segments, then calculate the correlations between the corresponding segments, and finally plot the correlation against the time index associated with each segment. Let the segment of firing data of a neuron be denoted by $w(t)$, and let the hand trajectory (either the x, y, or z component) be denoted by $u(t)$. The cross-correlation, denoted by $C(w, u)$, can be calculated by the simple equation

$$C(w, u) = max_L < w(t)u(t - L) >,$$

where $<>$ denotes the average within the segment and L is a small time (not necessarily positive) chosen in such a way that $< w(t)u(t - L) >$ is maximized. When $C(w, u)$ is normalized by the standard deviations of $w(t)$ and $u(t)$, one obtains the correlation coefficient. Before taking the average within each segment, one could remove the mean values of $w(t)$ and $u(t)$ first. The mutual information

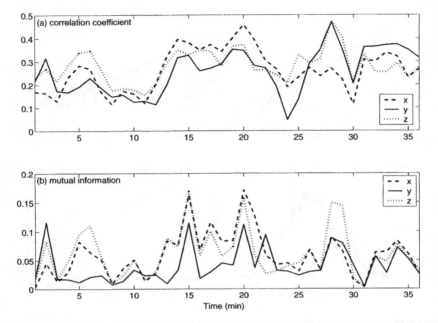

Figure 8.18.   Time-varying correlations between spike counting data and hand movement data. (a) correlation coefficient; (b) mutual information.

of $w(t)$ and $u(t)$, written as $I(W, U)$, is the amount of information gained about u when w is learned, and vice versa. Let the probability distribution for $w(t)$ and $u(t)$ be denoted by $P(W = w_i)$, $i = 1, \cdots, N_w$, and $P(U = u_i)$, $i = 1, \cdots, N_u$, respectively. Then

$$
\begin{aligned}
I(W, U) &= H(U) - H(W|U) = H(W) - H(U|W) \\
&= H(W) + H(U) - H(W, U) \\
&= \sum_{i=1}^{N_w} \sum_{j=1}^{N_u} P(W = w_i, U = u_j) \ln \frac{P(W = w_i, U = u_j)}{P(W = w_i)P(U = u_j)},
\end{aligned}
\tag{8.28}
$$

where $H()$ denotes entropy. $I(W, U) = 0$ if and only if W and U are independent. The variations of correlation coefficient and mutual information with time are shown in Figs. 8.18(a) and 8.18(b), respectively. Evidently, both types of correlations vary considerably with time. Since hand trajectory is stationary, this indicates that neuronal firing patterns are highly nonstationary.

8.9.2.4 *Long-range correlations in neuronal firings*   To gain further insights into the features defining the set of neurons that have strong correlations with hand trajectory (i.e., neurons similar to those shown in Figs. 8.16(c–f)), DFA is used to analyze spike-counting processes. The counting processes are obtained with the time window size $\Delta t = 0.1$ s. A few typical results are plotted in Fig. 8.19 for six neurons. It is observed that the lines are quite straight; therefore, the power-law relation of Eq. (8.22) is well defined. More interestingly, it is observed that three neurons, shown in Figs. 8.19(a–c), are characterized by a single fractal scaling, with the Hurst parameter ranging from about 0.5 to about 0.72, while three other neurons, shown in Figs. 8.19(d–f), have fractal scaling breaks around $m = 2^6 \sim 2^7$. Note that $m = 2^6 \sim 2^7$ corresponds to about $6.4 \sim 12.8$ s. For the range of m from 2^2 to about 2^7, the neurons shown in Figs. 8.19(d–f) have Hurst parameters all larger than 0.8. Interestingly, the time scale of $6.4 \sim 12.8$ s is comparable to the average time of 8 s between two successive reaching tasks (see Sec. A.3 of Appendix A). The neurons shown in Figs. 8.19(a–c) do not have much correlation with hand trajectory, just like the one shown in Fig. 8.16(b), while the neurons shown in Figs. 8.19(d–f) have very strong correlations with hand trajectory, like those plotted in Figs. 8.16(c–f). The large Hurst parameter and the time scale of $6.4 \sim 12.8$ s, therefore, strongly suggest that neurons well correlated with hand trajectory experienced a "resetting" effect at the start of each reaching task.

8.9.2.5 *Dynamic coalition of neurons*

The above analyses clearly indicate that, in executing a movement task, different neurons have different degrees of importance and that each neuron's importance can also change with time. In other words, the execution of a movement task is carried out by a subset of "important" neurons in a specific cortical area but not by all the neurons. These neurons may be termed a *dynamic coalition of neurons*. A coalition of neurons means that many types of excitatory and inhibitory interconnected neurons are involved, which support one another, directly or indirectly by increasing the activity of their fellow members. "Dynamic" means that neurons within the coalition may leave the coalition and not participate in neural information processing, such as controlling hand movements. After they leave the coalition, they may rejoin it later or new neurons can join. The process of leaving and joining the coalition makes the structure of the coalition network vary greatly over time.

Intuitively, the concept of a coalition of neurons is very attractive, since it generalizes the concept of synchronization, a treasured one in neuroscience. The picture becomes overwhelming if one analyzes the neuronal firing patterns of a large number of neurons simultaneously measured, such as those shown in Fig. 8.16: In the time interval of three hand movements shown in Fig. 8.16, *if we define the dynamic coalition of neurons by their correlation with the hand trajectory, then neuron 1 does not belong to the coalition, neuron 5 belongs to the coalition, and neurons 2 to 4 are transient members — they do not belong to the coalition at the first, second, and third hand movement, respectively.* We emphasize that few neurons were active in all the hand movements. This resulted in highly nonstationary neuronal firing patterns.

It is interesting to note that a similar concept, termed dynamic core, has been put forward by the Nobel laureate Edelman and his colleague Tononi. The concept

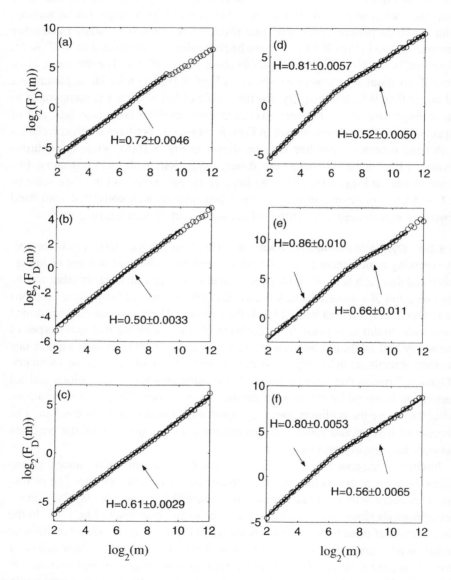

Figure 8.19.   DFA for the spike count data: (a–c) neurons with one scaling range and (d–f) neurons with two scaling ranges.

of dynamic core has recently been rephrased as coalition of neurons by Crick and Koch. Recently, this concept has been enriched by a study of Gao et al. [151] on long-range correlations in multistable (or ambiguous) visual perception (see Fig. 8.20). The basic picture in that study is the following. During multistable visual perception, a sustained winning coalition of neurons may be responsible for one of the possible percepts formed. The winning coalition, however, may break down and subsequently be either replaced by a competing coalition or itself transformed into a different coalition. This corresponds to the switching of percepts. The memory may come from the inertia of the coalitions: a strong and stable coalition has to be won over by another similarly stable and strong coalition, resulting in long switching times. The stochasticity comes from the observation that each coalition is composed of a dynamic group of neurons. Being dynamic, the structure of the network of the coalitions of neurons, especially the connectivity of the network, must be highly transient. Such a coalition may involve a few or many groups of neurons in one or a few regions of the brain, while each group in a specific region is only a fraction of all of the neurons in that region.

The above arguments suggest that the simplest model of neural information processing has to contain two levels. One level defines a subset of significant neurons, whose firings are well correlated with movements. The other level defines a signal component through the collective behavior of the neurons belonging to the dynamic coalition of neurons. This is a largely uncharted territory. For relevant references, see Sec. 8.10.

8.9.3   Protein coding region identification

We now consider gene finding, one of the most important problems in the study of genomes. Current computational approaches for finding genes are based either on comparative search or Markov (or hidden Markov) models. Both approaches require considerable knowledge of the genome sequence under investigation. In order to be successful, a gene-finding algorithm has to incorporate good indices that best discriminate coding and noncoding regions. While a number of good codon indices have been proposed, most of them are in one way or another related to the period-3 (P3) feature of coding sequences. The P3 feature is due to the fact that three nucleotide bases encode an amino acid and that the usage of codons is highly biased. This feature can be manifested in a number of similar ways. For example, if one maps a DNA sequence of length N to a numerical sequence and then takes the Fourier transform, typically one observes·a strong peak at (or around, if N is not a multiple of 3) $N/3$ in the magnitude of the Fourier transform if the sequence is a coding one. However, such a peak is either very weak or missing if the sequence is noncoding.

P3 characterizes certain regularity of a DNA sequence. A DNA sequence is, however, also random, with entropy per nucleotide base of around 1.9 bit. How can one simultaneously characterize both the P3 feature and the randomness of a DNA

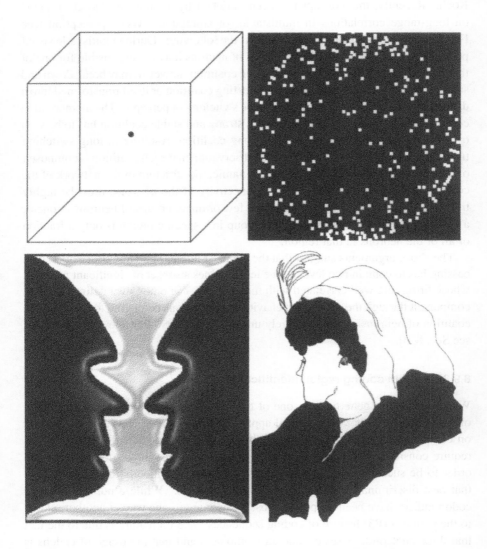

Figure 8.20.   Well-known ambiguous figures.   Upper left: Necker cube. When viewed steadily, the lower left side sometimes appears in the front of the cube, while at other times it appears to be in back.   Upper right: turning ball. It may appear to turn horizontally or vertically. It is an example of binocular rivalry, a form of interocular competition that arises when the patterns in the two eyes cannot be fused stereoscopically.   Lower left: Rubin's illustration of a vase and faces.   Lower right: Boring's classic picture of a young girl and an old lady.

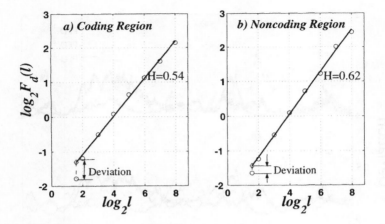

Figure 8.21.   Representative P3 fractal deviation for coding (a) and noncoding (b) DNA.

sequence? One simple and effective method is to characterize the fractal scaling deviation at the specific scale of P3. To perform fractal analysis, one can first map a DNA sequence to a numerical sequence, then take partial summation to form a random walk, and finally apply multifractal DFA. As shown in Figs. 8.21(a,b), P3-induced deviation from fractal behavior is typically much larger for a coding sequence than for a noncoding sequence. When one employs the sliding window technique and partitions the P3 fractal-deviation curve into three subsets (i.e., $1, 4, 7, \cdots$; $2, 5, 8, \cdots$; $3, 6, 9, \cdots$, which correspond to three reading frames), one obtains Fig. 8.22, from which it is observed that in coding regions, the three reading frame–specific deviation curves separate, while in noncoding regions they mix together. By quantifying the separation of the three reading frame–specific deviation curves, two very effective codon indices can be devised. The percentage accuracy of one of them (FD) is shown in Table 8.1, where the parameters N_i, $i = 1, 2$ denote the number of coding and noncoding sequences of all 16 yeast chromosomes with length equal to or larger than n_i, $i = 1, 2$, respectively. Compared with methods based on P3 only (obtained by Fourier transform, denoted by FM in the table) or fractal only (denoted by H in the table), simultaneous characterization of P3 and fractal behavior is far more accurate.

8.10   BIBLIOGRAPHIC NOTES

The basic concept for the study of LRD time series was formulated by the distinguished statistician Cox [87] in 1984. The concept was later introduced to the study of network traffic by Leland et al. [281] and Beran et al. [44] and made more rigorous by Tsybakov and Georganas [452]. An early good reference on the many estimators of the Hurst parameters (excluding the wavelet based ones) is [424]. A

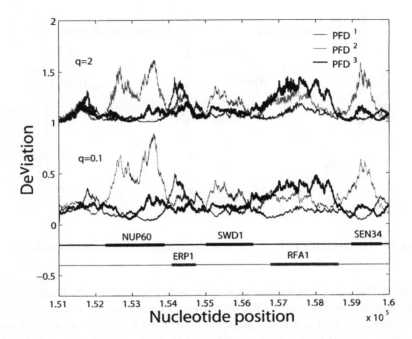

Figure 8.22.   The reading frame–specific fractal deviation curves for a segment of DNA in yeast chromosome I (from nucleotide 151000 to 160000). Horizontal bars on the two lines below the deviation curves are the open reading frames on the two strands of the chromosome (first line: positive strand; second line: reverse strand).

more recent one is [157]. The wavelet-based estimator in traffic engineering was first proposed by Abry and Veitch [4]. The structure-function–based multifractal was initially developed for the study of turbulence [143]. DFA was originally proposed by Peng et al. [341]; see [422] for a discussion on the relation between DFA and PSD; for further developments of DFA, see [72, 73, 227, 245]. The l_1 norm formulation of multifractal DFA was developed in [162], while the l_2 norm formulation was developed in [246]. The AR(1) model as a pseudo-LRD traffic model was explored in [5]. It was (erroneously) claimed in [253] as an exact model for the $1/f$ process.

Readers interested in applications are referred to [235] on human heartbeat dynamics, to [388] on analysis of earthquake data, to [474] on human gait data, to [422, 454, 487] on analysis of geophysical data, to [214, 228] on sea clutter, to [70, 186, 259, 382, 383, 393, 396, 428, 471] on BMI, to [88, 117, 496] on dynamic coalition of neurons, and to [154, 159, 160] on protein-coding sequence identification (In order not to overwhelm readers with too many references, only a few are cited here. Readers may also refer to references cited by those works as well as those that have cited the references included here.)

Coding / Noncoding	Sensitivity/Specificity		
	FD	FM	H
$(n_1, N_1) / (n_2, N_2)$	w=64	w=63	w=64
(1, 4125) / (1, 5993)	82.6%	70.9%	43.9%
(256, 4067) / (256, 4164)	84.3%	71.5%	45.2%
(512, 3756) / (512, 1939)	87.3%	71.2%	45.5%
(1026, 2674) / (512, 1939)	89.2%	72.0%	45.1%
(1026, 2674) / (1026, 638)	92.4%	71.1%	43.7%

Table 8.1   Accuracy of the gene-finding algorithm

Readers interested in PCA of fractal processes are referred to [153]. For spectral analysis of networks, see [79, 110, 125, 126, 191, 193, 309]. Finally, for information on the four microarray datasets, see [9, 78, 196, 374].

8.11 EXERCISES

1. Write simple computer programs for the variance-time method, R/S statistic, FA, and DFA; apply them to the fGn/fBm data generated by you. If you are unable to generate the data, you may download some of them at http://www.gao.ece.ufl.edu/book_data/. Check out the methods by integrating/differencing the data.

2. Implement FA- and DFA-based multifractal analysis; apply your programs to your own data or those downloaded from the book's website.

3. Compute PSD for the AR(1) model by taking the Fourier transform of its autocorrelation function. Numerically generate a time series from the AR(1) model and compute its PSD. Check to determine if your simulation is consistent with Eq. (8.25).

4. Repeat the results plotted in Figs.8.5, 8.6, and 8.8 (hint: choose a suitable sampling time; do not make it either too small or too large).

5. Explain why the H from wavelet analysis of the u_i process (Fig. 8.15(b)) should be compared to the H from DFA of the v_i process (Fig. 8.14).

CHAPTER 9

MULTIPLICATIVE MULTIFRACTALS

We have discussed $1/f^\beta$ processes and Levy motions. Now we will study the third type of random fractal model, the random cascade multifractal. This type of multifractal was initially developed to understand the intermittent features of turbulence. Mandelbrot was among the first to introduce this concept. Parisi and Frisch's work has made it widely known. It has been applied to the study of various phenomena, such as rainfall and liquid water distributions inside marine stratocumuli. It has also been applied to the study of finance by Mandelbrot and to network traffic modeling by numerous researchers.

Below, we first present the definition of measure-based multifractals and explain a simple construction procedure for generating a random cascade multifractal. After presenting examples and proving a number of properties of such fractal processes, we explain how the model can be used to understand turbulent motions, to analyze sea clutter and neuronal firing patterns, and to model network traffic.

9.1 DEFINITION

Consider a unit interval. Associate it with a unit mass. Partition the unit interval into a series of small intervals, each of linear length ϵ. Also, partition the unit mass into a series of weights or probabilities $\{w_i\}$ and associate w_i with the ith interval.

Now consider the moments

$$M_q(\epsilon) = \sum_i w_i^q, \tag{9.1}$$

where q is real. We follow the convention that whenever w_j is zero, the term w_j^q is set to zero as well. We also note that a positive q value emphasizes large weights, while a negative q value emphasizes small weights. If we have, for a real function $\tau(q)$ of q,

$$M_q(\epsilon) \sim \epsilon^{\tau(q)}, \quad as \ \epsilon \to 0 \tag{9.2}$$

for every q, and the weights $\{w_i\}$ are nonuniform, then the weights $w_i(\epsilon)$ are said to form a multifractal measure. Note that the normalization $\sum_i w_i = 1$ implies that $\tau(1) = 0$.

Note that if $\{w_i\}$ are uniform, then $\tau(q)$ is linear in q. When $\{w_i\}$ are weakly nonuniform, visually $\tau(q)$ may still be approximately linear in q. The nonuniformity in $\{w_i\}$ is better characterized by the so-called generalized dimensions D_q, defined as

$$D_q = \frac{\tau(q)}{q-1}. \tag{9.3}$$

D_q is a monotonically decreasing function of q. It exhibits a nontrivial dependence on q when the weights $\{w_i\}$ are nonuniform.

9.2 CONSTRUCTION OF MULTIPLICATIVE MULTIFRACTALS

To better appreciate the construction rules, we point out that these rules essentially involve dyadic partitions.

Consider a unit interval. Associate it with a unit mass. Divide the unit interval into two, say, left and right segments of equal length. Also, partition the associated mass into two fractions, r and $1 - r$, and assign them to the left and right segments, respectively. The parameter r is in general a random variable, governed by a PDF $P(r), 0 \leq r \leq 1$. The fraction r is called the multiplier. Each new subinterval and its associated weight are further divided into two parts following the same rule. This procedure is schematically shown in Fig. 9.1, where the multiplier r is written as r_{ij}, with i indicating the stage number and j (assuming only odd numbers, leaving even numbers for $1 - r_{ij}$) indicating the positions of a weight on that stage. Note that the scale (i.e., the interval length) associated with stage i is 2^{-i}. We assume that $P(r)$ is symmetric about $r = 1/2$ and has successive moments $\mu_1, \mu_2, \cdots$. Hence r_{ij} and $1 - r_{ij}$ both have marginal distribution $P(r)$. The weights at the stage N, $\{w_n, n = 1, ..., 2^N\}$, can be expressed as

$$w_n = u_1 u_2 \cdots u_N,$$

where $u_l, l = 1, \cdots, N$, are either r_{ij} or $1 - r_{ij}$. Thus, $\{u_i, i \geq 1\}$ are iid random variables having PDF $P(r)$.

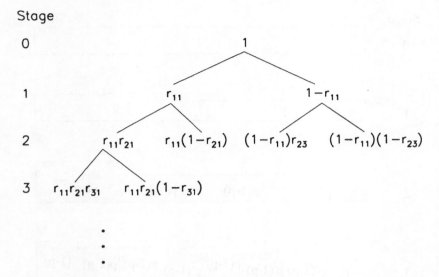

Figure 9.1.   Schematic illustrating the construction rule of a multiplicative multifractal.

Note that the rules described above generate a conservative process. To generate a nonconservative process, one can simply require that a mass m_i at stage i is split into two parts, $W_1 m_i$ and $W_2 m_i$, where W_i, $i = 1, 2$ are independent, follow the same PDF, and both have mean 1/2. Since the latter does not ensure that $W_1 + W_2 = 1$, the sum of the weights at stage n is not necessarily equal to 1.

In the following, we illustrate this process by selecting a specific pdf $P(r)$.

• Deterministic binomial multiplicative process

In this case, the PDF is set to be equal to $P(r) = \delta(r - p)$, where $\delta(x)$ is the Kronecker delta function. Thus, $r = p$ with probability 1, where $0 < p < 1$ is a fixed number. The weights obtained for the first several stages are schematically shown in Fig. 9.2.

For this process, at stage n, we have

$$M_q(\epsilon) = \sum_{i=0}^{n} C_n^i p^{qi} (1 - p)^{q(n-i)} = [p^q + (1 - p)^q]^n. \qquad (9.4)$$

Since, at stage n, $\epsilon = 2^{-n}$, we obtain

$$\tau(q) = -\ln[p^q + (1 - p)^q]/\ln 2, \qquad (9.5)$$

which is independent of n (or ϵ). Hence, this weight process constitutes a multifractal.

It is interesting to note that a similar model has been proposed by meteorologists to describe the stochastic growth process of raindrops. Fig. 9.3 schematically shows that from one time step to another, 90% of raindrops remain their original size, while

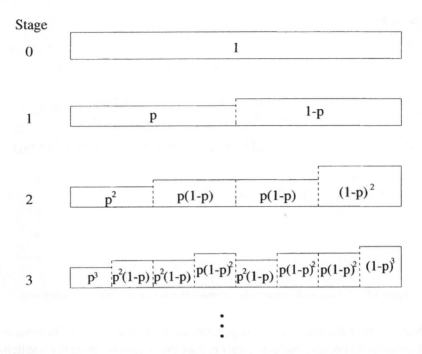

Figure 9.2.   A schematic showing the weights at the first several stages of the binomial multiplicative process ($r = p$ w.p.1).

10% of them increase their size by a fixed ratio. The physics behind this picture is that a raindrop attracts the water vapor in the surrounding air in an amount proportional to its surface area with probability 0.1 but fails to attract any water vapor with probability 0.9. Although simple, this model is able to account for the observed rapid growth of raindrops.

- **Random binomial multiplicative process**

To make the weight series random, we modify $P(r)$ to become

$$P(r) = [\delta(r - p) + \delta(r - (1 - p))]/2 \qquad (9.6)$$

so that $P(r = p) = P(r = 1-p) = 1/2$. Hence, $P(r)$ is symmetric about $r = 1/2$. A realization of the weights at stage 12 (with $p = 0.3$) is shown in Fig. 9.4(a). It is quite obvious that the $\tau(q)$ spectrum for this process is identical to that for the deterministic binomial process, since the weight sequence for this process is simply a shuffled version of that for the deterministic case.

- **Random multiplicative process**

The function $P(r)$ can be selected to follow any functional form. The following piecewise linear $P(r)$ function is used to generate the weight realization (at stage 12) shown in Fig. 9.4(b):

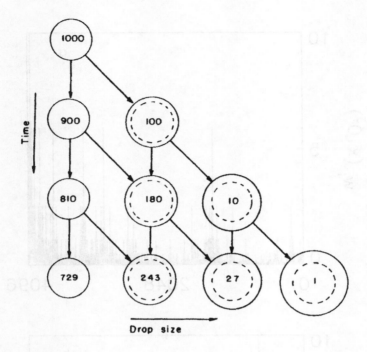

Figure 9.3.   Schematic of raindrop growth. After [47, 224].

$$P(r) = \begin{cases} 2r + 0.5 & \text{if } 0 \le r \le 0.5 \\ -2r + 2.5 & \text{if } 0.5 \le r \le 1. \end{cases} \tag{9.7}$$

9.3   PROPERTIES OF MULTIPLICATIVE MULTIFRACTALS

For the weights of a conservative cascade model at stage N, we prove the following properties.

(1)

$$M_q(\epsilon) \sim \epsilon^{\tau(q)}, \tag{9.8}$$

with $\epsilon = 2^{-N}$ and

$$\tau(q) = -\ln(2\mu_q)/\ln 2, \tag{9.9}$$

where μ_q is the qth moment of the iid random variables u_i. This follows from the observation that at stage N,

$$M_q(\epsilon) = \sum_{n=1}^{2^N} (w_n)^q \approx 2^N E(w^q) = 2^N E((u_1 u_2 \cdots u_N)^q) = 2^N \mu_q^N.$$

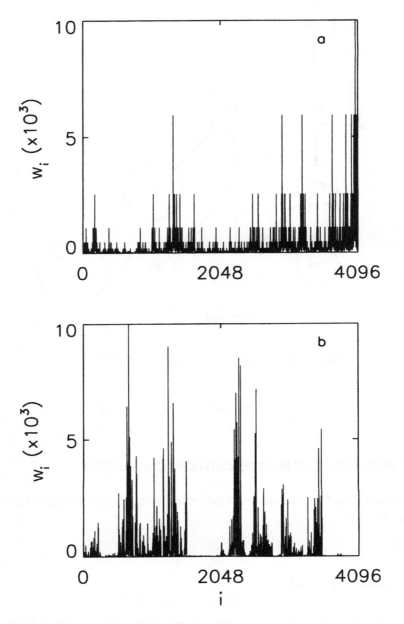

Figure 9.4.   Weight series at stage 12 for (a) the modified binomial multiplicative process ($p = 0.3$) and (b) the random multiplicative process with the multiplier PDF given in Eq. (9.7). The scale of w_i is multiplied by 10^3, as indicated in the parentheses.

This property indicates that a multiplicative process is a multifractal and relates the $\tau(q)$ spectrum to the moments of the multiplier distribution.

Note that for $P(r) = [\delta(r-p) + \delta(r-(1-p))]/2$, we have $\mu_q = [p^q + (1-p)^q]/2$. Hence, $\tau(q) = -\ln[p^q + (1-p)^q]/\ln 2$. This is identical to Eq. (9.5). However, Eq. (9.9) is much more general and, hence, more powerful.

(2)

$$E(w) = E(w_n) = E(u_1 u_2 \cdots u_N) = 2^{-N}, \quad n = 1, \cdots, 2^N. \quad (9.10)$$

(3)

$$E(w^q) = E((u_1 u_2 \cdots u_N)^q) = \mu_q^N. \quad (9.11)$$

In particular,

$$E(w^2) = \mu_2^N \quad (9.12)$$

and

$$Var(w) = Var(w_n) = \mu_2^N - 2^{-2N}, \quad n = 1, \cdots, 2^N. \quad (9.13)$$

(4) When $N >> 1$, the weights at stage N have a log-normal distribution. This is deduced directly by taking the logarithm of $w_n = u_1 u_2 \cdots u_N$ and applying the central limit theorem.

Note that log-normality is a salient feature that has been observed extensively in many different types of network traffic processes, including VBR video traffic, the call duration and dwell time in a wireless network, data connection and messages, and page size, page request frequency, and user's think time in WWW traffic. Hence this property can be used to determine if measured traffic trace data are consistent with a multiplicative multifractal process model.

(5)

$$E[(w_n - E(w))(w_{n+m} - E(w))] = (1/2 - \mu_2)\mu_2^{N-1}(4\mu_2)^{-k} - 2^{-2N} \quad (9.14)$$

for $m = 2^k$, where k is an integer. Hence, the covariance function decays with time lag m in a power-law manner.

To prove assertion (5), we consider two weights, w_{n_1} and w_{n_2}, at stage N. Assume that they share the same ancestor weight x at stage $N - k$, i.e.,

$$w_{n_1} = x \cdot r \cdot \prod_{l=1}^{k-1} r_{1l},$$

$$w_{n_2} = x \cdot (1 - r) \cdot \prod_{l=1}^{k-1} r_{2l},$$

where r and $\{r_{il}, i = 1, 2, l = 1, ..., k-1\}$ are independent random variables with distribution $P(r)$. Then

$$E[(w_{n_1} - 2^{-N})(w_{n_2} - 2^{-N})] = E(x^2)E[r(1-r)]E[\prod_{l=1}^{k-1} r_{1l}r_{2l}] - 2^{-2N}$$
$$= 2^{-2(k-1)}\mu_2^{N-k}(1/2 - \mu_2) - 2^{-2N}.$$

For $m = 2^k$, all pairs of $\{w_n, w_{n+m}\}$, for $n \geq 1$, share an ancestor at stage $N - k - 1$. Hence,

$$E[(w_n - E(w))(w_{n+m} - E(w))] = 2^{-2k}\mu_2^{N-k-1}(1/2 - \mu_2) - 2^{-2N}$$
$$= (1/2 - \mu_2)\mu_2^{N-1}(4\mu_2)^{-k} - 2^{-2N}.$$

(6)
$$Var(W^{(m)}) = \mu_2^N(4\mu_2)^{-k} - 2^{-2N}, \tag{9.15}$$

where

$$W^{(m)} = (w_{im-m+1} + \cdots + w_{im})/m, \quad m = 2^k, \quad k = 1, 2, \cdots, \text{ and } i \geq 1.$$

This is proved by expressing $W^{(m)} = 2^{-k}x$, where x is a weight at stage $N - k$.

Equation (9.15) expresses a variance-time relation. For time series with long memory, $Var(W^{(m)}) \sim m^{2H-2}$, where $1/2 < H < 1$ is the Hurst parameter. For multiplicative multifractal processes, when N is large and $\mu_2 > 0$, the term $\mu_2^N(4\mu_2)^{-k}$ dominates. When the term 2^{-2N} in Eq. (9.15) is dropped, the functional variation of $\log Var(W^{(m)})$ vs. $\log m$ is linear. The resulting slope, $-\log(4\mu_2)/\log 2$, provides an estimate of $2H - 2$. A moment's thought will convince us that this slope is an upper bound for $2H - 2$. Hence,

$$H \leq -\frac{1}{2}\log_2 \mu_2. \tag{9.16}$$

Since the multiplier distribution $P(r)$ is defined for $0 \leq r \leq 1$ and is symmetric about 1/2, its mean is 1/2 and its variance is upper bounded by 1/4. We thus have

$$\left(\frac{1}{2}\right)^2 \leq \mu_2 \leq \left(\frac{1}{2}\right)^2 + \frac{1}{4}.$$

Therefore $1/2 \leq H \leq 1$, with $H = 1$ corresponding to deterministic time series (i.e., $P(r) = \delta(r - 1/2)$). We thus observe that a multiplicative multifractal process also possesses a long-range-dependent property, especially for time scales not too close to the total time span of observation.

Let us see how good the linearity defined by Eq. (9.15) is. For this purpose, we consider three different functional forms for the multiplier function, namely, double exponential with parameter α_e,

$$P(r) \sim e^{-\alpha_e|r-1/2|} \tag{9.17}$$

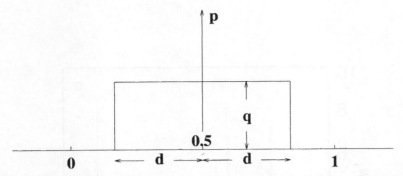

Figure 9.5. A schematic showing the functional form for the multiplier distribution described by Eq. (9.19).

Gaussian with parameter α_g,

$$P(r) \sim e^{-\alpha_g(r-1/2)^2} \tag{9.18}$$

and a function of the form

$$P(r) = \begin{cases} q + p\delta(r - 1/2) & 1/2 - d \leq r \leq 1/2 + d \\ 0 & \text{otherwise,} \end{cases} \tag{9.19}$$

where $0 \leq d \leq 1/2$. The last function is schematically shown in Fig. 9.5. Note that the three parameters d, p, and q are related by the equation $p + 2qd = 1$. Hence, the function contains two independent parameters. We shall choose p and d as the two basic parameters. Note that one may introduce a parameter equivalent to d for the functions characterized by Eqs. (9.17) and (9.18). Due to exponential decay of Eqs. (9.17) and (9.18), however, such a parameter is not too interesting. Also note that we may rewrite Eqs. (9.17) and (9.18) as $P(r) \sim e^{-\alpha|r-1/2|^\beta}$, with $\beta = 1$ for the double exponential and 2 for the Gaussian. Hence, Eqs. (9.17) and (9.18) really contain two parameters with a prefixed parameter β.

We generate a number of realizations of multiplicative processes with $P(r)$ given by one of the above three forms. Two examples are shown in Fig. 9.6. We then compute the variance-time relation from the generated time series. Some examples of the variance-time curves are shown in Fig. 9.7, with $P(r)$ given by Eq. (9.18) and $N = 18$. We observe that the variance-time curves are indeed approximately linear, with the linearity better defined for larger μ_2.

We can furthermore check how tight the bound determined by Inequality (9.16) is by estimating H from the variance-time curves and comparing it with the right-hand side of Inequality (9.16). Figure 9.8 shows such a comparison, where the dashed curve is generated from Eq. $H = -\frac{1}{2}\log_2 \mu_2$. The points denoted by the diamonds, triangles, and squares are estimated from multiplicative processes with $P(r)$ given by Eqs. (9.17), (9.18), and (9.19), respectively. We observe that the bound given by Inequality (9.16) is very tight, especially for not too small values of μ_2.

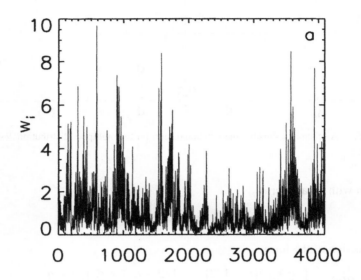

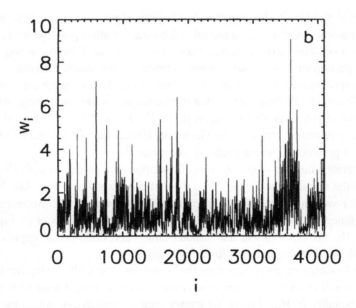

Figure 9.6. Weight series at stage 12 with the multiplier distributions given by (a) Eq. (9.18) with $\alpha_g = 30$ and (b) Eq. (9.19) with $(p, d) = (0.7, 0.4)$.

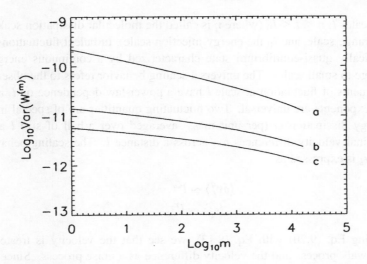

Figure 9.7. $\log Var(W^{(m)})$ vs. $\log m$ curves for (a) $\alpha_g = 10$, (b) $\alpha_g = 50$, and (c) $\alpha_g = 100$.

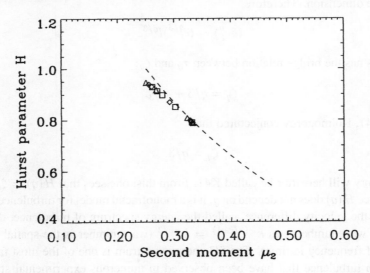

Figure 9.8.   Hurst parameter vs. the second moment μ_2. The dashed line is computed according to the right-hand side of Inequality (9.16), while the points designated by squares, diamonds and triangles are directly estimated from multiplicative processes with their multiplier distributions given by Eqs. (9.17), (9.18), and (9.19), respectively.

9.4   INTERMITTENCY IN FULLY DEVELOPED TURBULENCE

Understanding turbulent flows is one of the greatest challenges in physics. An intriguing aspect of fully developed turbulence is the possible *existence of universal* scaling behavior of small-scale fluctuations. In large Reynolds number flows, at spatial scales l, $\eta \leq l \leq l_0$ (where η is called the molecular dissipation scale, l the inertial-range scale, and l_0 the energy-injection scale), turbulent fluctuations reach a statistically quasi-equilibrium state characterized by a continuous energy flux from large to small scales. The universal scaling behavior refers to the observation that moments of fluctuation at scale l have a power-law dependence on l, and the scaling exponents are universal. Two fluctuating quantities are of special interest: the energy dissipation ϵ_l (per unit mass) averaged over a ball of size l and the longitudinal velocity differences δv_l across a distance l. The scaling behavior of δv_l and ϵ_l is expressed as

$$\langle \delta v_l^q \rangle \sim l^{\zeta_q}, \tag{9.20}$$

$$\langle \epsilon_l^q \rangle \sim l^{\tau_q}. \tag{9.21}$$

Comparing Eq. (9.20) with Eq. (8.13), we see that the velocity is treated as a random walk process and the velocity difference as a noise process. Since ϵ_l has the dimension $L^2 T^{-3}$, where L denotes length and T time, $(l\epsilon_l)^{q/3}$ and δv_l^q have the same dimension. Therefore,

$$\langle \delta v_l^q \rangle \sim \langle \epsilon_l^{q/3} \rangle l^{q/3}. \tag{9.22}$$

One thus has the bridge relation between τ_q and ζ_q:

$$\zeta_q = q/3 + \tau_{q/3}. \tag{9.23}$$

In 1941, Kolmogorov conjectured that

$$\zeta_q = q/3. \tag{9.24}$$

This theory will hereafter be called K41. From this, one sees that $H(q) = \zeta_q/q = 1/3$. Since $H(q)$ does not depend on q, it is a monofractal model for turbulence. The PSD for the velocity difference, called the energy spectrum of turbulence, decays with the wavenumber k as $k^{-(2H+1)} = k^{-5/3}$ (wavenumber is the spatial counterpart of frequency in time). The 5/3 energy spectrum is one of the most famous results in turbulence that have been observed in numerous experimental studies. However, Eq. (9.24) for arbitrary q is not supported by experiments and numerical simulations. See the sixth column of Table 9.1. This is called the intermittency phenomenon of turbulence. Many models have been proposed to explain intermittency. Random cascade multifractal models are among the most successful. The basic idea is

$$\epsilon_{l_2} = W_{l_1 l_2} \epsilon_{l_1}, \tag{9.25}$$

where $W_{l_1 l_2}$ is the cascade multiplicative factor for any arbitrary pair of length scales l_1, l_2, $l_1 > l_2$. Energy conservation requires that

$$\langle W_{l_1 l_2} \rangle = 1. \tag{9.26}$$

| Order | Experiment | Logstable | P model | SL model | K41 |
q	ζ_q	ζ_q	ζ_q	ζ_q	$\zeta_q = q/3$
1	0.37	0.362	0.361	0.364	0.333
2	0.70	0.694	0.694	0.696	0.667
3	1.00	1.000	1.000	1.000	1.000
4	1.28	1.285	1.282	1.279	1.333
5	1.54	1.551	1.543	1.538	1.667
6	1.78	1.800	1.786	1.778	2.000
7	2.00	2.032	2.014	2.001	2.333
8	2.23	2.249	2.229	2.211	2.667
10	2.60	2.638	2.632	2.593	3.333

Table 9.1   Comparison of the experimentally measured scaling exponents ζ_q and the theoretical predictions

Unlike conservative cascade models discussed earlier, this is a nonconservative model. Using Eq. (9.20), one has $\epsilon_{l_i} \sim l_i^{\tau_q}$, $i = 1, 2$. Therefore,

$$\ln\langle W_{l_1 l_2}^q \rangle = \tau_q(\ln l_2 - \ln l_1). \tag{9.27}$$

Below we discuss the concept of extended self-similarity, which facilitates experimental determination of ζ_q, and describe a few of the better-known models.

9.4.1 Extended self-similarity

To observe fully developed turbulence, the Reynolds number, $R = LV/\nu$, where L, V are the characteristic scale and velocity of the flow and ν is its (kinetic) viscosity, has to be very large. When R is not large, the inertial range is very short, and the scaling behavior described by Eq. (9.21) is either absent or difficult to observe. A few examples are shown at the top of Fig. 9.9. This hinders accurate experimental determination of ζ_q. The concept of extended self-similarity provides a nice solution to this problem. Benzi et al. [42] find that the scaling properties of the velocity increments can be extended up to the dissipation range in the form

$$\langle \delta v_l^q \rangle \sim \langle \delta v_l^3 \rangle^{\zeta_q/\zeta_3}, \quad l \geq 5\eta. \tag{9.28}$$

That is, if one plots $\langle \delta v_l^q \rangle$ against $\langle \delta v_l^3 \rangle$, then the scaling behavior is much better defined, and the scaling exponent is given by ζ_q/ζ_3. For an example of $q = 6$, see the bottom of Fig. 9.9. The fact that $\zeta_3 = 1$ makes this simple procedure even more attractive.

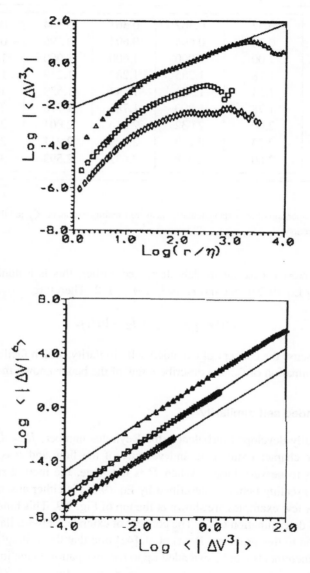

Figure 9.9. Top: Log-log plot of $\langle \delta v_l^3 \rangle$ against r/η for flows with Reynolds number 300,000 (triangles), 18,000 (squares), and 6000 (diamonds). The solid line correspond to a slope equal to 1. Bottom: Log-log plot of $\langle |\delta v_l^6| \rangle$ against $\langle |\delta v_l^3| \rangle$. Note that when $\log \langle |\delta v_l^3| \rangle$ is plotted against $\log |\langle \delta v_l^3 \rangle|$, one finds an excellent linear scaling relation with a slope very close to 1. From Benzi et al. [42].

9.4.2   The log-normal model

This is the first and perhaps the most famous (but not necessarily the best) model. The idea is that the scales between l_1 and l_2 can be further partitioned so that $W_{l_1 l_2}$ can be expressed as multiplication of a sequence of iid random variables. Taking the log, then, $\ln W_{l_1 l_2}$ becomes a summation of a sequence of iid random variables. When the central limit theorem is applicable, $W_{l_1 l_2}$ follows the log-normal distribution. More concretely, let $\delta = l_2/l_1$; then

$$W_{l_1 l_2} \overset{d}{=} \delta^X, \tag{9.29}$$

where X is $N(\mu_x, \sigma^2)$.

Let us now compute $\langle W_{l_1 l_2}^q \rangle$. It is simply

$$\langle \delta^{qX} \rangle = \langle e^{q \ln \delta X} \rangle.$$

The right-hand side of the equation can be obtained by replacing ju in the characteristic function of $N(\mu_x, \sigma^2)$ (see Eq. (3.18)) by $q \ln \delta$. Therefore,

$$\langle W_{l_1 l_2}^q \rangle = e^{q \ln \delta \mu_x + \sigma^2 q^2 (\ln \delta)^2 / 2} = \delta^{q \mu_x + \sigma^2 q^2 \ln \delta / 2}.$$

When $q = 1$, $\langle W_{l_1 l_2} \rangle = 1$. Therefore, $\mu_x + \sigma^2 \ln \delta / 2 = 0$. Using Eq. (9.27), we then have $\tau_q = \mu_x(q - q^2)$. It is common to write $\mu_x = \mu/2$. Hence,

$$\tau_q = \frac{\mu}{2}(q - q^2), \quad \zeta_q = q/3 + \frac{\mu}{2}[q/3 - (q/3)^2]. \tag{9.30}$$

A good value for μ is 0.3288.

There are a number of problems with the log-normal model. Two are pointed out here: (1) ζ_q is a decreasing function of q when $q > 3/2 + 3/\mu$. Frisch has proven that this causes a velocity to diverge (i.e., supersonic flow); and (2) The log-normal model violates Novikov's inequality, which states that, for three dimensions,

$$\tau_q + 3q \geq 0, \quad \text{for } q \geq 0 \quad \text{and} \quad \tau_q + 3q \leq 0, \quad \text{for } q \leq 0$$

and for one dimension,

$$\tau_q + q \geq 0, \quad \text{for } q \geq 0 \quad \text{and} \quad \tau_q + q \leq 0, \quad \text{for } q \leq 0.$$

Novikov's inequality is a constraint on energy: $l^3 \epsilon_l$ (in three dimensions) or $l \epsilon_l$ (in one dimension) is the total unaveraged energy and, hence, should be a nondecreasing function of l. When either one is raised to the power of q, it should be a nondecreasing function of l when $q \geq 0$ and a nonincreasing function of l when $q \leq 0$.

The log-normal model and many other random cascade models violate Novikov's inequality because of the nonconservative character of the cascade.

9.4.3 The log-stable model

This model applies the generalized central limit theorem to the random variable X in Eq. (9.29). $\langle W_{l_1 l_2}^q \rangle$ can again be simply obtained by replacing ju in the characteristic function of X (see Eq. 7.6) by $q \ln \delta$. Upon simplification, it becomes $\delta^{bq - \beta q^\alpha}$, where β is a complicated coefficient containing the term $|\ln \delta|^\alpha / \ln \delta$. When the condition $\langle W_{l_1 l_2} \rangle = 1$ is used, β is found to be b. Therefore, $\tau_q = b(q - q^\alpha)$. Using Kida's notation, $b = \mu/(2^\alpha - 2)$, we have

$$\tau_q = -\mu \frac{q^\alpha - q}{2^\alpha - 2},$$

$$\zeta_q = q/3 - \mu \frac{(q/3)^\alpha - q/3}{2^\alpha - 2}. \tag{9.31}$$

Kida found that the model gives an excellent fit to experimental data when $\alpha = 1.65$, $\mu = 0.2$. See the third column of Table 9.1. It is clear that the log-normal model is a special case of this model, and, as expected, the log-stable model has the same two problems as the log-normal model discussed in the previous subsection.

9.4.4 The β-model

The β-model is one of the earliest multifractal models proposed to describe intermittency in turbulence. Let a mother eddy at stage i generate N daughter eddies at stage $i + 1$. The size of the mother eddy is $l_i = r^i l_0$, while the size of the daughter eddies is $l_{i+1} = r l_i = r^{i+1} l_0$. It is clear that the fractal dimension of this cascade is

$$D = -\ln N / \ln r. \tag{9.32}$$

Let β denote the space-filling factor, $1 > \beta = N r^3$. Then

$$3 - D = \ln \beta / \ln r. \tag{9.33}$$

In the β-model, the random cascade factor between scales $i + 1$ and i, denoted by $W = \epsilon_{i+1}/\epsilon_i$, takes only two values, $1/\beta$ with probability β and 0 with probability $1 - \beta$. Using Eq. (9.27), we have

$$\tau_q = \ln(\beta^{1-q}) / \ln r. \tag{9.34}$$

9.4.5 The random β-model

In the random β-model, the space-filling factor β takes two possible values, β_1 and β_2, with probabilities x and $1 - x$, respectively. Therefore, the model is described by the following:

$$W = \begin{cases} 0 & \text{with probability } 1 - \beta_1 x - \beta_2(1 - x), \\ 1/\beta_1 & \text{with probability } \beta_1 x, \\ 1/\beta_2 & \text{with probability } \beta_2(1 - x). \end{cases} \tag{9.35}$$

Using Eq. (9.27), we have

$$\tau_q = \ln\left[x\beta_1^{1-q} + (1-x)\beta_2^{1-q}\right]\Big/\ln r. \qquad (9.36)$$

9.4.6 The p model

The p model is a binomial cascade model. Suppose that an eddy of size l breaks down into 2^d eddies of size $l/2$, d being the dimensionality of the space we are analyzing. Furthermore, suppose that the flux of energy to these smaller eddies proceeds as follows: a fraction p is distributed equally among one half of the 2^d new eddies, and a fraction $1 - p$ is distributed similarly among the other half. At this point, the model is the same as the binomial random cascade model discussed earlier. To obtain the relation between $\epsilon_{l/2}$ and ϵ_l, we note that $\epsilon_l \propto E_l/l^d$, $\epsilon_{l/2} \propto E_{l/2}/(l/2)^d$. Therefore, $\epsilon_{l/2} = W\epsilon_l$, where $W = 2p$ and $2(1 - p)$, both with probability $1/2$. Using Eq. (9.27), we have

$$\tau_q = 1 - q - \log_2[p^q + (1-p)^q]. \qquad (9.37)$$

Using Eq. (9.23), we have

$$\zeta_q = 1 - \log_2\left[p^{q/3} + (1-p)^{q/3}\right]. \qquad (9.38)$$

Note that when $p = 1/2$, the model gives the K41 theory (Eq. (9.24)). When $p = 0.7$, Meneveau and Sreenivasan [304] found that the model fit their experimental data very well. In fact, the model with the same p also fits other experimental data amazingly well. For an example, see Table 9.1, fourth column.

9.4.7 The SL model and log-Poisson statistics of turbulence

The SL model, proposed by She and Leveque [399], is one of the most remarkable models for intermittency in turbulence now available. With a few physically very appealing assumptions, it predicts that

$$\tau_q = -2q/3 + 2[1 - (2/3)^q]. \qquad (9.39)$$

Using Eq. (9.23), one finds

$$\zeta_q = q/9 + 2\left[1 - (2/3)^{q/3}\right]. \qquad (9.40)$$

The prediction fits experimental data amazingly well. See Table 9.1, fifth column. We show that Eq. (9.39) can be derived by assuming log-Poisson statistics for the cascade multiplier factor $W_{l_1 l_2}$.

To see the idea, let the range of the scales between l_1 and l_2 be further divided into n subranges, $l_1, \delta l_1, \cdots, l_2 = \delta^n l_1$, $n \to \infty$. From scale $\delta^i l_1$ to $\delta^{i+1}l_1$, the cascade multiplier factor W_δ takes two values. The first is $\alpha_\delta = \delta^{-\gamma} > 1$,

$\gamma = 0(1)$, characterizing the rate of amplification of the dissipation event ϵ_{l_1}. This event, assumed to be positive, is called a singularity structure event. The most singular structures in turbulence are related to filaments. Filaments exist in localized, rare, nondissipative regions in the flow. See Fig. 9.10. The second value is $\alpha_\delta \beta$ with $\beta < 1$ and characterizes modulation of the singular structure by the factor β. This event is called a modulation-defect event. In the limit, $\delta \to 1$, $\alpha_\delta \to 1$. In order to make $\langle W_\delta \rangle = 1$, the probability of observing the modulation-defect event must go to zero. At this point, the model is similar to the random-β model, taking certain limiting processes. It is clear that over a finite separation of scales l_1 to l_2, the probability of observing $X = m$ modulation defects obeys a Poisson distribution (recall that the binomial distribution, with the probability of success being x, approaches the normal distribution if x stays finite and the number of trials n tends to infinity. However, if $nx \to \lambda$ when $n \to \infty$, then the binomial distribution approaches a Poisson distribution of expectation λ).

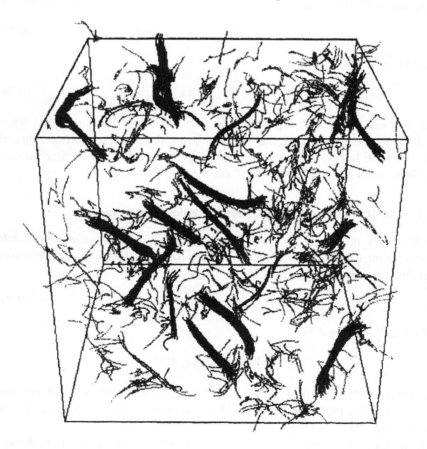

Figure 9.10. Intermittent vortex filaments in a three-dimensional turbulent fluid simulated on a computer (She et al. [398].)

Now $W_{l_1 l_2}$ can be written as

$$W_{l_1 l_2} = (\beta \alpha_\delta)^X \alpha_\delta^{n-X} = (l_1/l_2)^\gamma \beta^X, \tag{9.41}$$

where γ and β are constants and X is a Poisson random variable with mean $\lambda_{l_1 l_2}$. Since $\langle W_{l_1 l_2} \rangle = 1$, $\lambda_{l_1 l_2} = -\gamma(\ln l_1/l_2)/(\beta - 1)$, the qth moment of β^X is given by

$$\langle (\beta^X)^q \rangle = \sum_{m=1}^{\infty} \beta^{qm} e^{-\lambda_{l_1 l_2}} \frac{\lambda_{l_1 l_2}^m}{m!} = e^{-\lambda_{l_1 l_2}} e^{\lambda_{l_1 l_2} \beta^q}.$$

Therefore,

$$\ln \langle W_{l_1 l_2}^q \rangle = \gamma[q - (\beta^q - 1)/(\beta - 1)] \ln l_1/l_2. \tag{9.42}$$

Taking $\gamma = \beta = 2/3$, one obtains Eq. (9.39).

9.5   APPLICATIONS

To illustrate the usefulness of cascade models for real-world signal processing applications, in this section we consider three difficult problems: target detection within sea clutter, analysis of neuronal firing patterns, and analysis and modeling of network traffic. For details of these data, we refer to Secs. A.1–A.3 of Appendix A.

For simplicity, we assume the length of the time series to be 2^N. When the original time series is not positive, we can transform it into a positive one by squaring it, or taking the absolute value of it, or adding a sufficiently large positive value to all elements of the time series. Denote the resulting time series by $\{X_i, i = 1, 2, \cdots, 2^N\}$. There are two ways to analyze $\{X_i\}$. One method is to compute the moments $M_q(\epsilon)$ at different stages and check whether Eq. (9.2) is valid for certain ϵ ranges. The other method is to compute the multiplier distributions at different stages and check whether they are stage independent. We note that the latter method is typically more useful when constructing a multiplicative process for the time series of interest. In the following, we describe in detail a general procedure for obtaining weight sequences at different stages needed for computing the moments $M_q(\epsilon)$ and the multiplier distributions.

The basic idea in analyzing $\{X_i\}$ is to view $\{X_i, i = 1, \cdots, 2^N\}$ as the weight series of a certain multiplicative process at stage N. Under this scenario, the total weight $\sum_{i=1}^{2^N} X_i$ is set equal to 1 unit, and the scale associated with stage N is $\epsilon = 2^{-N}$. This is the smallest time scale resolvable by the measured data.

Given the weight sequence at stage N, the weights at stage $N - 1$, $\{X_i^{(2^1)}, i = 1, \cdots, 2^{N-1}\}$, are obtained by simply adding the consecutive weights at stage N over nonoverlapping blocks of size 2, i.e., $X_i^{(2^1)} = X_{2i-1} + X_{2i}$, for $i = 1, \cdots, 2^{N-1}$, where the superscript 2^1 for $X_i^{(2^1)}$ is used to indicate that the block size used for the summation at stage $N - 1$ is 2^1. This follows directly from the construction of a multiplicative multifractal process schematized in Fig. 9.1.

Stage

$\vdots$

N - 3 $X_1 + X_2 + X_3 + X_4 + X_5 + X_6 + X_7 + X_8$ $\cdots$

N - 2 $X_1 + X_2 + X_3 + X_4$ $X_5 + X_6 + X_7 + X_8$ $\cdots$

N - 1 $X_1 + X_2$ $X_3 + X_4$ $X_5 + X_6$ $X_7 + X_8$ $\cdots$

N X_1 X_2 X_3 X_4 X_5 X_6 X_7 X_8 $\cdots$

Figure 9.11.   A schematic showing the weights at the last several stages for the analysis procedure described in the text. It is instructive to compare this figure with Fig. 9.1, where it was shown that a weight at stage i is the sum of the two "daughter" weights at stages $i+1$.

Associated with this stage is the scale $\epsilon = 2^{-(N-1)}$. This procedure is carried out recursively. That is, given the weights at stage $j+1$, $\{X_i^{(2^{N-j-1})}, i = 1, \cdots, 2^{j+1}\}$, we obtain the weights at stage j, $\{X_i^{(2^{N-j})}, i = 1, \cdots, 2^j\}$, by adding consecutive weights at stage $j + 1$ over nonoverlapping blocks of size 2, i.e.,

$$X_i^{(2^{N-j})} = X_{2i-1}^{(2^{N-j-1})} + X_{2i}^{(2^{N-j-1})} \tag{9.43}$$

for $i = 1, \cdots, 2^j$. Here the superscript 2^{N-j} for $X_i^{(2^{N-j})}$ is used to indicate that the weights at stage j can be equivalently obtained by adding consecutive weights at stage N over nonoverlapping blocks of size 2^{N-j}. Associated with stage j is the scale $\epsilon = 2^{-j}$. This procedure stops at stage 0, where we have a single unit weight, $\sum_{i=1}^{2^N} X_i$, and $\epsilon = 2^0$. The latter is the largest time scale associated with the measured data. Figure 9.11 shows this procedure schematically.

After we have obtained all the weights from stages 0 to N, we compute the moments $M_q(\epsilon)$ according to Eq. (9.1) for different values of q. We then plot $\log M_q(\epsilon)$ vs. $\log \epsilon$ for different values of q. If these curves are linear over wide ranges of ϵ, then these weights are consistent with a multifractal measure. Note that, according to Eq. (9.2), the slopes of the linear part of $\log M_q(\epsilon)$ vs. $\log \epsilon$ curves provide an estimate of $\tau(q)$ for different values of q.

Next, we explain how to compute the multiplier distributions at different stages. From stage j to $j + 1$, the multipliers are defined by the following equation, based on Eq. (9.43):

$$r_i^{(j)} = \frac{X_{2i-1}^{(2^{N-j-1})}}{X_i^{(2^{N-j})}} \tag{9.44}$$

for $i = 1, \cdots, 2^j$. We view $\{r_i^{(j)}, i = 1, \cdots, 2^j\}$ as sampling points of the multiplier distribution at stage j, $P_j = \{P_j(r), 0 \leq r \leq 1\}$. Hence P_j can be determined from its histogram based on $\{r_i^{(j)}, i = 1, \cdots, 2^j\}$. We then plot $P_j(r)$ vs. r for different stages (j) together. If these curves collapse together so that $P_j \sim P$, then the multiplier distributions are stage independent and the weights form a multifractal measure P.

9.5.1   Target detection within sea clutter

In Sec. 8.9.1, we discussed target detection within sea clutter using structure-function–based multifractal analysis. Here we show that the cascade model provides another simple and effective solution.

Since sea clutter amplitude data are positive, we can simply follow the procedure detailed at the beginning of this section to analyze the data. Figure 9.12 plots $\log_2 M_q(\epsilon)$ vs. $-\log_2 \epsilon$ for the primary target bin and a bin without a target for two polarizations, HH and VV. We observe that the scaling between $M_q(\epsilon)$ and ϵ (i.e., the degree of linearity between $\log_2 M_q(\epsilon)$ and $-\log_2 \epsilon$) for all datasets is quite good up to the 15th stage. To further check whether these datasets are truly multifractals, we compute D_q for a certain ϵ range. The results are shown in Fig. 9.13. We observe that in all cases D_q has a nontrivial dependence on q. Therefore, we conclude that these time series are consistent with multifractals. Furthermore, when q is large, D_q can be used to effectively distinguish the data with and without a target.

9.5.2   Modeling and discrimination of human neuronal activity

Understanding neuronal firing patterns is one of the most important problems in theoretical neuroscience. It is also very important for clinical neurosurgery. Events in extracellular neuronal recording generate two types of time series: (1) the time interval between successive firings, called the interspike interval data, and (2) a counting process representing the number of firings in a chosen time period. In the past few decades, a lot of effort has been dedicated to the analysis of these two types of time series, using Fano factor analysis and wavelet analysis to characterize the difference between firing data and Poisson-based models, to quantify long-range-correlations, and to characterize neuronal dynamics.

Advances in surgical treatment of Parkinson's disease have stimulated much interest in the research of deep brain stimulation (DBS). The globus pallidus (GP) has been targeted in neurosurgical treatment of Parkinson's disease and dystonia. The GP is a complex structure in the basal ganglia and can be divided into two parts: the globus pallidus externa (GPe) and the globus pallidus interna (GPi). Both receive input from the caudate and putamen and communicate with the subthalamic nucleus. The GPi is thought to send the major inhibitory output from the basal ganglia back to the thalamus and also to send a few projections to an area of the midbrain (the PPPA), presumably to assist postural control. Distinguishing GPe from GPi is very important for surgical treatment of Parkinson's disease.

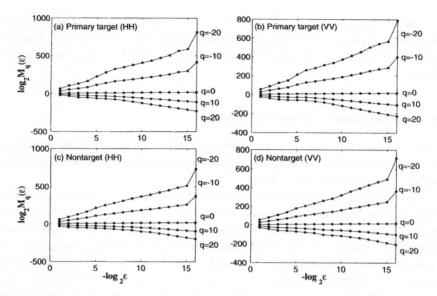

Figure 9.12. $\log_2 M_q(\epsilon)$ vs. $-\log_2 \epsilon$ for sea clutter amplitude data for several different q's.

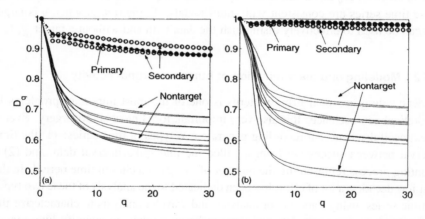

Figure 9.13.   The D_q spectrum for (a) HH and (b) VV datasets of one measurement.

In his pioneering work, DeLong [103] observed that discharge patterns of neurons in the GPi and GPe in an awake monkey (*Macaca mulatta*) at rest appeared to be very different. For GPe neurons, two types of firing patterns were observed: one had recurrent periods of high-frequency discharge with relatively silent periods in between, and the other had a low-frequency discharge with bursts. In contrast, only one firing pattern was found for GPi neurons: a continuous discharge without long periods of silence. DeLong's work has triggered much research on the characterization of different cell types in specific brain regions so that the accuracy of target acquisition of stereotactic electrode placement in the human brain can be automated

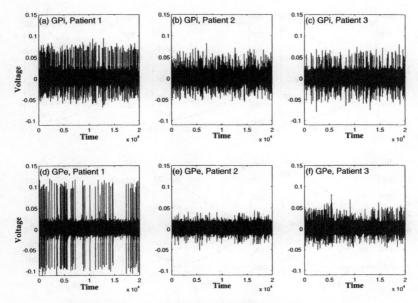

Figure 9.14. Human deep brain recordings from GPe and GPi.

with high accuracy. It is worth noting that in practice, a well-trained, experienced movement disorders specialist is able to do so by listening to neuronal activity over a loudspeaker through trial and error. We shall show below that the generalized dimension spectrum D_q can effectively differentiate the two brain areas, both within and between patients. Note that the method is effective not only on the spike train data, but also on the raw voltage recording data. The latter is very appealing, since when the raw recording data are very noisy, spike detection/sorting can be difficult or even impossible.

The following discussion is based on the analysis of raw voltage recordings from three patients. See Fig. 9.14. The data were sampled with a frequency of 20 KHz. The mean interspike interval is about 0.01 s, i.e., 200 sample points. Another shorter time scale is the duration of each spike, which is about 0.0008 s, i.e., 16 sample points. It should be emphasized that fractal analysis is meaningful only for time scales greater than the mean interspike interval, since in shorter time scales, only noise and individual spikes can be characterized.

Since the raw data contain both positive and negative values, they are squared first. Then, following the procedures detailed at the beginning of Sec. 9.5, we can obtain the weights at all the stages and compute $M_q(\epsilon)$. Figure 9.15 plots $\log_2 M_q(\epsilon)$ vs. $- \log_2 \epsilon$. It is observed that the scaling between $M_q(\epsilon)$ and ϵ (i.e., the degree of linearity between $\log_2 M_q(\epsilon)$ and $- \log_2 \epsilon$) for all datasets is quite good up to the ninth stage, which corresponds to the mean interspike interval. By fitting straight lines to $\log_2 M_q(\epsilon)$ vs. $- \log_2 \epsilon$, from stages 9 to 4, which correspond to $t_1 = 0.01$ to $t_2 = 0.40$ s, one can readily obtain τ_q and hence D_q. The latter is shown in

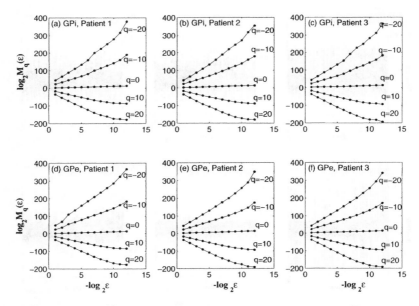

Figure 9.15. $\log_2 M_q(\epsilon)$ vs. $-\log_2 \epsilon$ for human brain deep recordings for several different q's.

Fig. 9.16. It is evident that for very negative q, D_q is larger for GPi than for GPe, both within and between patients. This feature is an effective discriminator between GPe and GPi.

It should be noted that the effectiveness of the method in distinguishing between GPe and GPi is due largely to the identification of the two time scales. While we have explained the meaning of t_1, we note that $t_2 = 0.40$ s may be linked to the local and global neuronal interconnectivity of the brain. That is, due to finite propagation of neural signals, time scales longer than about 0.4 s might already correspond to neural interactions beyond GPi and GPe and hence may also be irrelevant to distinguish between the two structures. Indeed, based on these time scales, one may wonder why many traditional signal processing approaches, such as those based on the total spectral density of the signal, fail to distinguish GPe from GPi. The reason is simple: the majority of the energy in the spectral density comes from the noise between spikes and individual spikes. The useful information — the spacing between spikes — contributes little to the total energy.

9.5.3   Analysis and modeling of network traffic

In Sec. 6.7.1, we explained four different representations of network traffic. The purpose of multifractal analysis of network traffic is to check whether the interarrival time series $\{T_i\}$, packet length sequences $\{B_i\}$, and the counting processes $\{\overline{B}_i\}$ and $\{\overline{P}_i\}$ of the traffic data can be viewed as realizations of multiplicative processes.

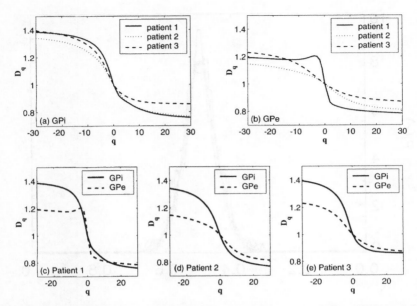

Figure 9.16. The generalized dimension spectrum for recording data of GPi and GPe area of three patients.

If they are only approximate multifractals, an equivalent multifractal model may still be constructed. To illustrate the idea, we consider video traffic modeling here.

Owing to the increasing demand on video services, variable bit rate (VBR) video traffic modeling has become a critical issue. Simple and accurate video traffic models are needed to solve problems arising from multimedia traffic management, such as traffic policing, shaping, and call admission control. Many services (such as video on demand and broadcast TV) impose on video traffic special variability through the evolution of scenes of different complexity and degrees of motion. We show that a video traffic trace of "star wars" can be very effectively modeled by a cascade process. The trace consists of 174,136 integers, where each integer represents the number of bits per video frame (24 frames/second for approximately 2 hrs). Our analysis uses the first 2^{17} points.

Now that we have explained the usefulness of the $\log_2 M_q(\epsilon)$ vs. $-\log_2 \epsilon$ curves and the D_q spectrum, we work with the multiplier distributions. Following the procedures detailed at the beginning of the section, the multiplier distributions P_j for the video traffic data are shown in Fig. 9.17, where the asterisk curve is generated from $P(r) \sim e^{-\alpha_g |r-1/2|^2}$, with $\alpha_g = 200$. Collapsed on it are P_j curves with $j = 8, \cdots, 11$. We thus see that the video traffic forms a multifractal process over certain time scales.

After the multiplier distributions are found, we can readily generate the multifractal video traffic and compare it with the actual trace by feeding both modeled and real traffic to a queuing system. Figure 9.18 shows the comparison of the

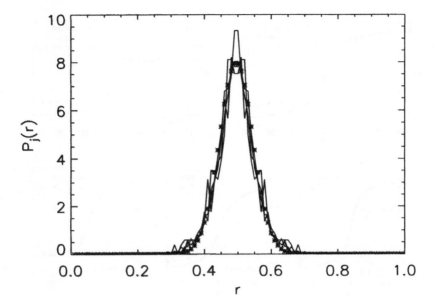

Figure 9.17.   Multiplier distributions P_j for the frame size sequence data MPEG.data. See the text for more detail.

system size tail distributions for a single-server FIFO queuing system loaded on the one hand by the measured traffic (dashed lines) and on the other hand by the multifractal traffic (solid lines). The system size is represented by the queue length, which measures the total number of queued frames normalized by the average frame size. The three curves, from top to bottom, correspond to three different utilization (i.e., normalized loading) levels, $\rho = 0.7$, 0.5, and 0.3. Clearly, the model yields an excellent fit of the system size tail distribution for a queuing system loaded by the measured video traffic.

9.6   BIBLIOGRAPHIC NOTES

The multiplicative cascade multifractal was initially developed for understanding the intermittent features of turbulence. The modern theory of turbulence started with Kolmogorov's theory of 1941 [265]. Since then, various models have been proposed. Early works include [293, 336]. For the various models discussed in Sec. 9.4, in order, we refer readers to [113, 258, 266, 304, 323, 399, 400]. See also [55] for a comparison of intermittency models of turbulence, [141] on experimental data analysis, [42, 43] on extended self-similarity, [398] on filaments in turbulence, and [47, 224] on raindrop formation. Since the discussions on turbulence here are brief, we have to omit many relevant references. Interested readers should read the exquisite books by Frisch [143] and Chorin [75] (in fact, we believe many readers

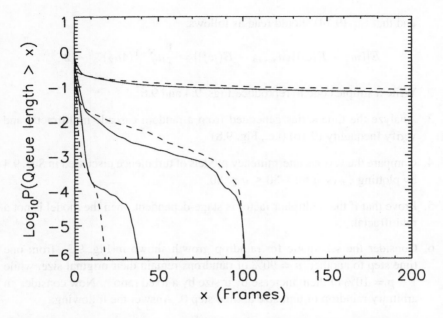

Figure 9.18.   System size tail probabilities obtained when VBR video traffic MPEG.data (solid curves) and its corresponding multifractal traffic process (dashed curves) are used to drive the queuing system. Three curves, from top to bottom, correspond to $\rho = 0.7, 0.5$, and 0.3, respectively.

have done so.) Also, it should be emphasized that besides the cascade models discussed here, there exist other ways to tackle the problem of intermittency in turbulence. See, for example, [31, 32].

It is interesting to note that the cascade model has been applied to the study of various phenomena such as rainfall [333], tropical convection [454], liquid water distributions inside marine stratocumulus [100, 101], and finance [296]. The generalized dimensions spectrum D_q was defined by Hentschel and Procaccia [219]. Systematic studies of the properties of the multiplicative processes can be found in [168]. Interesting sources on log-normality in network traffic are [30, 54, 63, 311].

Readers interested in applications of cascade models on sea clutter are referred to the references noted in the end of Chapter 8, to [103, 390, 495] and many references cited therein on analysis of neuronal firing patterns, and to [44, 89, 134, 148, 165– 173, 281, 368, 425] on traffic modeling.

9.7 EXERCISES

1. In Sec. 9.3, the properties of conservative cascade models are derived. Show that for a nonconservative model, all the properties except Eq. (9.14) still hold

and that Eq. (9.14) should read as follows:

$$E[(w_n - E(w))(w_{n+m} - E(w))] = \frac{1}{4}\mu_2^{N-1}(4\mu_2)^{-k} - 2^{-2N}.$$

2. Write a simple code to reproduce Figs. 9.4 and 9.6.

3. Analyze the time series generated from a random cascade multifractal and verify Inequality (9.16) (i.e., Fig. 9.8).

4. Compare the various intermittency models of turbulence discussed in Sec. 9.4 by plotting ζ_q vs. q for $-30 \le q \le 30$.

5. Prove that if the multiplier factor is stage-dependent, then the model is not a multifractal.

6. Consider the schematic for raindrop growth shown in Fig. 9.3: from one time step to another, $p = 90\%$ of raindrops remain their original size, while $1 - p = 10\%$ of them increase their size by a fixed ratio r. Now consider an arbitrary raindrop of unit size at time step 0. Answer the following:

   (a) At time step 1, the size of the raindrop can be either 1 or $1 + r$. What are their corresponding probabilities?

   (b) At time step 2, the size of the raindrop can be 1, $1 + r$, or $(1 + r)^2$. What are their corresponding probabilities?

   (c) Determine the sample space of the raindrop size at time step n, and determine the probabilities.

   (d) Prove that at step n, the k-th moment for drop size is

   $$[p + (1 + r)^k(1 - p)]^n$$

CHAPTER 10

STAGE-DEPENDENT MULTIPLICATIVE PROCESSES

We have discussed three different types of fractal models: $1/f^\beta$ processes with long memory, stable laws and Levy motions, and random cascade multifractals. To see better the relations among these models, in this chapter we describe a stage-dependent multiplicative process model.

10.1 DESCRIPTION OF THE MODEL

The stage-dependent multiplicative process is best described in a recursive manner (see the schematic of Fig. 10.1). It is a conservative model and can be readily modified to be a nonconservative model.

Step 1: At stage 0, we have a unit interval and a unit mass (or weight). The unit time interval is the total time span of interest. It could be 1 min, 1 hr, or even 1 year. The unit mass could be the total energy of a turbulent field under investigation or the total traffic loading to a network in the entire time span. We divide the unit interval into two (say, left and right) segments of equal length, partition the mass into two fractions, $r_{1,1}$ and $1 - r_{1,1}$, and assign them to the left and right segments, respectively. The first subscript "1" is used to indicate the stage number 1. The second subscript "1" denotes the

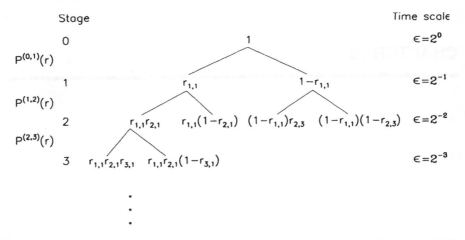

Figure 10.1.   Schematic illustrating the construction rule of the stage-dependent multiplicative process model.

position of the weight at that stage, starting from the left. We only assign odd positive integers, leaving the even integers to denote positions occupied by the "complementary" weights such as $1 - r_{1,1}$. The parameter $r_{1,1}$, called the multiplier, is a random variable governed by a PDF $P^{(0,1)}(r)$, $0 \leq r \leq 1$, where the superscript "(0,1)" is used to indicate the transition from stage 0 to stage 1. $P^{(0,1)}(r)$ is assumed to be symmetric about $r = 1/2$, so that $1 - r_{1,1}$ also has the distribution $P^{(0,1)}(r)$.

Step 2: Divide each interval at stage i into two segments of equal length. Also, partition each weight, say $r_{i,1}$, at stage i, into two fractions, $r_{i,1}r_{i+1,1}$ and $r_{i,1}(1 - r_{i+1,1})$, where $r_{i+1,1}$ is a random variable governed by a PDF $P^{(i,i+1)}(r)$, $0 \leq r \leq 1$, which is also symmetric about $r = 1/2$. Note that the random variables, $r_{i,j}$'s, are independent of all the other random variables used in the multiplicative process.

Step 3: A complex time series may be modeled with arbitrary precision by simply making the variances stage independent for a suitably chosen parameterized density function $P^{(i,i+1)}(r)$. However, while modeling a time series of length n, this would amount to estimating $\log n$ parameters, and given the universality result, one would suffer from all the problems inherent in global optimization problems, such as overfitting and being stuck in local minima. Hence, a more relevant question is how to constrain the parameter space and still retain the ability to model different types of stochastic processes. We require that the ratio of the variances satisfy a simple constraint,

$$\sigma^2(i, i+1) = \gamma\sigma^2(i-1, i), \qquad (10.1)$$

for a constant $\gamma \geq 1$. As we show in our analysis, it is indeed the case that by varying the single parameter γ, one can generate a wide range of processes including processes with long memory, pure multiplicative multifractals, and SRD processes. The implications of these results are further elaborated upon in Sec. 10.3.

Step 4: Interpret the weights at stage N, $\{w_j, j = 1, \cdots, 2^N\}$, as the process being modeled (such as a counting traffic process).

By comparing the above model with the pure multiplicative multifractal model described in Chapter 9, we immediately obtain the following properties of the model:

- When the PDFs $P^{(i,i+1)}(r)$ are stage independent, the model reduces to the ideal multiplicative multifractal model. This is the case when $\gamma = 1$ and all the $P^{(i,i+1)}(r)$ are of the same functional form.

- When $P^{(i,i+1)}(r)$ are of the same functional form but γ is slightly larger than 1, the process is not multifractal. However, the weights at stage N will have an approximately log-normal distribution.

- The variance parameter $\sigma^2(i, i+1)$ describes the burstiness of the process at the particular time scale $2^{-(i+1)}T$, where T is the total time span of interest. Hence, when $\gamma = 1$, the burstiness level remains unchanged for all time scales. When $\gamma > 1$, the burstiness level of the process decreases from shorter to longer time scales.

In the next section, we shall show that the γ parameter and the Hurst parameter are related by a simple relation (when $\gamma > 1$; the case of $\gamma = 1$ has been dealt with in Chapter 9),

$$2^{2-2H} \approx \gamma. \tag{10.2}$$

Hence, when $\gamma = 2$, the decrease in the burstiness level of the process is such that the process effectively becomes a short-range-dependent process such as a Poisson process.

We now comment on Eq. (10.1). Suppose that $\sigma^2(0,1) > 0$ is given. Then $\sigma^2(i, i+1) \to \infty$ when $i \to \infty$. This contradicts the restrictions made on the random variables, i.e., any distribution is nonzero only between 0 and 1 and is symmetric about $1/2$, and $0 \leq \sigma^2(i-1, i) \leq \frac{1}{2}$ for any i. The contradiction can be resolved by noticing that after a certain stage i^*, the stage-dependent multiplier distributions can no longer change with i (for example, if the shape of the multiplier distribution is initially Gaussian, when the variance gets larger and larger, within the unit interval the distribution eventually settles to the uniform distribution). Beyond stage i^*, the model reduces to the cascade multifractal discussed in Chapter 9. In this chapter, we shall focus on the time scales above that of stage i^*.

10.2 CASCADE REPRESENTATION OF $1/F^\beta$ PROCESSES

Since $E(w_j) = 2^{-N}$, the process, $\{w_j - 2^{-N}, j = 1, \cdots, 2^N\}$, is thus a zero mean time series. Below we shall prove that the random walk process, defined by

$$y_n = \sum_{j=1}^{n}(w_j - 2^{-N}),\tag{10.3}$$

is a fractal process.

Theorem 1 The random walk process y_n is asymptotically self-similar, characterized by a power spectral density $E(f) \sim f^{-(2H+1)}$, with $2^{2-2H} \approx \gamma$, when $\gamma > 1$.

Proof First, we note that a $1/f^{(2H+1)}$ process is characterized by

$$F(m) = E[(y_{n+m} - y_n)^2] \sim m^{2H}.\tag{10.4}$$

Hence,

$$F(2m)/F(m) = 2^{2H}.\tag{10.5}$$

It thus would suffice for us to prove that the ratio between the variance for the weights at stage $i-1$ and stage i is given by $4/\gamma$. A weight at stage i can be written as $r_1 r_2 \cdots r_i$, where r_k, $k = 1, ..., i$, is governed by a PDF $P^{(k-1,k)}(r)$. Thus, the variance for a weight at stage i is

$$var(i) = [1/4 + \sigma^2(0,1)][1/4 + \gamma\sigma^2(0,1)]\cdots[1/4 + \gamma^{i-1}\sigma^2(0,1)] - 2^{-2i}.$$

For simplicity, let us define

$$c^2 = \frac{\sigma^2(0,1)}{(1/2)^2}.$$

Hence,

$$var(i) = (\frac{1}{2})^{2i}\{\Pi_{j=1}^{i}(1 + \gamma^{j-1}c^2) - 1\}.$$

Since any $P^{(k-1,k)}(r)$ is defined in the unit interval and is symmetric about $r = 1/2$, its variance must be in the interval $[0, 1/4]$. This means that

$$\gamma^{k-1}c^2 < 1$$

for any k. Hence,

$$var(i) \approx (\frac{1}{2})^{2i}(\gamma^0 + \gamma^1 + \gamma^2 + \cdots + \gamma^{i-1})c^2 = (\frac{1}{2})^{2i}(\gamma^i - 1)/(\gamma - 1)c^2.$$

We can similarly find $var(i-1)$, and obtain the ratio between $var(i-1)$ and $var(i)$ to be

$$\frac{var(i-1)}{var(i)} = 4\frac{\gamma^{i-1} - 1}{\gamma^i - 1}.$$

Since $\gamma > 1$, for reasonably large stage number i, we thus have

$$\frac{var(i-1)}{var(i)} \approx \frac{4}{\gamma}.$$

Equating

$$\frac{4}{\gamma} = 2^{2H}$$

then completes our proof.

To better appreciate how the stage-dependent cascade model generates $1/f$ processes, we choose the stage-dependent PDFs to be Gaussian, as described by Eq. (9.18), and vary γ from 1.1 to 3.0 to obtain a number of random walk processes. Four examples for $\gamma = 1.4, 1.6, 2.0$, and 2.5 are shown in Fig. 10.2, where the variance-time plots are depicted. We observe excellent scaling laws in all four cases. The estimated Hurst parameters for these and other processes with different γ parameters are shown in Fig. 10.3 as triangles. Also shown in Fig. 10.3 is the variation of H against γ based on Eq. (10.2). We observe that Eq. (10.2) accurately estimates the Hurst parameter. Note that different estimators for the Hurst parameter give the same results.

A word on the numerical simulations of such processes is in order. One should always guarantee that

$$0 < \sigma^2(i, i+1) < 1/4$$

for all i. Again, this means that i cannot go to infinity, as we pointed out at the end of the previous section. In relation to fractal scaling law, this means that a power-law scaling relation cannot be valid for an infinite scaling regime. Indeed, physically observed fractal scaling relations are always truncated.

Our next theorem states that any $1/f$ process can be represented by such a cascade model.

Theorem 2 A $1/f^{(2H+1)}$ process can be represented as a stage-dependent random cascade process with $\gamma = 2^{2-2H}$.

Proof Suppose that the $1/f^{(2H+1)}$ process is represented by a time series y_n, $n = 1, 2, \cdots$. First, we obtain the increment process $x_n = y_n - y_{n-1}$. Then we add to the x_n time series a positive value *level* so that the time series $\{x_n + level\}$ is positive. Denote this time series as X_n. For simplicity, assume that n runs from 1 to 2^N. We put X_j, $j = 1, \cdots, 2^N$, at stage N, according to our stage-dependent random cascade model. The weights at stages $N - 1$, $N - 2$, etc., can then be obtained by forming nonoverlapping running summations of length 2, 4, etc., as shown schematically in Fig. 10.4. Let two weights at stage $i - 1$ and stage i be $X^{(i-1)}$ and $X^{(i)}$, respectively. Let $EX^{(i)} = ave(i)$ and $varX^{(i)} = \sigma^2(i)$. Then, $EX^{(i-1)} = 2ave(i)$, $varX^{(i-1)} = 2^{2H}\sigma^2(i)$. Now suppose that $X^{(i-1)}$ and $X^{(i)}$ are connected by the following simple relation,

$$X^{(i)} = r(i - 1 \to i)X^{(i-1)},$$

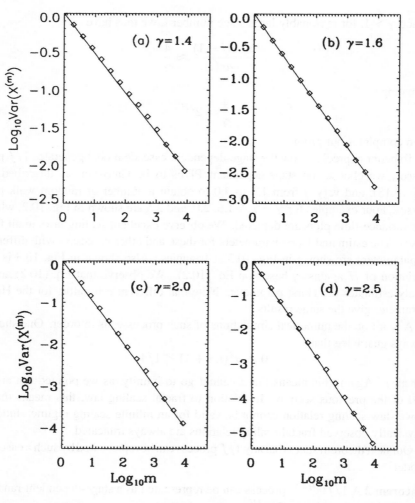

Figure 10.2.   Variance-time plots for the time series generated by the stage-dependent cascade model with $\gamma = 1.4$, 1.6, 2.0, and 2.5.

where by construction, $r(i - 1 \to i)$ and $X^{(i-1)}$ are independent. Then we have

$$ave^2(i) + \sigma^2(i) = [4ave^2(i) + 2^{2H}\sigma^2(i)][\frac{1}{4} + \sigma^2_{r(i-1 \to i)}].$$

Hence,

$$\sigma^2_{r(i-1 \to i)} = \frac{[1 - 2^{2H-2}]\sigma^2(i)}{4[ave^2(i) + 2^{2H-2}\sigma^2(i)]}.$$

Therefore,

$$\frac{\sigma^2_{r(i \to i+1)}}{\sigma^2_{r(i-1 \to i)}} = \frac{2^{-2H}ave^2(i) + 2^{-2}\sigma^2(i)}{2^{-2}ave^2(i) + 2^{-2}\sigma^2(i)}. \tag{10.6}$$

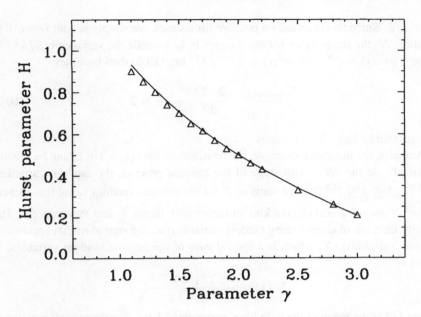

Figure 10.3. Variation of H vs. γ. The smooth curve is computed based on Eq. (10.2). Triangles are calculated based on variance-time plots such as shown in Fig. 10.2.

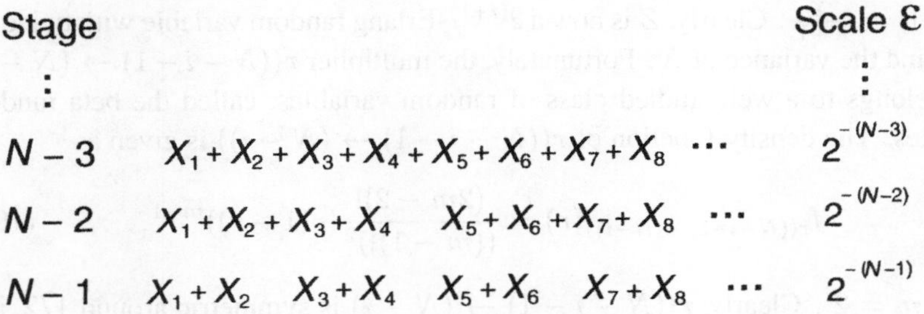

Stage   **Scale ε**

⋮   ⋮

$N - 3$ $X_1 + X_2 + X_3 + X_4 + X_5 + X_6 + X_7 + X_8$ ⋯ $2^{-(N-3)}$

$N - 2$ $X_1 + X_2 + X_3 + X_4$ $X_5 + X_6 + X_7 + X_8$ ⋯ $2^{-(N-2)}$

$N - 1$ $X_1 + X_2$ $X_3 + X_4$ $X_5 + X_6$ $X_7 + X_8$ ⋯ $2^{-(N-1)}$

N X_1 X_2 X_3 X_4 X_5 X_6 X_7 X_8 ⋯ 2^{-N}

Figure 10.4. Schematic for obtaining weights at stages $N - 1$, $N - 2$, etc.

We can select *level* to be large enough so that $\sigma^2(i)$ is small compared to $ave^2(i)$. Hence

$$\frac{\sigma^2_{r(i \to i+1)}}{\sigma^2_{r(i-1 \to i)}} = 2^{2-2H}.$$

Equating $\gamma = 2^{2-2H}$ then completes our proof.

As an example, let us consider a sequence of independent RVs with *exponential distribution*, $P(x) = \lambda e^{-\lambda x}$, $x \geq 0$. Such a time series has a Hurst parameter

$H = 1/2$. Since this is already a positive time series, we simply set our *level* $= 0$. At stage N, the mean value for the weights is $1/\lambda$, while the variance is $1/\lambda^2$. At stage i, $ave(i) = 2^{N-i}/\lambda$, $\sigma^2(i) = 2^{N-i}/\lambda^2$. Eq. (10.6) then becomes

$$\frac{\sigma^2_{r(i\rightarrow i+1)}}{\sigma^2_{r(i-1\rightarrow i)}} = \frac{2 \cdot 2^{N-i} + 1}{2^{N-i} + 1} \approx 2 \tag{10.7}$$

for reasonably large $N - i$ values.

Actually, for the above example, the distribution for $r(i \rightarrow i + 1)$ can be readily obtained. At the $(N - i)$th stage of the cascade process, the random variables, $w_j^{(N-i)}$, $1 \le j \le 2^{(N-i)}$, are sums of 2^i iid exponential random variables. Hence, $w_j^{(N-i)}$'s are iid 2^i-Erlang random variables with mean $\frac{2^i}{\lambda}$ and variance $\frac{2^i}{\lambda^2}$. The density function of an m-Erlang random variable (i.e., the sum of m iid exponential random variables), X, which is a special case of the gamma random variables, is given as

$$f_X(x) = \frac{\lambda e^{-\lambda x}(\lambda x)^{m-1}}{(m-1)!},$$

where $1/\lambda$ is the mean of the individual exponential RVs. For the sake of notational convenience, let us represent two consecutive odd- and even-numbered RVs at the $(N - i)$th level as X and Y, respectively. Then, the corresponding weight at the $(N - i - 1)$th stage is given by $Z = X + Y$ and the multiplier $r((N - i) - 1 \rightarrow (N-i)) = \frac{X}{X+Y}$. Clearly, Z is now a $2^{(i+1)}$-Erlang random variable with twice the mean and the variance of X. Fortunately, the multiplier $r((N - i - 1) \rightarrow (N - i))$ also belongs to a well-studied class of random variables, called the beta random variables. The density function of $r((N - i - 1) \rightarrow (N - i))$ is given as

$$f_{r((N-i-1)\rightarrow(N-i))}(r) = \frac{(2m - 2)!}{((m - 1)!)^2}(r(1 - r))^{m-1}, \tag{10.8}$$

where $m = 2^i$. Clearly, $r((N - i - 1) \rightarrow (N - i)$ is symmetric around $1/2$, and its variance is given by

$$\sigma^2((N - i - 1) \rightarrow (N - i)) = \frac{1}{4(2^{i+1} + 1)}.$$

Thus, the multiplicative parameter $\gamma_{(N-i)}$ (for $i \ge 1$) is given by

$$\gamma_{(N-i)} = \frac{\sigma^2((N - i) \rightarrow (N - i + 1))}{\sigma^2((N - i - 1) \rightarrow (N - i))} = \frac{2^{i+1} + 1}{2^i + 1} \tag{10.9}$$

and is identical to Eq. (10.7).

Moreover, one can show that the random variables Z and $r((N-i-1) \rightarrow (N-i))$ are statistically independent. Hence, the stage-dependent cascade model can *exactly generate an SRD process comprising a sequence of independent RVs with identical exponential distributions*. In the model, choose the multiplier random variables connecting consecutive stages $(N-i)-1$ and $(N-i)$ $(0 \leq i \leq N-1)$ independently and identically from the distribution given by Eq. (10.8).

As another example, let us consider independent, uniformly distributed RVs in the unit interval. Here again we have $H = 1/2$. Set $level = 0$. At stage N, the mean and variance of the time series are 1/2 and 1/12, respectively. At stage i, $ave(i) = 2^{-1} \cdot 2^{N-i}$, $\sigma^2(i) = 2^{N-i}/12$. Thus, Eq. (10.6) becomes

$$\frac{\sigma^2_{r(i \to i+1)}}{\sigma^2_{r(i-1 \to i)}} = \frac{6 \cdot 2^{N-i} + 1}{3 \cdot 2^{N-i} + 1} \approx 2 \qquad (10.10)$$

for reasonably large $N - i$ values.

To assess how well a cascade model represents a $1/f$ process with different H parameters, we numerically study fGn processes with $H = 0.75$ and 0.25, as well as a sequence of independent RVs with exponential distribution and uniform distribution. Following the procedures described in the proof, we compute $\sigma^2_{r(i \to i+1)}$ and plot $\log_2 \sigma^2_{r(i \to i+1)}$ vs. the stage number. The results are shown in Fig. 10.5. We observe excellent scaling relations in all these cases, with the slopes of those straight lines given by $2 - 2H$, thus γ given by Eq. (10.2).

10.3   APPLICATION: MODELING HETEROGENEOUS INTERNET TRAFFIC

10.3.1   General considerations

For simplicity, in this section we only consider the modeling of counting traffic processes. The discussion will, of course, also be pertinent to the modeling of any other interesting time series.

Suppose that we have a positive time series of length 2^N. We want to model it using our stage-dependent model. For simplicity, we may choose all the PDFs to be of the same functional form, say the truncated Gaussian (Eq. (9.18)). In the ideal setting of the stage-dependent cascade model, we constrain the model by Eq. (10.1). So the model contains two independent parameters, γ and $\sigma^2(0, 1)$ or $\sigma^2(N-1, N)$. What are the physical meanings of these two parameters? Following the arguments in Sec. 10.1, we can readily understand that they are the burstiness indicators: when γ is fixed, the burstiness of the modeled traffic increases when $\sigma^2(0, 1)$ or $\sigma^2(N - 1, N)$ is increased. When the latter is fixed, the burstiness of the modeled traffic decreases when γ is increased. Since γ is related to the Hurst parameter by Eq.(10.2), it is clear that the Hurst parameter alone is not the burstiness indicator. More interestingly, this model generates an ideal multiplicative multifractal ($\gamma = 1$), conventional LRD processes ($1 < \gamma < 2$), and short-range-dependent processes such as Poisson processes ($\gamma = 2$) and antipersistent processes ($2 < \gamma < 3$) as special cases. Equally important is that this process makes us realize that the boundary between conventional LRD traffic models and Markovian traffic models is far more vague than one might have thought.

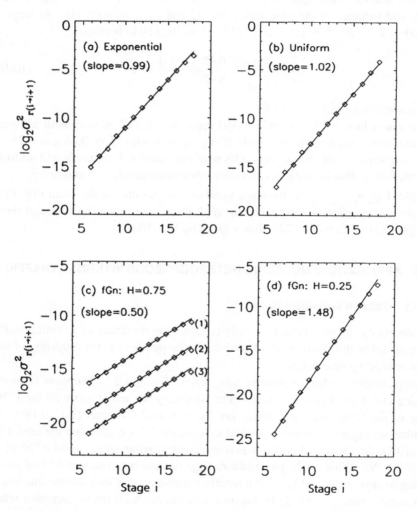

Figure 10.5. $\log_2 \sigma^2_{r(i \rightarrow i+1)}$ vs. the stage number for a sequence of independent RVs with exponential distribution (a) and with uniform distribution (b), and for fGn processes with $H = 0.75$ (c) and $H = 0.25$ (d). The three lines in (c) correspond to three different choices of $level$. We see that $level$ does not affect the slope of these straight lines.

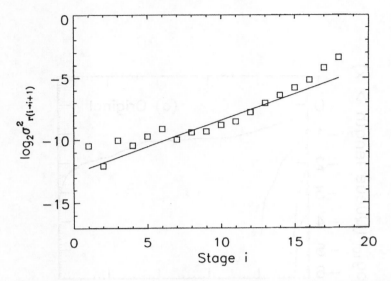

Figure 10.6. $\log_2 \sigma^2_{r(i \to i+1)}$ vs. stage i for the counting process of OSU (square symbols). The straight line is a least linear squares fit of the square symbols.

There are two specific important advantages of such modeling over the fBm processes: these processes have a non-Gaussian nature and are easier to simulate.

At the next level of complexity, one may wish the model to be constrained by two conditions similar to Eq. (10.1). That is to say, the modeled traffic has two distinct fractal scaling laws. We need four parameters to specify the model — γ_1, γ_2, $\sigma^2(0, 1)$, or $\sigma^2(N - 1, N)$ — and a parameter specifying where the transition from one type of fractal to another occurs. When one of the γ is one, then the number of parameters in the model is reduced to three, the model is an ideal multiplicative multifractal in one scaling region and a $1/f$ process in another scaling region. Following this line of argument, it is clear that, under the constraint that all PDFs $P^{(i,i+1)}(r)$ are truncated Gaussian (Eq. (9.18)), the most general stage-dependent cascade model will be specified by N variances, $\sigma^2(i - 1, i)$, $i = 0, \cdots, N - 1$. It is this type of model (in the wavelet domain) that has been used by Feldmann et al. [134], Gilbert et al. [187], and Riedi et al. [368]. It should now be clear why their models are not truly multifractals. Nevertheless, Feldmann et al. still bravely went on to explain the mechanism for their model based on the layered structure of a network. We should emphasize here that any random function may be approximated by an unconstrained stage-dependent model with any precision.

10.3.2   An example

As an example, we model the very-high-speed Backbone Network Service (vBNS) traffic trace OSU (see Sec. A.1 of Appendix A for the description of the trace data). We analyze the data using the same procedure for the analysis of the data presented

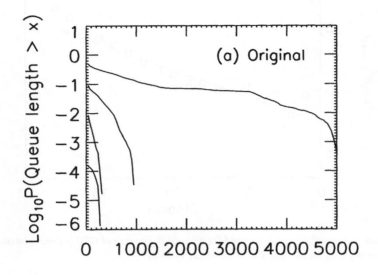

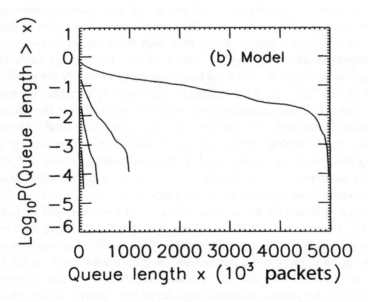

Figure 10.7.   System size tail probabilities obtained when the traffic trace OSU (a) and modeled stage-dependent multiplicative process traffic (b) are used to drive the queuing system. Four curves, from top to bottom, correspond to $\rho = 0.9, 0.7, 0.5$, and 0.3, respectively.

in Fig. 10.5. The result is shown in Fig. 10.6 as the square symbols for the counting process of the OSU data. Note that when i is small, $\sigma^2(i, i + 1)$ may not be well estimated due to the sparsity of data points. So we conclude that at least for $i > 7$, $\log_2 \sigma^2_{r(i \rightarrow i+1)}$ vs. i follows approximately a straight line. However, we start from a "global" (i.e., for all i) least linear squares fitting of those square symbols. The fitted line is shown in Fig. 10.6 as the solid line. Next, we generate a modeled counting traffic process according to that fitted line, and drive a FIFO single-server queuing system by the modeled traffic as well as by the original OSU traffic trace. The results are shown in Fig. 10.7. We note that for utilization levels $\rho = 0.9, 0.7$, and 0.5, such a model is already excellent. The deviation observed in Fig. 10.7 when $\rho = 0.3$ is due to the fact that we have actually fitted a straight line for all i in Fig. 10.6. Thus, if we are willing to increase the number of parameters slightly, the model will be more accurate.

10.4 BIBLIOGRAPHIC NOTES

The model discussed here was first proposed in [174]. For a proof on the independence between Z and $r((N - i - 1) \rightarrow (N - i))$, see pp. 60–61 of [375]. For traffic modeling, see [134, 187, 368].

10.5 EXERCISES

1. Reproduce Fig. 10.2.

2. Reproduce Fig. 10.3.

3. Reproduce Fig. 10.5.

CHAPTER 11

MODELS OF POWER-LAW-TYPE BEHAVIOR

Due to the ubiquity of heavy-tailed distribution as well as $1/f^\beta$ noise, modeling of such power-law-type behavior is important for two basic reasons: (1) A good model may shed new light on the mechanism for such power-law-type behavior and (2) an explicit, succinct account of such behavior may greatly simplify a pattern recognition/classification problem. In this chapter, we describe a few simple models for power-law-type behavior.

11.1 MODELS FOR HEAVY-TAILED DISTRIBUTION

In Chapter 7, we discussed stable distributions. We explained that each stable distribution can be considered an attractor, with the normal distribution being at the edge of the family of stable distributions. Any stable distribution can be a good model for a heavy-tailed distribution. In this section, we describe five more simple models.

11.1.1 Power law through queuing

In queuing theory, the simplest and most fundamental model is the Poisson traffic model, or the M/M/1 queuing system, where "M" stands for exponential distribu-

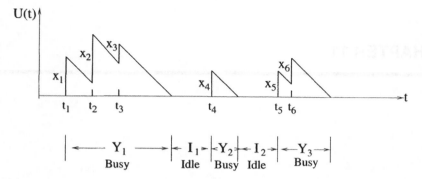

Figure 11.1.   Schematic of alternating busy and idle periods for a queuing system with a constant service rate. $U(t)$ denotes an unfinished load of the system.

tion; the interarrival time follows an exponential distribution with rate λ, while the service time follows another exponential distribution with rate μ. More specifically, assume that at time instants t_1, $t_2, \cdots$, packets arrive at a queuing system. The interarrival times, defined as $T_i = t_{i+1} - t_i$, are exponentially distributed with CDF $1 - e^{-\lambda t}$. The packet size x_i, $i = 1, 2, \cdots$, in units of time measuring how soon the packet can be transmitted, follows another exponential distribution with CDF $1 - e^{-\mu t}$. The utilization level ρ is defined as λ/μ. It is well known that the busy period (which is schematically depicted in Fig. 11.1) of such a queuing system has the following distribution:

$$p(t) = \frac{\rho^{-1/2}}{t} e^{-(\lambda+\mu)t} I_1[2t\sqrt{\lambda\mu}], \qquad (11.1)$$

where I_1 is the modified Bessel function of the first kind (of order 1). When $z \to \infty$, $I_1(z) \approx \frac{1}{\sqrt{2\pi}} z^{-1/2} e^z$. Therefore,

$$p(t) \sim t^{-3/2} e^{-\sqrt{\mu}(1-\sqrt{\rho})^2 t}, \quad t \to \infty. \qquad (11.2)$$

That is, the tail of the distribution is a power law, followed by an exponential truncation. The truncation becomes less and less relevant when $\rho \to 1$. In fact, when $\rho = 1$, the tail becomes a pure power law. Comparing Eq. (11.2) with Eqs. (3.23) and (3.24), we note that $\alpha = 0.5$. The tail is in fact the same as that of the Levy distribution (see Eq. (7.4)).

11.1.2   Power law through approximation by log-normal distribution

Montroll and Shlesinger [316] have found that the power-law distribution with the parameter $\alpha = 0$ can be approximated by the log-normal distribution. Let $\log x$ have a normal distribution, with mean $\log \overline{x}$ and variance σ^2. Their idea is to find the distribution for $x/\overline{x}$. Denote the PDF for $x/\overline{x}$ by $g(x/\overline{x})$. We have

$$\frac{e^{-[\log(x/\overline{x})]^2/2\sigma^2}}{(2\pi\sigma^2)^{1/2}}d\left(\log\frac{x}{\overline{x}}\right) = \frac{e^{-[\log(x/\overline{x})]^2/2\sigma^2}}{(2\pi\sigma^2)^{1/2}}\frac{dx/\overline{x}}{x/\overline{x}} = g\left(\frac{x}{\overline{x}}\right)d\left(\frac{x}{\overline{x}}\right). \quad (11.3)$$

Therefore,

$$\log[g(x/\overline{x})] = -\log(x/\overline{x}) - [\log(x/\overline{x})]^2/2\sigma^2 - \log(2\pi\sigma^2)/2. \quad (11.4)$$

When x is measured in multiples f of its mean $\overline{x}$, $x = f\overline{x}$,

$$\log[g(f)] = -\log f - (\log f)^2/2\sigma^2 - \log(2\pi\sigma^2)/2. \quad (11.5)$$

When the second term is negligible compared with the first term, the distribution is a power law, $g(f) \sim f^{-1}$. Comparing Eq. (11.5) with Eq. (3.23), we see that $\alpha = 0$. To neglect the second term, we may let $\sigma^2 \to \infty$. When σ^2 is large but finite, the power-law relation is valid in a finite range of f.

11.1.3   Power law through transformation of exponential distribution

As we have seen in Chapter 3, many basic models of stochastic processes involve exponential distributions. For example, a Poisson process is defined through exponential interarrival times, while the sojourn times of Markov processes follow exponential distributions. Exponential distributions play an even more fundamental role in physics, since the basic laws in statistical mechanics and quantum mechanics are expressed as exponential distributions, while finite spin glass systems are equivalent to Markov chains. Given the prevalence of power-law-type behavior in complex interconnected systems, it is of fundamental importance to find a way to derive power-law distributions from exponential distributions. We develop a simple and elegant scheme here.

Recall that when a random variable T has an exponential distribution with parameter $\lambda > 0$, its complementary cumulative distribution function (CCDF) or tail probability is

$$P(T \geq t) = e^{-\lambda t}, \quad t \geq 0, \quad \lambda > 0.$$

In many situations, the parameter λ may also be a random variable with a PDF $p(\lambda)$. Then the tail probability for T becomes

$$P(T \geq t) = \int_0^\infty e^{-\lambda t}p(\lambda)d\lambda. \quad (11.6)$$

This is the very one-sided Laplace transform of $p(\lambda)$! With Eq. (11.6), the problem of transforming an exponential distribution into a power-law distribution involves finding Laplace transforms of PDFs $p(\lambda)$ with power-law-type behavior. This can be easily accomplished by looking up tables of Laplace transforms. There are many functions whose Laplace transform has a power-law tail. Such power-law behavior may be said to come from the heterogeneity of the constituents of a complicated system.

For concreteness, in the following, we examine four simple functional forms for $p(\lambda)$.

1. λ is uniformly distributed within $[\lambda_{min}, \lambda_{max}]$:

$$
\begin{aligned}
p(T \geq t) &= \int_{\lambda_{min}}^{\lambda_{max}} e^{-\lambda t} \frac{1}{\lambda_{max} - \lambda_{min}} d\lambda \\
&= \frac{1}{t} \frac{e^{-\lambda_{min}t} - e^{-\lambda_{max}t}}{\lambda_{max} - \lambda_{min}}.
\end{aligned} \tag{11.7}
$$

We now consider three regimes of t:

(a) $t < 1/\lambda_{min}$: using Taylor series expansion, one readily finds that $p(T \geq t) \approx 1$. This corresponds to the power law with the exponent $\alpha = 0$.

(b) $1/\lambda_{max} < t < 1/\lambda_{min}$: in this case, we may drop the term $e^{-\lambda_{max}t}$ and approximate the term $e^{-\lambda_{min}t}$ by 1. Then $p(T \geq t) \sim t^{-1}$.

(c) $t > 1/\lambda_{min}$: in this regime, exponential decay dominates.

In conclusion, $p(T \geq t)$ follows a truncated power law.

2. λ is exponentially distributed within $[\lambda_{min}, \lambda_{max}]$:

$$
\begin{aligned}
p(T \geq t) &= c \int_{\lambda_{min}}^{\lambda_{max}} e^{-\lambda t} \gamma e^{-\gamma \lambda} d\lambda \\
&= \frac{c\gamma}{t + \gamma} [e^{-(t+\gamma)\lambda_{min}} - e^{-(t+\gamma)\lambda_{max}}].
\end{aligned} \tag{11.8}
$$

c is the normalization constant satisfying

$$
c \int_{\lambda_{min}}^{\lambda_{max}} \gamma e^{-\gamma \lambda} d\lambda = 1.
$$

Thus

$$
c = \frac{1}{e^{-\gamma \lambda_{min}} - e^{-\gamma \lambda_{max}}}.
$$

Equation (11.8) then becomes

$$
p(T \geq t) = \frac{\gamma}{t + \gamma} \frac{e^{-(t+\gamma)\lambda_{min}} - e^{-(t+\gamma)\lambda_{max}}}{e^{-\gamma \lambda_{min}} - e^{-\gamma \lambda_{max}}}. \tag{11.9}
$$

Note that case 1 can be obtained from this case by taking $\gamma = 0$. Following the discussion of case 1, it is then clear that $p(T \geq t)$ here also follows a truncated power law.

3. λ follows the gamma distribution:

The PDF of the gamma distribution is given as

$$f(\lambda) = \frac{1}{\Gamma(\alpha)}\beta^\alpha \lambda^{\alpha-1} e^{-\beta\lambda}, \quad \lambda \geq 0, \ \alpha > 0, \ \beta > 0,$$

where $\Gamma(t)$ is the gamma function:

$$\Gamma(t) = \int_0^\infty y^{t-1} e^{-y} dy.$$

The Laplace transform for the gamma distribution can be readily found from a mathematics handbook. It is given by

$$P(T \geq t) = \left(1 + \frac{t}{\beta}\right)^{-\alpha}. \tag{11.10}$$

It is clear that when t is large, $p(T \geq t)$ follows a power law.

4. $1/\lambda$, which is the mean of T, follows the Pareto distribution of Eq. (3.24). Here it is more convenient to work with the PDF, which is given by

$$f(t) = \alpha b^\alpha t^{-\alpha-1}, \quad t \geq b > 0, \tag{11.11}$$

where α is the shape parameter and b is the location parameter. Recall that when $0 < \alpha < 2$, the variance is unbounded. When $0 < \alpha < 1$, the mean also diverges. With the distribution for $1/\lambda$ given by Eq. (11.11), one can easily find the distribution for λ to be

$$p(\lambda) = \alpha b^\alpha \lambda^{\alpha-1}, \quad \lambda \leq 1/b. \tag{11.12}$$

Then

$$p(T \geq t) = \int_0^{1/b} e^{-\lambda t} \alpha b^\alpha \lambda^{\alpha-1} d\lambda. \tag{11.13}$$

In simulations, we may choose b very close to 0; then Eq. (11.13) can be approximated as

$$\begin{aligned} p(T \geq t) &= \int_0^\infty e^{-\lambda t} \alpha b^\alpha \lambda^{\alpha-1} d\lambda \\ &= \alpha b^\alpha t^{-\alpha} \Gamma(\alpha), \end{aligned} \tag{11.14}$$

where $\Gamma(t)$ is the gamma function. It is interesting to note that T and $1/\lambda$ have the same type of heavy-tailed behavior (i.e., the exponent α is the same).

11.1.4 Power law through maximization of Tsallis nonextensive entropy

An interesting way of obtaining a heavy-tailed distribution is to maximize the Tsallis entropy under a few simple constraints. The Tsallis entropy is a generalization of the Shannon (or Boltzmann-Gibbs) entropy. Mathematically, it is closely related to the Renyi entropy. Let us define all these entropies first.

Assume that there are m distinctive events, each occurring with probability p_i. The Shannon entropy is defined by

$$H = -\sum_{i=1}^{m} p_i \log p_i, \tag{11.15}$$

where the unit for H is a bit or baud corresponding to base 2 or e in the logarithm. Without loss of generality, we shall choose base e.

Next, we consider the Renyi entropy, defined by

$$H_q^R = \frac{1}{1-q} \log\Big(\sum_{i=1}^{m} p_i{}^q\Big). \tag{11.16}$$

H_q^R has a number of interesting properties:

- When $q = 1$, H_1^R is the Shannon entropy: $H_1^R = H$.

- $H_0^R = \log(m)$ is the topological entropy.

- If $p_1 = p_2 = \cdots = p_m = \frac{1}{m}$, then for all real valued q, $H_q^R = \log(m)$.

- In the case of unequal probability, H_q^R is a nonincreasing function of q. In particular, if we denote

$$p_{max} = \max_{1 \le i \le m} (p_i), \quad p_{min} = \min_{1 \le i \le m} (p_i),$$

then

$$\lim_{q \to -\infty} H_q^R = -\log(p_{min}), \quad \lim_{q \to \infty} H_q^R = -\log(p_{max}).$$

Finally, let us define the Tsallis entropy. It is given by

$$H_q^T = \frac{1}{q-1}\Big(1 - \sum_{i=1}^{m} p_i{}^q\Big). \tag{11.17}$$

In the continuous case, it can be written as

$$H_q^T = \frac{1}{q-1}\Big(1 - \int_{-\infty}^{\infty} d\Big(\frac{x}{\sigma}\Big)[\sigma p(x)]^q\Big). \tag{11.18}$$

The Tsallis entropy is often called extended entropy formalism. *Tsallis and co-workers assert that this formalism has a number of remarkable mathematical* properties. For example, it preserves the Legendre transformation structure of thermodynamics and keeps the Ehrenfest theorem, von Neumann equation, and Onsgaer reciprocity theorem in the same form with all values of q.

The Renyi entropy and the Tsallis entropy are related by the following simple equation:

$$H_q^R = \frac{\ln\left[1 + (1-q)H_q^T\right]}{1-q}, \qquad \lim_{q\to 1} H_q^R = \lim_{q\to 1} H_q^T = -\sum_{i=1}^{m} p_i \ln p_i.$$

Although mathematically the Renyi and Tsallis entropies seem to be equivalent, we note an important distinction. Under the Shannon entropy formalism, the object or situation under consideration is assumed to be fully random. The Renyi entropy aims to better characterize the situation by resorting to a spectrum of index q to emphasize the difference in p_i. While the Tsallis entropy also has this function, it focuses more on characterizing a type of motion whose complexity is neither regular nor fully chaotic/random by finding a specific q, often different than 1, that best describes the motion. Motions that are neither regular nor fully chaotic/random are ubiquitous. A distinguished class of examples are the fractal processes with long memory.

We now derive the Tsallis distribution under the following two constraints:

$$\int_{-\infty}^{\infty} p(x)dx = 1, \tag{11.19}$$

$$\sigma^2 = \frac{\int_{-\infty}^{\infty} x^2[p(x)]^q dx}{\int_{-\infty}^{\infty} [p(x)]^q dx}. \tag{11.20}$$

The first constraint simply says that the total probability is 1. The second constraint defines the second normalized moment in the framework of extended formalism. It can be written as

$$\int_{-\infty}^{\infty} [x^2 - \sigma^2][p(x)]^q dx = 0.$$

To maximize the Tsallis entropy, we introduce two Lagrange multipliers, λ_1, λ_2, and write $y = p(x)$. We then have

$$S(y) = \frac{1}{q-1}\left(1 - \int_{-\infty}^{\infty} d(\frac{x}{\sigma})[\sigma y]^q\right)$$

$$+ \lambda_1 \int_{-\infty}^{\infty} y dx + \lambda_2 \int_{-\infty}^{\infty} (x^2 - \sigma^2)y^q dx$$

$$= \frac{1}{q-1} + \int_{-\infty}^{\infty} \left[-\frac{\sigma^{q-1}}{q-1}y^q + \lambda_1 y + \lambda_2(x^2 - \sigma^2)y^q\right] dx.$$

For simplicity, let us denote

$$L(x,y) = -\frac{\sigma^{q-1}}{q-1}y^q + \lambda_1 y + \lambda_2(x^2 - \sigma^2)y^q.$$

Note that $S(y + \delta y) - S(y) = \int_{-\infty}^{\infty} \partial L / \partial y \delta y dx$. To maximize S, $\partial L / \partial y = 0$. The Tsallis distribution is then found to be

$$p(x) = \frac{1}{Z_q}[1 + \beta(q - 1)x^2]^{1/(1-q)} \tag{11.21}$$

for $1 < q < 3$. When $q < 1$, the same formula holds when $|x| \le [\beta(1 - q)]^{-1/2}$ and $p(x) = 0$ for any other x. It is clear that Z_q is a normalization constant and β is related to the second moment.

Note that when $q = 1$, the derivation gives the normal distribution. When $q = 2$, the distribution reduces to the Cauchy distribution and, hence, is a stable distribution. When $5/3 < q < 3$, the distribution is heavy-tailed, with

$$0 \le \alpha = 2/(q - 1) - 1 = (3 - q)/(q - 1) \le 2.$$

We may generalize the Tsallis distribution by replacing constraint expressed by Eq. (11.20) by

$$\sigma^\alpha = \frac{\int_{-\infty}^{\infty} x^\alpha [p(x)]^q dx}{\int_{-\infty}^{\infty} [p(x)]^q dx}. \tag{11.22}$$

Following the same procedure leading to Eq. (11.21), we have

$$p(x) = \frac{1}{Z_q}[1 + \beta(q - 1)x^\alpha]^{1/(1-q)}, \tag{11.23}$$

where again Z_q is a normalization constant and β is related to the second moment. Equation (11.23) contains three parameters and is more flexible. To illustrate the power of Eq. (11.23), we apply it to describe sea clutter radar returns in Sec. 11.3.2.

11.1.5   Power law through optimization

Engineering systems are usually designed through constraint optimization. Doyle and Carlson [111] have proposed an interesting means of obtaining power-law behavior by minimizing a cost function of the form

$$J = \left\{ \sum p_i l_i \mid l_i = f(r_i), \quad \sum r_i \le R \right\}, \tag{11.24}$$

where p_i is the probability of events with index i, $1 \le i \le N$, r_i is the resource allocation per unit loss, normalized to lie in the unit interval $0 \le r_i \le 1$, the relationship $l_i = f(r_i)$ describes how allocation of resources r_i limits the size l_i of events, and $\sum r_i \le R$ is an overall constraint coarsely accounting for the connectivity and spatial structure of a real design problem. By assuming resource vs. loss function $l_i = f(r_i)$ having the following form,

$$f_\beta(r_i) = \begin{cases} -\log(r_i), & \beta = 0 \\ \frac{c}{\beta}(r_i^{-\beta} - 1), & \beta > 0, \end{cases} \tag{11.25}$$

they find the optimal solution to be

$$r_i = Rp_i^{\frac{1}{1+\beta}} \left(\sum_j p_j^{\frac{1}{1+\beta}} \right)^{-1} \tag{11.26}$$

$$l_i = \begin{cases} -\log(Rp_i) + \log(\sum_j p_j), & \beta = 0 \\ \frac{c}{\beta}\left[\left(Rp_i^{\frac{1}{1+\beta}} \right)^{-\beta} \left(\sum_j p_j^{\frac{1}{1+\beta}} \right)^{\beta} - 1 \right], & \beta > 0 \end{cases} \tag{11.27}$$

$$l_i = \begin{cases} -\sum_i p_i \log(Rp_i) + (\sum_i p_i)\log(\sum_i p_i), & \beta = 0 \\ \beta^{-1}\left[R^{-\beta}\left(\sum_i p_i^{\frac{1}{1+\beta}} \right)^{1+\beta} - \sum_i p_i \right], & \beta > 0. \end{cases} \tag{11.28}$$

Assuming $r_i < 1$ and inverting Eq. (11.27), they obtain a power-law relation between p_i and l_i:

$$p_i(l_i) = c_1(l_i + c_2)^{-(1+1/\beta)}. \tag{11.29}$$

11.2 MODELS FOR $1/F^\beta$ PROCESSES

In the past few decades, considerable efforts have been made to find universal mechanisms for $1/f^\beta$ noise. While many excellent models have been proposed, it does not seem that a single universal mechanism exists for such noise. To provide a glimpse of this exciting field, we describe a few simple models here. Their use or abuse will also be briefly touched upon.

11.2.1 $1/f^\beta$ processes from superposition of relaxation processes

Superposition of relaxation processes was originally proposed to explain the $1/f^\beta$ noise in vacuum tubes. It can be considered one of the earliest mechanisms proposed for $1/f^\beta$ noise. The basic idea is that a random process

$$x(t) = \sum_i N_0 e^{-\lambda(t-t_i)},$$

where N_0 is a constant, t_i, $i = 1, 2, \cdots$ are random time instances when a relaxation process is initiated, and λ is another random variable having a PDF $p(\lambda)$, may possess a power-law decaying PSD when $p(\lambda)$ is chosen appropriately. In the language of vacuum tube noise, each relaxation process $N(t) = N_0 e^{-\lambda(t-t_i)}$ for $t \geq t_i$ and $N(t) = 0$ for $t < t_i$, which originates from cathode surface trapping sites, contributes to the vacuum tube current. The Fourier transform of a single exponential relaxation process is

$$F(\omega) = \int_{-\infty}^{\infty} N(t)e^{-j\omega t}dt = N_0 \int_{0}^{\infty} e^{-(\lambda+j\omega)t}dt = \frac{N_0}{\lambda + j\omega}. \tag{11.30}$$

For a train of such pulses $N(t, t_k) = N_0 e^{-\lambda(t-t_k)}$ for $t \geq t_k$ and $N(t, t_k) = 0$ for $t < t_k$, we have

$$
\begin{aligned}
F(\omega) &= \int_{-\infty}^{\infty} \sum_k N(t, t_k) e^{-j\omega t} dt \\
&= N_0 \sum_k e^{j\omega t_k} \int_0^{\infty} e^{-(\lambda+j\omega)t} dt = \frac{N_0}{\lambda + j\omega} \sum_k e^{j\omega t_k} \quad (11.31)
\end{aligned}
$$

and the spectrum is

$$
\begin{aligned}
S(\omega) &= \lim_{T \to \infty} \frac{1}{T} \langle |F(\omega)|^2 \rangle = \frac{N_0^2}{\lambda^2 + \omega^2} \lim_{T \to \infty} \frac{1}{T} \left\langle \left| \sum_k e^{j\omega t_k} \right|^2 \right\rangle \\
&= \frac{N_0^2 n}{\lambda^2 + \omega^2}, \quad\quad\quad (11.32)
\end{aligned}
$$

where n is the average pulse rate and the triangle brackets denote an ensemble average. This spectrum is nearly flat near the origin, and after a transition region it becomes proportional to $1/\omega^2$ at high frequency. $1/\omega^\beta$-type behavior can be obtained by choosing appropriate forms of the PDF $p(\lambda)$. For example, if the relaxation rate is uniformly distributed between two values λ_1 and λ_2 and the amplitude of each pulse remains constant, we find the spectrum

$$
\begin{aligned}
S(\omega) &= \frac{1}{\lambda_2 - \lambda_1} \int_{\lambda_1}^{\lambda_2} \frac{N_0^2 n}{\lambda^2 + \omega^2} d\lambda = \frac{N_0^2 n}{\omega(\lambda_2 - \lambda_1)} \left[\arctan \frac{\lambda_2}{\omega} - \arctan \frac{\lambda_1}{\omega} \right] \\
&= \begin{cases} N_0^2 n, & 0 < \omega \ll \lambda_1 \ll \lambda_2 \\[2mm] \frac{N_0^2 n \pi}{2\omega(\lambda_2 - \lambda_1)}, & \lambda_1 \ll \omega \ll \lambda_2 \\[2mm] \frac{N_0^2 n}{\omega^2}, & \lambda_1 \ll \lambda_2 \ll \omega. \end{cases} \quad (11.33)
\end{aligned}
$$

As another example, assume that $dP(\lambda) = \frac{A}{\lambda^\gamma} d\lambda$ occurs in the range $\lambda_1 < \lambda < \lambda_2$. In this case, we obtain

$$
\begin{aligned}
S(\omega) &\propto \int_{\lambda_1}^{\lambda_2} \frac{1}{\lambda^2 + \omega^2} \frac{d\lambda}{\lambda^\gamma} = \frac{1}{\omega^{1+\gamma}} \int_{\lambda_1}^{\lambda_2} \frac{1}{1 + \lambda^2/\omega^2} \frac{d(\lambda/\omega)}{(\lambda/\omega)^\gamma} \\
&= \frac{1}{\omega^{1+\gamma}} \int_{\lambda_1/\omega}^{\lambda_2/\omega} \frac{1}{1 + x^2} \frac{dx}{x^\gamma}.
\end{aligned}
$$

The above integral can be approximated as

$$
S(\omega) \approx \frac{1}{\omega^{1+\gamma}} \int_0^{\infty} \frac{1}{1 + x^2} \frac{dx}{x^\gamma} \propto \frac{1}{\omega^{1+\gamma}}. \quad (11.34)
$$

Note that the superposition of relaxation processes may be transformed into an ON/OFF train. The question of whether the noise may have a spectrum of $1/\omega^\beta$

instead of $1/\omega^2$ is then equivalent to the question of whether ON/OFF trains may have power-law tails. Our discussions in Sec. 11.1.3 provide the general framework for finding a suitable PDF $p(\lambda)$ such that the spectrum is $1/\omega^\beta$.

11.2.2 $1/f^\beta$ processes modeled by ON/OFF trains

In Sec. 8.7, we saw that power-law distributed ON/OFF trains possess $1/f^{2H+1}$ spectra. When $1 \leq \alpha \leq 2$, it can be rigorously proven that the parameter H and the parameter α are related by Eq. (8.27). In numerical simulations, we have found that when ON and OFF periods have the same heavy-tailed distribution, Eq. (8.27) is still valid even if $0 \leq \alpha \leq 1$. In this subsection, we discuss other means of obtaining ON/OFF trains without prescribing their distributions.

One simple way is to start from ON/OFF trains with exponential distributions with rate λ. As we have seen in Sec. 11.1.3, when λ follows a suitable distribution, the ON and OFF periods can become heavy-tailed. While a power-law distributed λ or $1/\lambda$ is an interesting case, to obtain heavy-tailed ON/OFF periods, the distribution of λ may be light-tailed. This is clear from the in-depth discussion in Sec. 11.1.3; therefore, we shall say no more about this here.

Another way of modeling the ON/OFF trains is through chaotic maps. Our in-depth discussion of chaos will be presented in Chapter 13. Here, we shall use the basic understanding of chaos as described in Chapter 2 to give a glimpse of the idea. For this reason, it would be helpful to review Sec. 2.2 before reading the following presentation.

One simple way of obtaining an ON/OFF train from a chaotic time series $x_1, x_2, \cdots$ is through thresholding,

$$y_n = \begin{cases} 0 & 0 \leq x_n < d_y, \\ 1, & d_y \leq x_n \leq 1, \end{cases} \tag{11.35}$$

where the chaotic time series $x_1, x_2, \cdots$ are generated by the following iterations:

$$x_{n+1} = f(x_n) = \begin{cases} f_1(x_n), & 0 \leq x_n < d, \\ f_2(x_n), & d \leq x_n \leq 1, \end{cases} \tag{11.36}$$

where $f_1(x_n)$ and $f_2(x_n)$ are two chaotic maps. The challenge is to find suitable $f_1(\cdot)$ and $f_2(\cdot)$ such that y_n have desired properties. One map that has proven to be useful in modeling long-range-dependent network traffic is the following:

$$x_{n+1} = \begin{cases} f_1(x_n) = x_n + \frac{1-d}{d^{m_1}}(x_n)^{m_1}, & 0 \leq x_n < d, \\ f_2(x_n) = x_n - \frac{d}{(1-d)^{m_2}}(1-x_n)^{m_2}, & d \leq x_n \leq 1, \end{cases} \tag{11.37}$$

where $0 < x_n, d < 1$ and $1 < m_1, m_2 < 2$. This system has a number of interesting properties. For example, the ON/OFF trains have heavy-tailed distributions, and the Hurst parameter is given by

$$H = \frac{3m_i - 4}{2(m_i - 1)},$$

where $m_i = \max(m_1, m_2)$.

It should be noted that in the physical literature, there are types of maps called ON/OFF intermittency maps. Thresholding a time series generated by such maps is straightforward. One generic form of such maps is the following:

$$x_{n+1} = z_n f(x_n), \tag{11.38}$$

with $f(0) = 0$, $f'(0) = 1$, and $z_n = a\eta_n \geq 0$ a random or chaotic process. Given $\{\eta_n\}$, there is a critical value of a denoted as a_c such that for $a < a_c$, the origin is asymptotically stable. If a is increased slightly above a_c, the variable x then exhibits ON/OFF intermittency. An example of such a map is given by

$$x_{n+1} = z_n x_n e^{-bx_n}, \tag{11.39}$$

where $x \geq 0$ and $b > 0$. When $z_n = a > 1$ is a constant, the fixed point is found to be $x^* = \ln a / b$. To understand this process, one should realize that when $x > x^*$, the system is in an ON state; otherwise, it is in an OFF state. When z_n is a random process (such as a fGn process) or chaotic process, the x_n signal is intermittent.

11.2.3 $1/f^\beta$ processes modeled by self-organized criticality

It is common to find that complex systems tend to organize or disorganize, depending on whether they start from highly disorganized or organized states. The ubiquity of $1/f^\beta$ processes in nature motivated Bak and co-workers [23, 24] to suggest that the preferred state in a complex system is neither regular nor fully chaotic/random, but something similar to the "edge of chaos" in the sense that there is no characteristic scale in space or time. This idea is so appealing to scientists in vastly different fields that now the term self-organized criticality (SOC) has become one of the most popular in the literature.

Technically, the theory of SOC refers to the behavior of the sand pile model. The model consists of an $N \times N$ square grid of cells, each characterized by a positive integer $Z(i, j)$, which can be considered the local slope (or height difference between adjacent cells) of the pile. Initial conditions are chosen randomly in the range $1 \leq Z \leq K$, where $K \geq 3$ is a threshold value. At each time step, a randomly chosen cell (i_0, j_0) is incremented by 1:

$$Z_{n+1}(i_0, j_0) = Z_n(i_0, j_0) + 1. \tag{11.40}$$

This amounts to adding a grain of sand to the sand pile. To simulate an avalanche phenomenon, a local instability condition is introduced, which indicates that any cells with $Z(i, j) > K$ and their four nearest neighbors are readjusted according to

$$Z_{n+1}(i, j) = Z_n(i, j) - 4 \tag{11.41}$$

$$Z_{n+1}(i \pm 1, j) = Z_n(i \pm 1, j) + 1 \tag{11.42}$$

$$Z_{n+1}(i, j \pm 1) = Z_n(i, j \pm 1) + 1. \tag{11.43}$$

The cells outside the boundary $Z(0, j)$, $Z(i, 0)$, $Z(N + 1, j)$, $Z(i, N + 1)$ are kept at zero. This condition makes the model an open system, allowing sand grains to leave the system. Equations (11.41)–(11.43) are iterated until no cells with $Z > K$ remain.

There are many types of power-law behavior in the sand pile model. For example, the avalanche size, defined as the number of cells that topple at each time step, follows a power-law distribution; the PSD of the total size of the pile

$$Z_{total} = \sum_{i=1}^{N} \sum_{j=1}^{N} Z(i, j)$$

also follows a power law.

Being discrete and nonlinear, the sand pile model is hard to tackle analytically. One therefore turns to numerical simulations. The drawback of such an approach is that when calculating the PSD, some researchers mistake the magnitude of the Fourier transform for the PSD. By doing so, they incorrectly interpret $1/f^2$ as $1/f$. It turns out that the sand pile model of Bak et al. really produces a $1/f^2$ spectrum instead of $1/f$. Therefore, the SOC model is not quite able to explain the $1/f^\beta$ process. Unfortunately, quite often, when the $1/f^\beta$ spectrum is found from certain data, some researchers attribute it to SOC.

11.3 APPLICATIONS

The materials presented in this chapter have found numerous applications in many areas of science and engineering. To illustrate their usefulness, in this section we consider two important problems, the mechanism for long-range-dependent network traffic and modeling of sea clutter.

11.3.1 Mechanism for long-range-dependent network traffic

Long-range dependence is a prevalent feature of network traffic. Willinger et al. [476] have modeled traffic between a single source and destination by an ON/OFF process, where ON and OFF periods are independent (see Fig. 11.2). The ON periods are identically distributed; so are the OFF periods. Furthermore, at least one of them follows heavy-tailed distributions, and the heavy tails come solely from user behavior. Willinger et al. have further proved that superposition of such processes can be approximated by fBm. This model has now been accepted as the most plausible mechanism for LRD network traffic. However, a number of fundamental questions remain unanswered. For example, where do ON/OFF trains come from? Can heavy-tailed distributions come from mechanisms other than user behavior? Amazingly, by addressing such questions, one arrives at a universal, protocol-independent mechanism for LRD traffic.

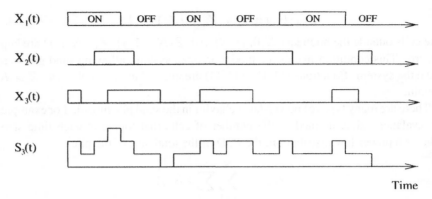

Figure 11.2. $N = 3$ ON/OFF sources, $X_1(t)$, $X_2(t)$, $X_3(t)$, and their summation $S_3(t)$.

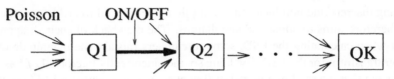

Figure 11.3. Schematic of a tandem of queuing system.

To address the above issues, we first note that network traffic processes represent aggregate traffic streams collected at some backbone network nodes. Before they are measured, most of them must have been transmitted across several links. This suggests that the most relevant scenario to study is the simple tandem network presented in Fig. 11.3. We now ask a question: If the input traffic to Q1 is the simplest Poisson traffic (i.e., the M/M/1 queuing system), can we observe heavy-tailed ON/OFF sources to Q2, Q3, $\cdots$, and so on? The answer is yes, as we shall explain below.

Our first crucial observation is that a busy period of Q1 is an ON period on the link to Q2, while an idle period of Q1 is an OFF period on the link to Q2, as shown in Fig. 11.3. This simple observation shows that ON/OFF processes naturally arise in a network. Realizing the equivalence of an ON episode and a busy period, and noticing (as discussed in Sec. 11.1.1) that a busy period has a heavy-tailed distribution when utilization approaches 1, we immediately see that the power-law behavior does not depend solely on user behavior. It can be generated through transformation of Poisson traffic in a tandem network.

Let us consider transmission of packets in a network. Assume that a long ON period is generated by Q1. Typically, it may not be passed through Q2 as a whole when the utilization level at Q2 is not too low. It has to be packetized and transmitted piece by piece so that the bandwidth can also be used by other users to ensure fairness. While the first piece would be fairly large, other pieces may be considerably smaller. The time intervals between the successive transmissions of these pieces may be quite random. These pieces may even be delivered out of order. In short, the output traffic component originating from this single long ON episode may constitute a very bursty traffic component to Q3. For this reason, the exponent α characterizing the heavy-tailed distributions of the ON periods may become quite different from that of a M/M/1 queuing system, which is 1/2. In fact, the OFF periods can also become heavy-tailed.

11.3.2   Distributional analysis of sea clutter

We have discussed sea clutter in some depth in a number of previous chapters. In this subsection, we consider distributional analysis of sea clutter. The details of sea clutter data are presented in Sec. A.2 of Appendix A.

As we noted in Chapter 1, sea clutter data are often highly non-Gaussian. Much effort has been made to fit various distributions to the observed amplitude data of sea clutter, including Weibull, log-normal, K, and compound-Gaussian distributions. However, the fitting of those distributions to real sea clutter data is not excellent, and quite often using parameters estimated from those distributions is not very effective for distinguishing sea clutter data with and without targets. Only recently have we realized that the ineffectiveness of conventional distributional analysis may be due to the fact that sea clutter data are highly nonstationary. This nonstationarity motivates us to perform distributional analysis on the data obtained by differencing amplitude data of sea clutter. Denote the sea clutter amplitude data by $y(n), n = 1, 2, \cdots$. The differenced data of sea clutter are denoted as

$$x(n) = y(n+1) - y(n), \quad n = 1, 2, \cdots .$$

Specifically, we fit the differenced data by the Tsallis distribution and compare the fitting with the best distribution for sea clutter, the K distribution,

$$f(x) = \frac{\sqrt{2\nu}}{\sqrt{\mu}\Gamma(\nu)2^{\nu-1}} \left(\sqrt{\frac{2\nu}{\mu}}x \right)^{\nu} K_{\nu-1} \left(\sqrt{\frac{2\nu}{\mu}}x \right), \quad x \geq 0 \qquad (11.44)$$

where ν and μ are parameters (μ is in fact equal to half of the second moment), $\Gamma(\nu)$ is the usual gamma function, and $K_{\nu-1}$ is the modified Bessel function of the third kind of order $\nu - 1$. Note, however, that conventionally, the K distribution is used to fit the original sea clutter data, not the differenced data.

Figures 11.4(a,b) show typical results of using the Tsallis distribution to fit the differenced sea clutter data without and with a target, respectively. We observe that the Tsallis distribution fits the differenced data very well. For comparison, Figs. 11.4(c,d) show the results of using the K distribution to fit the same amplitude data that were used to generate Figs. 11.4(a,b). We observe that the fitting to these data using the K distribution is worse than that using the Tsallis distribution.

Note that the goodness of fit of a distribution can be quantified by the Kolmogorov-Smirnov (KS) test. We have found that the Tsallis distribution indeed fits sea clutter data much better than the K distribution. We also note that for detecting low observ-

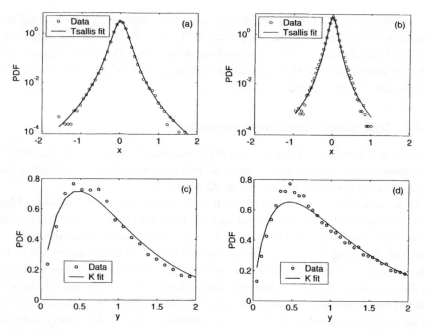

Figure 11.4.   Representative results of using (a,b) the Tsallis distribution to fit the differenced data and (c,d) the K distribution to fit the amplitude data of sea clutter. (a,c) are for the sea clutter data without a target, while (b,d) are for the data with a target. Circles and solid lines denote the raw and fitted PDFs, respectively.

able targets within sea clutter, the parameters estimated from the Tsallis distribution are more effective than the ones from K or other distributions.

To understand why the Tsallis distribution provides a better fit to the sea clutter data, we note two sources of complexity for sea clutter: the roughness of the sea surface, largely due to wave-turbulence interactions on the sea surface and ocean sprays, and the multipath propagation. Either source of complexity suggests that sea clutter may be considered as a superposition of signals massively reflected from the ocean surface. Therefore, the central limit theorem or the generalized central limit theorem has to play a crucial role. Consequently, the Tsallis distribution can be expected to fit sea clutter data well.

11.4 BIBLIOGRAPHIC NOTES

This chapter contains a lot of background material that cannot be fully explained here. We refer readers to the classic book by Kleinrock [263] on the background of queuing theory, to an exquisite paper by Montroll and Shlesinger on the relation between log-normal distribution and power-law distribution [315, 316], to a number of recent papers on the Tsallis distribution [38, 449] and the distribution obtained by Doyle and Carlson [62, 111], to a few classic papers [46, 60, 243, 391, 456] as well as an entertaining survey article [310] (downloadable at http://www.nslij-genetics.org/wli/1fnoise/1fnoise_review.html) on the superposition of relaxation processes, to [216,366,426,476] on the correlation structure of ON/OFF models, to [109, 120, 210, 215, 312–314, 350] on modeling ON/OFF trains by chaotic maps, to [23, 24, 98, 129, 217, 237, 240] on SOC, to [279, 457, 476] on the mechanism of LRD network traffic, and finally, to [69, 128, 189, 190, 238, 318, 320, 447, 466] on distributional analysis of sea clutter.

11.5 EXERCISES

1. Numerically verify
    (a) Eq. (11.7),
    (b) Eq. (11.9), and
    (c) Eq. (11.14).
    (Hint: when solving (a), be aware that in order to observe the power-law tail, λ_{min} and λ_{max} cannot be too close together. Similar care has to be taken for (b) and (c)).

2. Fill in the details in deriving Eq. (11.29).

3. Simulate $1/f^\beta$ processes by superposing relaxation processes.

4. Simulate exponentially distributed ON/OFF trains with the rate λ taking various distributions discussed in Sec. 11.1.3. Estimate the Hurst parameter of those simulated processes.

5. Generate time series by Eq. (11.39) and estimate the Hurst parameter.

CHAPTER 12

BIFURCATION THEORY

In Chapter 2, we briefly discussed elements of chaotic dynamics. Since in reality simple and complex motions coexist, it is important to consider the transitions from a simple motion to a complex one and vice versa. Insights into this issue can be gained by studying bifurcations and routes to chaos. The latter was briefly discussed in Chapter 2. In this chapter, we study bifurcation, which is the change in the qualitative character of a solution as a control parameter is varied. While the theory might seem simple, it can be extremely effective in solving difficult engineering problems. As an example, we shall discuss how to find the exact error threshold values for noisy NAND and majority gates to function reliably. This is a topic of considerable current interest, since in emerging nanotechnologies, reliable computation will have to be carried out with unreliable components as integral parts of computing systems.

12.1 BIFURCATIONS FROM A STEADY SOLUTION IN CONTINUOUS TIME SYSTEMS

12.1.1   General considerations

Assume that our dynamical system is described by a first-order ordinary differential equation (ODE),

$$\dot{y} = f(y, r),$$

where r is a control parameter. A fixed point $y = y^*$ is a solution to

$$f(y, r) = 0.$$

That is, the velocity at $y = y^*$ is zero. Typically, y^* is a function of r. To find the stability of $y = y^*$, we can consider an orbit $y(t) = y^* + u(t)$, where r is fixed and $u(t)$ is a small perturbation. We have

$$\frac{d}{dt}(y^* + u(t)) = f(y^* + u(t), r) \approx \left.\frac{\partial f}{\partial y}\right|_{(y^*, r)} u.$$

Therefore, $y = y^*$ is stable when $\left.\frac{\partial f}{\partial y}\right|_{(y^*, r)} < 0$, unstable when $\left.\frac{\partial f}{\partial y}\right|_{(y^*, r)} > 0$. At the bifurcation point $r = r_c$, $\left.\frac{\partial f}{\partial y}\right|_{(y^*, r_c)} = 0$.

We now consider classification of bifurcations at $r = r_c$. Using Taylor's expansion of $f(y, r)$ at (y^*, r_c), we have

$$\dot{y} = f(y, r) = f(y^*, r_c) + (y - y^*)\left.\frac{\partial f}{\partial y}\right|_{(y^*, r_c)}$$

$$+ (r - r_c)\left.\frac{\partial f}{\partial y}\right|_{(y^*, r_c)} + \frac{1}{2}(y - y_*)^2 \left.\frac{\partial^2 f}{\partial y^2}\right|_{(y^*, r_c)} + \cdots . \qquad (12.1)$$

The first term, $f(y^*, r_c)$ is zero, since at $r = r_c$, $y = y^*$ is a fixed point solution. Now rewrite terms such as $y - y^*$ as u, $r - r_c$ as a, $\dot{y}$ as $d(y - y^*)/dt = \dot{u}$, etc. We have $\dot{u}$ (which is equal to $\dot{y}$) as a summation of various terms of powers of u, a, etc.

To fix the idea, let us consider

$$\dot{u} = au + bu^3 - cu^5, \qquad (12.2)$$

where $b, c > 0$. Usually, we would like the coefficients before powers of x to be 1. For this example, we can let $x = u/U$, $\tau = t/T$, choose U, T, and r, and obtain

$$\frac{dx}{d\tau} = rx + x^3 - x^5. \qquad (12.3)$$

The various normal forms of bifurcations discussed below amount to a classification of this Taylor series expansion, with higher-order terms ignored.

12.1.2 Saddle-node bifurcation

The saddle-node bifurcation is the basic mechanism by which fixed points are created and destroyed. The prototypical example of a saddle-node bifurcation is given by the first-order system

$$\dot{x} = r + x^2, \tag{12.4}$$

where r is a parameter taking on real values. When $r < 0$, there are two fixed points: $-\sqrt{-r}$, which is stable, and $\sqrt{-r}$, which is unstable. The bifurcation point is at $r = 0$.

A bifurcation diagram is a plot of the fixed point solutions vs. the parameter. The bifurcation diagram for the saddle-node bifurcation is shown in Fig. 12.1(a). The solid branch, which is stable, means that if Eq. (12.4) is numerically solved, then eventually the solution will settle on the curve for the corresponding r. The dashed branch, which is unstable, cannot be observed by numerical simulations if the unstable solution is not chosen as the initial condition.

12.1.3 Transcritical bifurcation

There are situations where multiple fixed point solutions coexist. At least one of them must be stable for all values of the controlling parameter, yet the stability of a specific fixed point solution may change with the value of the parameter. Such situations are characterized by a transcritical bifurcation. The normal form is

$$\dot{x} = rx - x^2, \tag{12.5}$$

where r is a parameter taking on real values. In this situation, there are two fixed point solutions, $x = 0$ and $x = r$. When $r < 0$, $x = 0$ is stable but $x = r$ is not; when $r > 0$, $x = 0$ becomes unstable while $x = r$ becomes stable. The bifurcation diagram is shown in Fig. 12.1(b). Again, $r = 0$ is the bifurcation point.

12.1.4 Pitchfork bifurcation

Pitchfork bifurcation is common in physical problems that have symmetry. In such cases, fixed points tend to appear or disappear in symmetrical pairs. There are two different types of pitchfork bifurcation. Here $r = 0$ is also the bifurcation point. The simpler type is called supercritical. Its normal form is

$$\dot{x} = rx - x^3, \tag{12.6}$$

where r is a real parameter. When $r \leq 0$, there is only one fixed point solution, $x = 0$, which is stable. When $r > 0$, $x = 0$ becomes unstable. Two new stable fixed point solutions emerge, $x = \pm\sqrt{r}$. See Fig 12.1(c). In numerical simulations, depending on which initial conditions are used, one of the stable fixed point solutions will be observed.

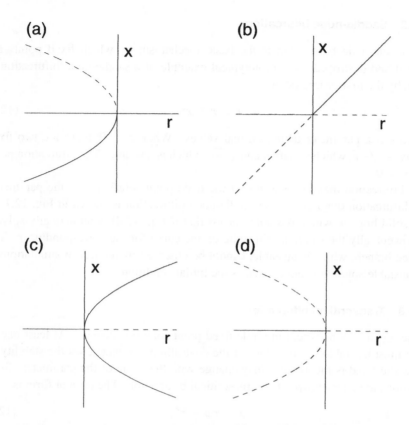

Figure 12.1. Normal forms for stationary bifurcations. Solid lines are stable solutions; dashed lines are unstable solutions. (a) Saddle-node; (b) transcritical; (c) supercritical pitchfork; (d) subcritical pitchfork.

A slightly more complicated situation is called subcritical. Its normal form is

$$\dot{x} = rx + x^3, \qquad (12.7)$$

where r is a real parameter. When $r > 0$, there is only one unstable fixed point solution, $x = 0$. When $r < 0$, three fixed point solutions coexist; $x = 0$ is stable, while $x = \pm\sqrt{-r}$ are unstable. See Fig 12.1(d).

When $r > 0$, a typical solution to Eq. (12.7) diverges to infinity very rapidly. To stabilize it, let us add a term x^5 to Eq. (12.7),

$$\dot{x} = rx + x^3 - x^5. \qquad (12.8)$$

The system is still symmetric under $x \to -x$. For $r < r_s = -1/4$, there is only one stable fixed point solution, $x = 0$. When $r_s \leq r < 0$, there are five fixed point solutions; three are stable, given by $x = 0$ and $x^2 = [1 + \sqrt{1 + 4r}]/2$, and two are unstable, given by $x^2 = [1 - \sqrt{1 + 4r}]/2$. When $r \geq 0$, there are three fixed point

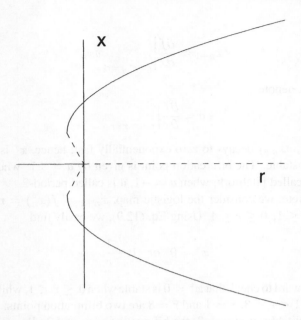

Figure 12.2. Subcritical pitchfork bifurcation with phenomenological quintic stabilizing terms.

solutions: $x = 0$, which is unstable, and $x^2 = [1 + \sqrt{1 + 4r}]/2$, which are stable. These solutions are summarized in Fig. 12.2.

Figure 12.2 provides a mechanism for hysteresis as the parameter r is varied. Suppose we start from the solution $x = 0$ and a negative r. When r is increased but remains negative, the solution will still be $x = 0$. When r passes zero, however, any tiny disturbance will cause the solution to switch to one of the stable fixed points. Let us assume that the upper branch is reached. Then the system will stay at that branch so long as r is kept equal to or larger than $r_s = -1/4$. When r is below r_s, the unique stable fixed point solution of $x = 0$ will again be reached.

12.2 BIFURCATIONS FROM A STEADY SOLUTION IN DISCRETE MAPS

We now consider a discrete map,

$$x_{n+1} = f(x_n, r).$$

Denote the fixed point solution corresponding to $r = r_c$ as x^*. It satisfies

$$x^* = f(x^*, r_c). \tag{12.9}$$

To determine the stability of x^*, we examine a trajectory, $x_0, x_1, \cdots, x_n$, where $x_n = \delta x_n + x^*$. We have

$$x_{n+1} = \delta x_{n+1} + x^* = f(\delta x_n + x^*, r_c) \approx f(x^*, r_c) + \left.\frac{\partial f}{\partial x}\right|_{x=x^*, r=r_c} \delta x_n.$$

Therefore,

$$\delta x_{n+1} \approx \left. \frac{\partial f}{\partial x} \right|_{x=x^*, r=r_c} \delta x_n.$$

For simplicity, denote

$$a = \left. \frac{\partial f}{\partial x} \right|_{x=x^*, r=r_c}.$$

When $|a| < 1$, δx_{n+1} decays to zero exponentially fast; hence, x^* is stable. Otherwise, it is unstable. The bifurcation point is given by $a = \pm 1$: when $a = 1$, the bifurcation is called pitchfork; when $a = -1$, it is called period-2.

To be concrete, we consider the logistic map, $x_{n+1} = f(x_n) = r x_n (1 - x_n)$, where $0 \le x_n \le 1$, $0 \le r \le 4$. Using Eq. (12.9), we easily find

$$x^* = 0 \quad \text{or} \quad 1 - \frac{1}{r}.$$

It is straightforward to check that $x^* = 0$ is stable when $0 \le r < 1$, while $x^* = 1 - \frac{1}{r}$ is stable when $1 < r < 3$. $r = 1$ and $r = 3$ are two bifurcation points: at $r = 1$, the bifurcation is pitchfork; at $r = 3$, the bifurcation is period-2. Beyond $r = 3$, the motion eventually becomes chaotic through the period-doubling route (see Fig. 2.4).

12.3   BIFURCATIONS IN HIGH-DIMENSIONAL SPACE

In general, a dynamical system may be described by a high-dimensional ODE or map and possibly by multiple control parameters. In the former case, the analysis would still be simple – interpreting x as a vector in the discussions of Secs. 12.1 and 12.2 would suffice. The situation is more complicated when there are multiple parameters. One good strategy is to first find fixed point solutions and then carry out numerical simulations of the dynamical system by systematically varying the control parameters.

12.4   BIFURCATIONS AND FUNDAMENTAL ERROR BOUNDS FOR FAULT-TOLERANT COMPUTATIONS

In the emerging nanotechnologies, faulty components may be an integral part of a system. For the system to be reliable, the error of the building blocks — the noisy gates — has to be smaller than a threshold. Therefore, finding exact error thresholds for noisy gates is one of the most challenging problems in fault-tolerant computations. Since von Neumann's foundational work in 1956, this issue has attracted a lot of attention, and many important results have been obtained using information theoretic approaches. Here we show that this difficult problem can be neatly solved by employing bifurcation theory.

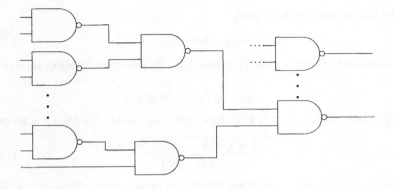

Figure 12.3. Schematic of a circuit built out of two-input NAND gates.

12.4.1   Error threshold values for arbitrary K-input NAND gates

In this subsection, we first derive the error threshold value for a two-input NAND gate and then extend the analysis to an arbitrary K-input NAND gate.

12.4.1.1 Two-input NAND gates   The basic computation in the von Neumann multiplexing scheme involves the circuit schematic shown in Fig. 12.3, where it is assumed that there are no feedback loops and the output of each gate is connected to an input of only one other gate in the circuit. For the basic gate error, we adopt the simple von Neumann model, which assumes that the gate flips the output with a probability $\epsilon \leq 1/2$, while the input and output lines function reliably. For a single NAND gate, let us denote the probabilities of the two inputs being 1 by X and Y. Since the circuits considered have neither closed loops nor fan-out, the two inputs can be treated as independent. Thus the probability Z of the output being a 1 is

$$Z = (1 - \epsilon)(1 - XY) + \epsilon XY = (1 - \epsilon) + (2\epsilon - 1)XY. \tag{12.10}$$

We observe a few interesting properties of Z: (1) When $\epsilon = 0$ and X, Y, Z take on values of either 1 or 0, Eq. (12.10) reduces to the standard definition of an error-free NAND gate, $Z = 1 - XY$. (2) $Z_{\max} = 1 - \epsilon$, $Z_{\min} = \epsilon$. (3) When X (or Y) is fixed, Z linearly decreases with Y (or X). (4) Z decreases most rapidly along $X = Y$, and hence $X = Y$ constitutes the worst-case scenario.

In von Neumann's multiplexing scheme, duplicates of each output randomly become the inputs to another NAND gate. This motivates us to first consider the case $X = Y$. Furthermore, we label a sequence of NAND gates by index i, $i = 1, 2, \cdots, n, \cdots$, where the output of gate i becomes the input to gate $i + 1$. Noticing that a NAND gate is a universal gate, in an actual computation i may *also* be considered equivalent to the ith step of the computation. Equation (12.10) thus reduces to a simple nonlinear map,

$$X_{n+1} = (1 - \epsilon) + (2\epsilon - 1)X_n^2. \tag{12.11}$$

Bifurcation analysis of Eq. (12.11) reveals that a period-doubling bifurcation occurs at

$$\epsilon_* = (3 - \sqrt{7})/4 = 0.08856. \tag{12.12}$$

See Fig. 12.4. When $\epsilon_* < \epsilon \leq 1/2$, the system has a stable fixed point solution,

$$x_0 = \frac{-1 + \sqrt{4(1 - \epsilon)(1 - 2\epsilon) + 1}}{2(1 - 2\epsilon)}. \tag{12.13}$$

Exactly at $\epsilon = \epsilon_*$, a period-doubling bifurcation occurs, since $f'(x_0) = -1$. From Eq. (12.13), one can readily find ϵ_*, as given by Eq. (12.12). When $0 \leq \epsilon < \epsilon_*$, x_0 loses stability, and the motion is periodic with period 2. These two periodic points have been labeled x_+ and x_- in Fig. 12.4. For convenience in our later discussions, we explicitly write down the following three equations:

$$x_0 = (1 - \epsilon) + (2\epsilon - 1)x_0^2, \tag{12.14}$$

$$x_+ = (1 - \epsilon) + (2\epsilon - 1)x_-^2, \tag{12.15}$$

$$x_- = (1 - \epsilon) + (2\epsilon - 1)x_+^2. \tag{12.16}$$

Noting that x_0 is also a solution to Eqs. (12.15) and (12.16), after factoring out the terms involving x_0, one obtains the two periodic points on the limit cycle,

$$x_\pm = \frac{1 \pm \sqrt{4(1 - \epsilon)(1 - 2\epsilon) - 3}}{2(1 - 2\epsilon)}. \tag{12.17}$$

For a NAND gate to function reliably, two identical inputs of 1 or 0 should output a 0 or 1, respectively. We thus see that $0 \leq \epsilon < \epsilon_*$ is the parameter interval where the NAND gate can function. When $\epsilon > \epsilon_*$, the output x_0 can be interpreted as either 1 or 0 and hence, is what von Neumann called a state of irrelevance. Equations (12.12), (12.13), and (12.17) are identical, with expressions derived by Pippenger [348] using a much more complicated approach. In what follows, it will become clear that bifurcation analysis provides additional insights and generalizes to K-input NAND gates.

12.4.1.2 Arbitrary K-input NAND gate

First, we note that with current technology, an arbitrary K-input NAND gate can be readily built, independent of a two-input NAND gate. A circuit built of arbitrary K-input NAND gates may be more reliable and less redundant than a circuit built of two-input NAND gates.

For a K-input NAND gate, we can generalize Eqs. (12.10) and (12.11) to come up with the following two basic equations:

$$\begin{aligned} Z &= (1 - \epsilon) + (2\epsilon - 1)Y_1 Y_2 \cdots Y_K, \tag{12.18} \\ X_{n+1} &= (1 - \epsilon) + (2\epsilon - 1)X_n^K \\ &= 1 - X_n^K + \epsilon(2X_n^K - 1) = f(X_n). \tag{12.19} \end{aligned}$$

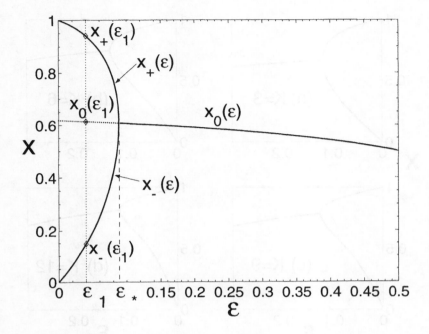

Figure 12.4. The bifurcation diagram for Eq. (12.11). ϵ is the individual gate error probability and X is the probability of a gate output being 1. $x_0(\epsilon)$, given by Eq. (12.13), is unstable when $0 < \epsilon < \epsilon_*$ and stable when $\epsilon_* < \epsilon < 1/2$. In the interval $0 < \epsilon < \epsilon_*$, $x_\pm(\epsilon)$ are given by Eq. (12.17) and form the upper and lower branches of the bifurcation.

The bifurcation diagrams for Eq. (12.19) with $K = 3, 6, 9$, and 12 are shown in Figs. 12.5(a–d). We observe that the fixed point x_0 monotonically increases with K and that the bifurcation involved is always period-doubling. The latter makes perfect sense, since for a sequence of NAND gates to function, the states of the outputs must flip. Such behavior can only be described by a period-2 motion.

Let us now find the bifurcation point ϵ_*. Let x_0 again be a fixed point of the map. We have

$$x_0 = 1 - x_0^K + \epsilon(2x_0^K - 1). \tag{12.20}$$

It is stable when $\epsilon > \epsilon_*$. When K is large, the explicit expression for x_0 may be hard to obtain. Let us not bother to do so. Instead, let us examine when x_0 may lose stability when ϵ is varied. This can be found by requiring the derivative of $f(x)$ evaluated at x_0 to be -1, $f'(x_0) = -1$. Hence

$$(2\epsilon_* - 1)Kx_0^{K-1} = -1. \tag{12.21}$$

Combining Eqs. (12.20) and (12.21), we have

$$\left(1 + \frac{1}{K}\right)\left[\frac{1}{K(1 - 2\epsilon_*)}\right]^{\frac{1}{K-1}} = 1 - \epsilon_*. \tag{12.22}$$

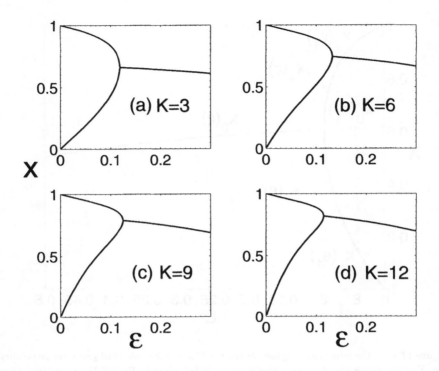

Figure 12.5.   Bifurcation diagrams for Eq. (12.19) with $K = 3, 6, 9$, and 12.

For example, let us solve Eq. (12.22) for $K = 2$ and 3. We obtain Eq. (12.12) for a two-input NAND gate and

$$\epsilon_* = \frac{5}{6} - \frac{1}{6(33 - 8\sqrt{17})^{\frac{1}{3}}} - \frac{1}{6}(33 - 8\sqrt{17})^{\frac{1}{3}} \approx 0.1186$$

for a three-input NAND gate. For $K \geq 4$, analytical expressions for ϵ_* are messy or hard to obtain; however, we can still obtain the threshold values by numerically solving Eq. (12.22). Figure 12.6 shows the variation of ϵ_* with K for $K \leq 20$. We observe that the error threshold value ϵ_* of 0.1186 for $K = 3$ is almost 34% larger than that for $K = 2$. We also note that ϵ_* assumes a maximum value of 0.1330 at $K = 5$, which is more than 50% larger than that for $K = 2$. We thus see that in situations where a two-input NAND gate is too noisy to compute reliably, it may be advantageous to use gates with more than two (but fewer than six) inputs.

12.4.2 Noisy majority gate

We now consider noisy majority gates with $2K + 1$ inputs. Let ϵ be the probability that the noisy gate makes a von Neumann error (flipping the output logic value). Assume that the inputs are independent and that their probabilities of being 1 are all

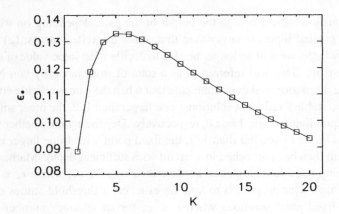

Figure 12.6.   Variation of ϵ_* vs. K.

the same, which we shall denote by x. Let θ represent the probability of the output being 1 when the gate is fault-free. Notice that the majority gate outputs 1 when the number of excited inputs is larger than the number of unexcited inputs. Otherwise, the gate outputs 0. Thus θ is given by

$$\theta = \sum_{i=K+1}^{2K+1} \binom{2K+1}{i} x^i (1-x)^{2K+1-i}. \tag{12.23}$$

It follows that the probability Z that the output is 1 is

$$
\begin{aligned}
Z &= \epsilon(1-\theta) + (1-\epsilon)\theta \\
&= \epsilon + (1-2\epsilon)\left[\sum_{i=K+1}^{2K+1} \binom{2K+1}{i} x^i (1-x)^{2K+1-i}\right]. \tag{12.24}
\end{aligned}
$$

Often there are situations (such as in the case of a network of gates) where one needs to consider a sequence of gates, where the output from one gate becomes the input to the next gate. This motivates us to denote the probability of the output from gate n being 1 as x_n, while the probability of the output from gate $n+1$ being 1 is x_{n+1}. We thus obtain the following map:

$$x_{n+1} = \epsilon + (1-2\epsilon)\left[\sum_{i=K+1}^{2K+1} \binom{2K+1}{i} \cdot x_n^i \cdot (1-x_n)^{2K+1-i}\right], \tag{12.25}$$

where ϵ is identified as a bifurcation parameter. Alternatively, we may interpret x_n as the probability of the input to gate n being 1 and x_{n+1} as the probability of the input to gate $n+1$ being 1.

To better understand the dynamic behavior of Eq. (12.25), let us recall a few facts on how faulty majority gates function. When the gate is faulty, there are two distinct cases: (1) when the error is small enough, we anticipate that we can assign unambiguously a logic 1 or 0 to the output of the gate, depending on whether the number of excited inputs is larger than that of the unexcited inputs; (2) when the error is very large, we will no longer be able to decide what logic value of the output of the gate is in. This was referred to as a state of irrelevance by von Neumann. This simple discussion makes us anticipate that when the gate is reliable enough, the map has two stable fixed point solutions: one larger than $1/2$, the other smaller than $1/2$, corresponding to logic 1 and 0, respectively. Depending on whether the initial condition is larger or smaller than $1/2$, the fixed point with value larger or smaller than $1/2$ will then be approached in a circuit with sufficient depth. Mathematically, the gate being reliable enough can be described by $\epsilon < \epsilon_*$, where ϵ_* is a certain threshold value. Our purpose is to find the exact error threshold values ϵ_* and the two stable fixed point solutions when $\epsilon < \epsilon_*$ for an arbitrary number of inputs $2K + 1$. To simplify the subsequent analysis, let us make a simple transformation:

$$x = \frac{1 - y}{2}.$$

Equation (12.25) then becomes

$$
\begin{aligned}
y_{n+1} &= f(y_n, \epsilon) \\
&= (1 - 2\epsilon)\left\{1 - 2^{-2K} \sum_{i=K+1}^{2K+1} \binom{2K + 1}{i} \cdot (1 - y_n)^i \cdot (1 + y_n)^{2K+1-i}\right\}.
\end{aligned}
$$

$$(12.26)$$

We have the following simple but important lemma.

Lemma 1 $y = 0$ is a fixed point solution to Eq. (12.26) for $0 \leq \epsilon \leq 1/2$. Equivalently, $x = 1/2$ is a fixed point solution to Eq. (12.25).

Proof: This amounts to proving that

$$1 - 2^{-2K} \sum_{i=K+1}^{2K+1} \binom{2K + 1}{i} = 0. \qquad (12.27)$$

This is indeed so, since, by symmetry,

$$\left(\frac{1}{2} + \frac{1}{2}\right)^{2K+1} = 1 = 2 \cdot \left\{\sum_{i=K+1}^{2K+1} \binom{2K + 1}{i} \cdot \left(\frac{1}{2}\right)^i \cdot \left(\frac{1}{2}\right)^{2K+1-i}\right\}.$$

Lemma 1 says that when $x = 1/2$ or $y = 0$, a meaningful logic 1 or 0 cannot be assigned to the output of the gate. The problem is thus to determine when this state dominates and when it cannot be reached. These two questions are answered by the following theorem.

Theorem 1 Equation (12.26) undergoes a pitchfork bifurcation at

$$\epsilon_* = \frac{1}{2} - \frac{2^{2K-1} \cdot (K!)^2}{(2K+1)!}. \tag{12.28}$$

When $\epsilon \geq \epsilon_*$, the fixed point solution $y = 0$ is stable, and the gate is in a state of irrelevance. When $0 < \epsilon < \epsilon_*$, however, $y = 0$ becomes unstable, and two new stable fixed point solutions are created, corresponding to the scenario in which the gate functions reliably.

Proof: Our discussion so far has pointed to the possibility that the bifurcation pertinent to Eq. (12.26) is a pitchfork bifurcation. Mathematically, a pitchfork bifurcation means that the derivative of the right-hand side of Eq. (12.26) evaluated at ϵ_* is 1, $\partial f(y, \epsilon)/\partial y|_{y=0, \epsilon=\epsilon_*} = 1$. We thus have a simple equation:

$$(1 - 2\epsilon_*)\left(-2^{-2K}\right)\left\{ \sum_{i=K+1}^{2K+1} (2K + 1 - 2i)\binom{2K+1}{i} \right\} = 1. \tag{12.29}$$

To simplify Eq. (12.29), we first note, using Eq. (12.27), that

$$\left\{ \sum_{i=K+1}^{2K+1} (2K + 1)\frac{(2K+1)!}{i! \cdot (2K+1-i)!} \right\} = (2K + 1)\, 2^{2K}.$$

The term

$$\left\{ \sum_{i=K+1}^{2K+1} (-2i)\frac{(2K+1)!}{i! \cdot (2K+1-i)!} \right\},$$

on the other hand, can be rewritten as

$$\left\{ \sum_{j=i-1=K}^{2K} (-2)\frac{(2K+1)\cdot(2K)!}{j! \cdot (2K-j)!} \right\}.$$

By expanding

$$\left(\frac{1}{2} + \frac{1}{2}\right)^{2K}$$

and again using symmetry, one finds that

$$\left\{ \sum_{j=K}^{2K} (-2)\frac{(2K+1)\cdot(2K)!}{j! \cdot (2K-j)!} \right\} = -(2K + 1)\left(2^{2K} + \frac{(2K)!}{K! \cdot K!} \right).$$

Equation (12.28) then follows. Next, let us prove our last statement: when $0 < \epsilon < \epsilon_*$, the gate functions reliably. For this purpose, we define the logic value of the output of the gate to be 1 if $1/2 < x \leq 1$, and 0 if $0 \leq x < 1/2$. Equivalently, this can be expressed as $-1 \leq y < 0$ and $0 < y \leq 1$. To prove the statement, it is sufficient to prove that when $-1 \leq y_0 < 0$ or $0 < y_0 \leq 1$, then $-1 \leq y_1 < 0$ or $0 < y_1 \leq 1$, where y_0 and y_1 are related by Eq. (12.26). For concreteness, let us focus on the case of $-1 \leq y_0 < 0$. We need to prove

$$\sum_{i=K+1}^{2K+1} \binom{2K+1}{i} \cdot (1 - y_0)^i \cdot (1 + y_0)^{2K+1-i} > 2^{2K}.$$

This is indeed the case, noting that

$$(1 - y_0 + 1 + y_0)^{2K+1} = 2^{2K+1}$$

and

$$(1 - y_0)^i \cdot (1 + y_0)^{2K+1-i} > (1 - y_0)^{2K+1-i} \cdot (1 + y_0)^i \qquad (i \geq K + 1).$$

Note that this part of the proof can be made more formal by defining the logic value of the output to be 1 if $1/2 + \eta \leq x \leq 1$ and 0 if $0 \leq x \leq 1/2 - \eta$, where η is an arbitrarily small positive number.

It is interesting to examine whether the gate can function reliably when $\epsilon = \epsilon_*$. In a circuit with sufficient depth, where n can be arbitrarily large, y_n can be arbitrarily close to 0. Hence, for any prescribed η, sooner or later, y_n will be within $(-\eta, \eta)$. This means $\epsilon = \epsilon_*$ has to be excluded.

To make the above discussions more informative, let us numerically investigate the bifurcations of Eq. (12.25) for a few fixed $2K + 1$. A few examples are shown in Fig. 12.7 for $2K + 1 = 3, 5, 15, 101$. We observe that the bifurcation involved is indeed a pitchfork bifurcation and that ϵ_* increases with the number of inputs $2K + 1$.

12.4.3   Analysis of von Neumann's multiplexing system

In emerging nanotechnologies, reliable computation will have to be carried out with unreliable components being integral parts of computing systems. One promising scheme for designing these systems using universal NAND gates is von Neumann's multiplexing technique. A multiplexed system consists of a chain of stages (in space or time) each of which consists of NAND gates whose outputs are duplicated and randomly connected to the inputs of the gates of the next stage in the chain. The NAND multiplexing unit schematically shown in Fig. 12.8 is comprised of a randomizing unit R and N copies of NAND gates each failing (flipping the output bit value) with gate error probability ϵ. It takes two bundles of N wires as inputs and generates a bundle of N wires as the output. Through a random permutation by the "randomizing unit," the inputs in one bundle are randomly paired with those from the other bundle to form input pairs to the duplicated NAND gates. In systems based on this construction, each signal is carried on a bundle of N wires instead of a single wire and every logic computation is done by N duplicated gates simultaneously. The

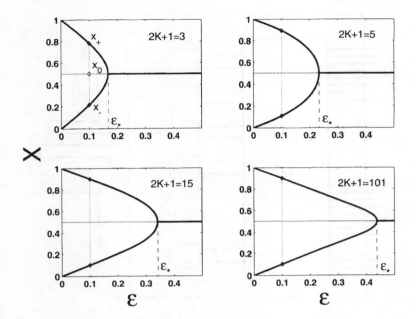

Figure 12.7. Bifurcation diagrams for $2K + 1 = 3$, 5, 15, 101.

logic state 1 or 0 is decided for a bundle when its excitation level, i.e., the fraction of excited wires, is above or below a preset threshold. In a multiplexed system, each computation node is comprised of three such multiplexing units connected in series, as illustrated in Fig. 12.9. The executive unit carries out logic computation, and the restorative unit (comprised of two NAND multiplexing units) restores the excitation level of the output bundle of the executive unit to its nominal level.

Recently, a number of research groups have used a Markov chain model to analyze the reliability of multiplexed systems. We shall show here that the bifurcation approach can effectively analyze the behavior of the multiplexing system, while the approach based on the Markov chain model is at best a very crude approximation.

12.4.3.1 Probability distributions for the excitation level of a multiplexing unit
First, we note that due to the randomization operation in the multiplexing system, starting from the second unit of the multiplexing system, it is not meaningful to talk about whether the state of a wire is either in state 1 or 0. It is only meaningful to talk about the state of wires probabilistically: i.e., with probability Z being excited and probability $1 - Z$ unexcited. Let us now consider one NAND multiplexing unit depicted in Fig. 12.8. Denote the number of excited wires in a bundle of size N by a random variable L and call L/N the bundle's excitation level. Since each wire in the bundle has a probability of Z to be excited, obviously L follows a binomial

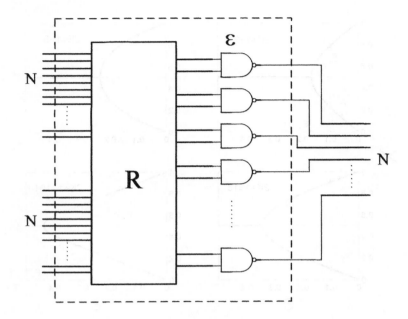

Figure 12.8. A NAND multiplexing unit.

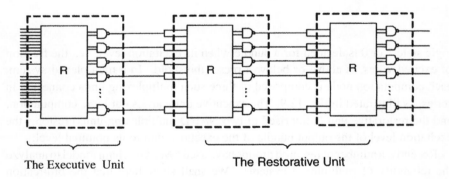

The Executive Unit   The Restorative Unit

Figure 12.9. Multistage NAND multiplexing system. Each computation node in the system consists of three NAND multiplexing units shown in Fig. 12.8. The executive unit does logic computation and two NAND multiplexing units act as a restorative unit.

distribution with parameters N and Z:

$$P(l) = P(L = l) = \binom{N}{l} Z^l (1 - Z)^{N-l}. \qquad (12.30)$$

The mean and variance of L are NZ and $NZ(1 - Z)$. The excitation level L/N is a normalized version of L with mean value Z. Von Neumann suggests that

Eq. (12.30) may be approximated by a normal distribution, $f(l)$:

$$f(l) = \frac{1}{\sqrt{2\pi}\sqrt{NZ(1-Z)}} e^{\frac{(l-NZ)^2}{2NZ(1-Z)}}. \tag{12.31}$$

Mathematically, by the central limit theorem, such an approximation is justified when Z is fixed and the bundle size N is very large. However, when building nanocomputers, we are concerned with small system error probabilities (e.g., on the order of 10^{-10} or smaller) and a small economic redundancy factor (say, $N < 1000$); hence, Gaussian approximation is not very useful. This point will be made clearer shortly. Below we show that the parameter Z can be easily found; hence, Eq. (12.30) can be readily computed without any approximations.

In order to evaluate Eq. (12.30), one only needs to find the parameter Z. To understand how Z can be identified, let us consider two consecutive multiplexing units, the ith and $(i + 1)$th units. It is then clear that

$$Z = X_i \text{ for the } i\text{th stage} \quad \text{and} \quad Z = X_{i+1} \text{ for the } (i + 1)\text{th stage,}$$

respectively, where X_i and X_{i+1} are related by Eq. (12.11) or Eq. (12.19), depending on whether the building blocks are two-input NAND gates or arbitrary K-input NAND gates. When N is very large, Z can be arbitrarily close to either x_+ or x_- (see Eq. (12.17) and Fig. 12.4). Let us denote the distributions expressed by Eq. (12.30) by

$$P(l) = \vec{\pi}_+ \quad \text{when} \quad Z = x_+$$

and by

$$P(l) = \vec{\pi}_- \quad \text{when} \quad Z = x_-.$$

Figure 12.10 shows three examples of $\vec{\pi}_+$ and $\vec{\pi}_-$ for three different gate error probabilities ϵ and a fixed value of $N = 100$. We note that when transients in a multiplexing system die out, the distribution for the excitation level of the output bundle of a multiplexing unit is either $\vec{\pi}_+$ or $\vec{\pi}_-$, while the distribution for the excitation level of the output of the multiplexing unit following the previous one is either $\vec{\pi}_-$ or $\vec{\pi}_+$. In other words, the distributions oscillate between $\vec{\pi}_+$ and $\vec{\pi}_-$ along the multiplexing stages. When the gate error probability ϵ exceeds the threshold value, however, x_+ and x_- become x_0, and we only have one stationary distribution, which we shall denote by $\vec{\pi}_0$, obtained by substituting Z in Eq. (12.30) by x_0. When this is the case, the multiplexing system no longer functions.

In practice, the number of cascaded multiplexing units (or logic depth) may not be very large. Then, how relevant are $\vec{\pi}_+$ and $\vec{\pi}_-$ to reality? The answer is easiest to obtain when system reliability is considered, as shown below.

12.4.3.2 Reliability of multiplexing systems
Characterizations of error and reliability are natural extensions of the notion of probabilistic computation and are particularly important for the design of fault-tolerant nanoelectronic systems

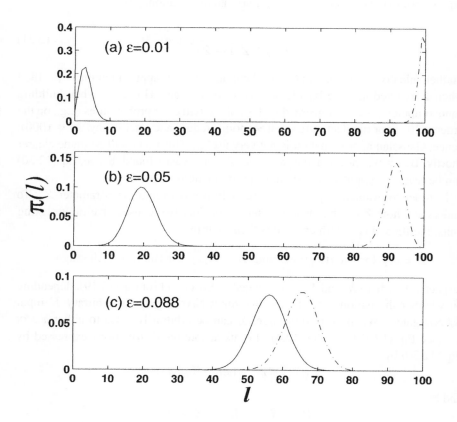

Figure 12.10. Distributions $\vec{\pi}_-$ (solid curve) and $\vec{\pi}_+$ (dash dot curve) for three different gate error probabilities for von Neumann's multiplexing system with $N = 100$. They amount to the distributions for two consecutive stages of the multiplexing system when the stage number is large.

with a large number of defective components. As pointed out, when $\epsilon > \epsilon_*$, the multiplexing system has a single stationary distribution and the logic states 1 and 0 of the system become indistinguishable. Thus only the case $\epsilon < \epsilon_*$ will be considered in the analysis that follows.

Let us now give our definitions of error and reliability. We start by noting that computation by multistage NAND multiplexing systems is essentially a binary communication problem. Errors occur at each stage as the signal propagates through the noisy channel — a cascade of NAND multiplexing units. Given the gate error probability ϵ, bundle size N, for each input instance and a particular stage n, we need to make a decision on the output logic value based on L_n, the observed number of excited wires in the bundle. Let us employ a simple decision rule: set a decision threshold $\gamma \in [0, 1]$; if the number of excited wires in the bundle exceeds γN, then 1 is decided; otherwise, 0 is decided. When the stage number n is odd, the desired state of its output is the opposite of the state of the input to the first stage. For instance, the desired state of the output of the nth stage is 1 when 0 is sent; hence, the output error is the total probability for it to be interpreted in state 0,

$$P_{e,0} = Pr(L_n < \gamma N) = \sum_{i=0}^{i=\lfloor N\gamma \rfloor} P_n(i), \tag{12.32}$$

where P_n denotes the distribution of the output excitation level for the nth stage given by Eq. (12.30). Similarly, when 1 is sent, the error probability is

$$P_{e,1} = Pr(L_n > \gamma N) = \sum_{i=\lceil N\gamma \rceil}^{i=N} P_n(i). \tag{12.33}$$

When n is even, the desired state of the output of the nth stage is the same as the state of the input to the first stage, and the output error probabilities can be defined similarly. With a logic depth of n, system error and reliability are then defined by

$$P_e = \max(P_{e,0}, P_{e,1}), \tag{12.34}$$

$$P_r = 1 - P_e. \tag{12.35}$$

There are two basic issues. One is to choose a suitable decision threshold $\gamma \in [0, 1]$. The other is to decide a suitable P_n to use in evaluating Eqs. (12.32) and (12.33). Let us discuss the latter first.

We claim that in Eq. (12.32), we can choose $P_n = \vec{\pi}_+$, i.e., Z in Eq. (12.30) is given by x_+. Similarly, in Eq. (12.33), we can choose $P_n = \vec{\pi}_-$, with Z in Eq. (12.30) substituted by x_-. Such a choice is natural when the logic depth, i.e., the number of stages, is very large. When the logic depth is small, the error probabilities given by Eq. (12.32) and Eq. (12.33) based on $P_n = \vec{\pi}_+$ and $P_n = \vec{\pi}_-$, respectively, can be considered proper upper bounds: when the first stage is in a state of 1, we require that at least Nx_+ wires are excited; when the state of the first stage is 0, we require that at most Nx_- wires are excited. This way, $\vec{\pi}_+$ will be approached from the right, and $\vec{\pi}_-$ from the left; hence, $P_{e,1}$ and $P_{e,0}$ based on $\vec{\pi}_-$ and $\vec{\pi}_+$ are upper bounds. The input to the first stage can, of course, be easily controlled.

Let us now discuss how to choose γ. First, we examine whether the commonsense choice $\gamma = 1/2$ would work or not. Since a NAND multiplexing unit is purported to work as a NAND gate, we only need to consider if the threshold $\gamma = 1/2$ would work for a NAND gate. If $\gamma = 1/2$ would work, then given an input $x_n < 1/2$, we should get an output $x_{n+1} > 1/2$, and vice versa. However, by Eqs. (12.11) and (12.19), this is not the case: when an input x_n satisfies $1/2 < x_n < x_0$, where x_0 is given by Eq. (12.13) (see also Fig. 12.4), we have an output $x_{n+1} > x_0 > 1/2$. This discussion makes it clear that a natural decision rule is to set $\gamma = x_0$. We

Bundle Size N	50	100	200	300
P_e	1.29×10^{-9}	9.6×10^{-18}	4.47×10^{-33}	1.76×10^{-47}
P_g	5.29×10^{-14}	3.79×10^{-26}	2.71×10^{-50}	2.23×10^{-74}

Table 12.1   Output error probability for a NAND multiplexing system built from 2–input NAND gates. P_e is computed using binomial distribution and P_g is computed using Gaussian approximation. The gate error probability is assumed to be $\epsilon = 0.05$.

emphasize, however, that so long as $\epsilon \ll \epsilon_*$, $N \gg 1$ and γ is somewhere in the middle of x_- and x_+, $P_{e,0}$ and $P_{e,1}$ given by Eqs. (12.32) and (12.33) will be fairly insensitive to the actual choice of γ. This is because the probability distributions are sharply peaked around x_- and x_+ (see Fig. 12.10).

Now we can explicitly derive the expression for P_e or P_r. Due to the asymmetry in the shape of the bifurcation diagram of Fig. 12.4 and the shape of the distributions for $\vec{\pi}_+$ and π_- (see Fig. 12.10), we have found that $\vec{\pi}_+$ should be used to evaluate P_e. That is,

$$P_e = \vec{\pi}_+(I < Nx_0) = \sum_{i=0}^{i=\lfloor Nx_0 \rfloor} \binom{N}{i} (x_+)^i (1 - x_+)^{N-i} . \qquad (12.36)$$

To appreciate how system error depends on redundancy, in Table 12.1 we have listed a few P_e for four different bundle sizes N when the gate error probability ϵ is fixed at 0.05. For convenience of comparison, the output error probabilities computed using Gaussian approximation are also included.

Now we are in a position to truly appreciate why normal approximation to Eq. (12.30) is not useful in practice. There are two basic reasons: (1) the distributions for the output of the multiplexing unit (see Fig. 12.10) peak around x_+ or x_- but are not symmetric about x_+ or x_-, especially when the gate error probability ϵ is close to 0; (2) the system error probability is basically contributed by the probabilities that are away from x_+ or x_-, not around x_+ or x_-. Those tail probabilities tend to be underestimated by Gaussian distribution, whose tail probabilities decay faster than binomially. As reflected by the error probabilities listed in Table 12.1, normal approximation underestimates system error probability by many orders of magnitude. The smaller the gate error ϵ or the larger the bundle size N, the more severely P_g underestimates P_e.

12.4.3.3   Failure of Markov chain modeling   When we model the multiplexed system by a Markov chain, stimulated wires in an output bundle of size N are described by a stochastic process $K = \{K_n; n \in M\}$ where n is the *stage number*, $K_n \in J = \{0, 1, ..., N\}$ is the state of the process K at stage n, and

$M = \{0, 1, \cdots\}$. The relation between the input and output distributions of each multiplexing unit (also referred to as the stage) is described by a first-order Markov chain:

$$P(K_{n+1} = j_{n+1} | K_0 = j_0, ..., K_n = j_n) = P(K_{n+1} = j_{n+1} | K_n = j_n). \quad (12.37)$$

The probability transition matrix $\mathbf{P}$ is given by

$$\mathbf{P} = [P(j|i)],$$

where the matrix element $P(j|i) = P(K_{n+1} = j | K_n = i)$ is given by the conditional probability

$$P(j|i) = \binom{N}{j} Z(i)^j (1 - Z(i))^{N-j}$$

and

$$Z(i) = (1 - \epsilon) - (1 - 2\epsilon)(i/N)^2. \quad (12.38)$$

Note that Eq. (12.38) is in fact used only to compute the distribution for the first stage of the Markov chain. Distributions for the later stages of the Markov chain are obtained through iteration of the probability transition matrix,

$$\vec{\pi}(k + 1) = \vec{\pi}(k)\mathbf{P},$$

where $\vec{\pi}(k)$ is the distribution for the kth stage of the Markov chain.

To understand why such a Markov chain model fails to describe the multiplexing system, it is sufficient to note that the stationary distribution for the Markov chain is simply $(\vec{\pi}_+ + \vec{\pi}_-)/2$. In other words, the stationary distribution is given by the average of the solid and dash-dot curves in Fig. 12.10. This is completely different from the analysis using the bifurcation approach: the solid and dash-dot curves are the distributions for two successive stages when the stage number of the multiplexing system is large. Now, by the definition of system reliability discussed earlier, we see that the error probability based on the Markov chain model eventually deteriorates to 1/2 when the stage number increases. Therefore, the Markov chain approach is not valid.

12.5 BIBLIOGRAPHIC NOTES

An excellent discussion on elementary bifurcation theory can be found in [415]. Readers interested in error thresholds of noisy gates are referred to [121, 122, 161, 211, 348, 359, 460].

12.6 EXERCISES

1. Show that the first-order system $\dot{x} = r - x - e^{-x}$ undergoes a saddle-node bifurcation as r is varied and find the value of r at the bifurcation point.

2. Show that the first-order system $\dot{x} = x(1 - x^2) - a(1 - e^{-bx})$ undergoes a transcritical bifurcation at $x = 0$. Find a combination of the parameters a and b as the control parameter.

3. Explicitly find U, T, and r that lead Eq. (12.2) to Eq. (12.3).

4. By making a simple transformation, prove that Eq. (12.11) is equivalent to the logistic map.

CHAPTER 13

CHAOTIC TIME SERIES ANALYSIS

In Chapter 2, we explained the essence of chaotic dynamics. In practice, one may only be able to measure a scalar time series, $x(t)$, which may be a function of the variables $V = (v_1, v_2, \cdots, v_n)$ describing the underlying dynamics (i.e., $dV/dt = f(V)$). If the dynamics are chaotic, then one can surely expect $x(t)$ to be complicated, resembling a random signal when the time scale examined is not too short. From $x(t)$, how much can we learn about the dynamics of the system? To gain insights into this fundamental problem, in Sec. 13.1 we explain how a suitable phase space can be constructed from $x(t)$ so that the dynamics of the system can be conveniently studied. This geometrical approach is among the most important methods developed in the study of chaotic dynamics. Using the problem of defending against Internet worms, we shall show that even if a dataset is not chaotic, this approach may still yield considerable insights into the complexity of the data and thus greatly simplify the problem of pattern classification. In Sec. 13.2, we discuss three most important measures for characterizing chaotic dynamics: dimension, Lyapunov exponents, and the Kolmogorov-Sinai (KS) entropy. We shall also explain how they can be computed from the $x(t)$ time series. In Sec. 13.3, we discuss a dynamical test for low-dimensional chaos. As we shall see, to determine whether a time series is chaotic or random, it is critical to incorporate the concept of scale. This point is further emphasized in Sec. 13.4, where we shall see that even if sea clutter is not chaotic, the concept of scale can uncover nonlinear structures in the data and thus aid in detecting targets within sea clutter. Our exploration of this topic will be picked up again in Chapter 15 and greatly deepened.

13.1   PHASE SPACE RECONSTRUCTION BY TIME DELAY EMBEDDING

Given a scalar time series data $x(t)$, can we determine the dynamics of the system without reference to other variables describing it? The answer is yes; it involves constructing a suitable phase space from $x(t)$. This procedure was first introduced by Packard et al. [334], and was given a rigorous mathematical basis by Takens [421], Mane [298], and more recently by Sauer et al. [385].

13.1.1   General considerations

Suppose that a dynamical system is described by n first-order ordinary differential equations (ODEs). The dynamical system can be equivalently described by a single ODE involving terms $d^n x/dt^n$, $d^{n-1}x/dt^{n-1}$, etc. It is then clear that one way to construct a phase space is to estimate the derivatives of $x(t)$ by finite differences:

$$\frac{dx}{dt} \approx \frac{x(t + \Delta t) - x(t)}{\Delta t},$$

$$\frac{d^2 x}{dt^2} \approx \frac{x(t + 2\Delta t) - 2x(t + \Delta t) + x(t)}{\Delta t^2},$$

and so on, where Δt is the sampling time for $x(t)$. Unfortunately, in practice, derivatives, especially high-order ones, are quite noisy. A better approach is to develop vectors of the form

$$V_i = [x(i), x(i + L), ..., x(i + (m - 1)L)], \tag{13.1}$$

where m is called the embedding dimension and L the delay time. More explicitly, we have

$$V_1 = [x(t_1), x(t_1 + \tau), x(t_1 + 2\tau), ..., x(t_1 + (m - 1)\tau],$$
$$V_2 = [x(t_2), x(t_2 + \tau), x(t_2 + 2\tau), ..., x(t_2 + (m - 1)\tau],$$
$$\vdots$$
$$V_j = [x(t_j), x(t_j + \tau), x(t_j + 2\tau), ..., x(t_j + (m - 1)\tau],$$
$$\vdots$$

where $t_{i+1} - t_i = \Delta t$ and $\tau = L\Delta t$. This procedure then defines a mapping (i.e., dynamics),

$$V_{n+1} = M(V_n). \tag{13.2}$$

Under the assumption that the dynamics of the system can be described by an attractor with boxing counting dimension D_F (to be defined in the next section), it can be proven that when $m > 2D_F$, the dynamics of the original system are topologically equivalent to those described by Eq. (13.2). When this is the case, the delayed reconstruction is called embedding. The basic ideas behind this fundamental theorem are that (1) given an initial condition, the solution to a set of ODEs is unique, and (2) the trajectory in the phase space does not intersect with itself. When m is not large enough, however, self-intersection may occur.

In practice, m and L have to be chosen properly. This is the issue of optimal embedding. Before taking on this issue, we consider an interesting application.

13.1.2   Defending against network intrusions and worms

Enterprise networks are facing ever-increasing security threats from distributed denial of service (DDoS) attacks, worms, viruses, intrusions, Trojans, port scans, and network misuses, and thus effective monitoring approaches to quickly detect these activities are greatly needed. Traffic-based approaches are among the most favored, because they can provide zero-hour protection against network threats if they can be executed quickly enough. To illustrate the usefulness of the geometrical approach based on the time delay embedding technique, we briefly discuss worm detection in this section.

An Internet worm is a self-propagating program that automatically replicates itself to vulnerable systems and spreads across the Internet. Most deployed worm-detection systems are signature-based. They look for specific byte sequences (called attack signatures) that are known to appear in the attack traffic. Conventionally, the signatures are manually identified by human experts through careful analysis of the byte sequence from captured attack traffic. To appreciate what attack signatures may look like, Fig. 13.1 shows two traffic traces, one normal, another for a worm. It is observed that near the left lower corner, the worm traffic is very different from the normal traffic. For this specific worm traffic, the signature sequence is indeed located there. For other worm traffic traces, the signature sequence can appear elsewhere.

In what sense may a phase space–based approach be useful? The rationale is that the signature sequence of a specific worm will occupy specific regions in the phase space. To see if this is the case, Fig. 13.2 shows the two-dimensional phase diagrams for the normal and worm traffic traces with delay time L being 1. Note that, visually, the phase diagrams remain the same when L is changed. We observe that the phase diagrams define a few threshold values. One is $TH_1 \approx 30$; another is $TH_2 \approx 120$. The exact values for TH_1 and TH_2 can be readily found by calculating the gradients of the distributions of the points in the phase plane. The phase diagrams for the normal and worm traffic differ significantly in four regions:

1. Lower left corner defined by $x(n) < TH_1$, $x(n + 1) < TH_1$.

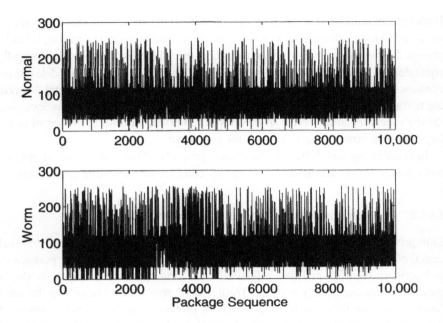

Figure 13.1. Normal and worm traffic traces.

2. Upper right corner defined by $x(n) > TH_2$, $x(n+1) > TH_2$.

3. Left vertical stripe defined by $x(n) < TH_1$, $x(n+1) > TH_2$.

4. Lower horizontal stripe defined by $x(n) > TH_2$, $x(n+1) < TH_1$.

Due to symmetry, regions 3 and 4 can be combined. Based on the percentage of the number of points in those regions, one can define three simple indices to distinguish normal and worm traffic. One result is shown in Fig. 13.3, where we observe that the histograms completely separate normal and worm data. Therefore, the accuracy is 100%. More importantly, this method is much faster than other methods, such as expectation maximization (EM) or hidden Markov model (HMM)-based approaches.

It is interesting to note that conventionally, a detection problem such as the one discussed here requires extensive training. This is no longer needed when one uses a phase space-based approach because this approach identifies a subspace that contains the signature sequence of the worm. Since the subspace depends on the rules (or dynamics) that have generated the worm signature sequence, but not on when the worm attack signatures are present, it may be appropriate to call the subspace an invariant subspace of the specific worm. While one might be concerned that byte sequences due to other network activities might occupy similar phase space regions, there is an effective solution: enlarge the embedding *dimension so that the* worm signature sequence can be located more precisely in the high-dimensional

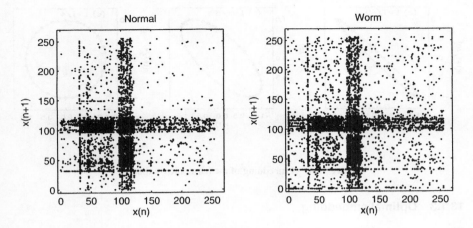

Figure 13.2.   Two-dimensional phase diagrams for normal and worm traffic traces.

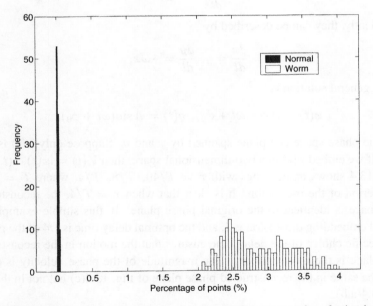

Figure 13.3.   Histograms for an index based on the percentage of the number of points in region 1. 100 normal trace data and 200 worm data are used in the evaluation.

phase space. The effectiveness of this approach lies in the fact that the chance for the small phase space region occupied by the signature sequence of the worm to be shared by byte sequences of other network activities is very small.

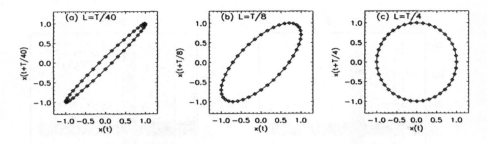

Figure 13.4.   Embedding of the harmonic oscillator.

13.1.3   Optimal embedding

First, we examine a simple example, the harmonic oscillator. Without loss of generality, the dynamics of the harmonic oscillator can be described by

$$\frac{d^2x}{dt^2} = -\omega x.$$

Equivalently, they can be described by

$$\frac{dx}{dt} = y, \quad \frac{dy}{dt} = -\omega x,$$

and the general solution is

$$x(t) = A\cos(\omega t + \phi_0), \quad y(t) = A\sin(\omega t + \phi_0).$$

Here, the phase space is a plane spanned by x and y. Suppose only $x(t)$ is measured. If we embed $x(t)$ in a two-dimensional space, then $V(t) = [x(t), x(t + \tau)]$. Figure 13.4 shows embeddings with $\tau = T/40$, $T/8$, $T/4$, where $T = 2\pi/\omega$ is the period of the oscillation. It is clear that when $\tau = T/4$, the reconstructed phase space is identical to the original phase plane. In this simple example, the minimal embedding dimension is 2, and the optimal delay time is 1/4 of the period. This specific choice of the delay time ensures that the motion in the reconstructed phase plane is the most uniform. The magnitude of the phase velocity is everywhere the same in the reconstructed phase plane of Fig. 13.4(c) but not in those of Figs. 13.4(a,b).

We now discuss the issue of optimal embedding. First, we make a few general comments. (1) The time span represented by the embedding vector V (Eq. (13.1)) is $(m-1)\tau$. It is called the embedding window and quite often is more important than m or τ alone. However, we shall focus on the selection of m and τ, since $(m-1)\tau$ can be easily calculated once m and τ are determined. (2) As we have remarked, when m is too small, the trajectory in the reconstructed phase space may intersect with itself. This violates the uniqueness of a solution described by a set of ODEs. Therefore, m should not be too small. However, with finite data, especially when data contain noise, m should not be too large either. (3) As illustrated in Fig. 13.4(a), when the delay time τ is small, the reconstructed phase diagram is clustered around the diagonal. This is true in an arbitrary space R^m when the embedding window is too small. Therefore, τ should not be too small. However, τ should not be too large either, since when τ is very large, successive elements of the reconstructed vector may already be almost independent, making the deterministic structure of the data difficult to assess. Below we discuss five different methods.

(1) **Visual inspection**: It is often instructive to check the data in a two-dimensional plane and see whether the chosen delay time makes the plot stretch in the phase plane uniformly, i.e., whether it is similar to that shown in Fig. 13.4(c). This is a qualitative method, however. Furthermore, no information on the embedding dimension m can be obtained. When one gradually increases m, it is better to decrease the delay time so that the embedding window $(m-1)\tau$ is kept approximately constant or increases more slowly than m.

(2) **Determining τ based on the autocorrelation function**: Empirically, it has been found that the time corresponding to the first zero of the autocorrelation function (Eq. (3.31)) of the signal is often a good estimate for τ. As with the first method, no information on m can be obtained.

(3) **Determining τ based on mutual information**: A refinement of the second method is based on mutual information. Let the data be partitioned into a number of bins. Denote the probability that the signal assumes a value inside the ith bin by p_i, and let $p_{ij}(\tau)$ be the probability that $x(t)$ is in bin i and $x(t+\tau)$ is in bin j. Then the mutual information for time delay τ is

$$I(\tau) = \sum_{i,j} p_{ij}(\tau) \ln p_{ij}(\tau) - 2 \sum_i p_i \ln p_i. \tag{13.3}$$

In the special case $\tau = 0$, $p_{ij} = p_i \delta_{ij}$, and I yields the Shannon entropy of the data distribution. When τ is large, $x(t)$ and $x(t+\tau)$ are independent and p_{ij} factorizes to $p_i p_j$; therefore, $I \approx 0$. It has been found that a good τ corresponds to the first minimum of $I(\tau)$, when $I(\tau)$ has a minimum. In practice, when calculating $I(\tau)$, it may be advantageous to use equal-probability bins instead of equal-size bins, i.e., $p_i = 1/n = const$, when n is the number of bins used to partition the data. It should be emphasized that $p_{ij}(\tau)$ is not a constant.

(4) **False nearest neighbor method**: This is a geometrical method. Consider the situation in which an m_0-dimensional delay reconstruction is embedding but an $(m_0 - 1)$-dimensional reconstruction is not. Passing from $m_0 - 1$ to m_0, self-intersection in the reconstructed trajectory is eliminated. This feature can be quantified by the sharp decrease in the number of nearest neighbors when m is increased from $m_0 - 1$ by 1. Therefore, the optimal value of m is m_0. More precisely, for each reconstructed vector $V_i^{(m)} = [x(t_i), x(t_i + \tau), x(t_i + 2\tau), \cdots, x(t_i + (m-1)\tau)]$, its nearest neighbor $V_j^{(m)}$ is found (for clarity, the superscript (m) is introduced

to explicitly indicate that this is an m-dimensional reconstruction). If m is not large enough, then $V_j^{(m)}$ may be a false neighbor of $V_i^{(m)}$. If embedding can be achieved by increasing m by 1, then the embedding vectors become $V_i^{(m+1)} = [x(t_i), x(t_i+\tau), x(t_i+2\tau), \cdots, x(t_i+(m-1)\tau), x(t_i+m\tau)] = [V_i^{(m)}, x(t_i+m\tau)]$ and $V_j^{(m+1)} = [V_j^{(m)}, x(t_j+m\tau)]$, and they will no longer be close neighbors. Instead, they will be far apart. The criterion for optimal embedding is then

$$R_f = \frac{|x(t_i+m\tau) - x(t_j+m\tau)|}{\|V_i^{(m)} - V_j^{(m)}\|} > R_T, \qquad (13.4)$$

where R_T is a heuristic threshold value. Abarbanel [1] recommends $R_T = 15$.

After m is chosen, τ can be found by minimizing R_f.

While this method is intuitively appealing, we should point out that it works less effectively in the noisy case. Partly, this is because the concept of nearest neighbors is not well defined when there is noise.

(5) Time-dependent exponent curves: This is a dynamical method developed by Gao and Zheng [178,179], and provides another convenient means of quantifying the effect of self-intersection when an m-dimensional delay reconstruction is not an embedding. Let the reconstructed trajectory be denoted by $V_1^{(m)}, V_2^{(m)}, \cdots$. Assume that $V_i^{(m)}$ and $V_j^{(m)}$ are false neighbors. It is unlikely that points $V_{i+k}^{(m)}, V_{j+k}^{(m)}$, where k is called the evolution time, will continue to be close neighbors. That is, the separation between $V_{i+k}^{(m)}$ and $V_{j+k}^{(m)}$ will be much larger than that between $V_i^{(m)}$ and $V_j^{(m)}$ if the delay reconstruction is not an embedding. The measure proposed by Gao and Zheng is

$$\Lambda(m, L, k) = \left\langle \ln \left(\frac{\|V_{i+k} - V_{j+k}\|}{\|V_i - V_j\|} \right) \right\rangle, \qquad (13.5)$$

where, for ease of later presentation, the superscript (m) in the reconstructed vectors has been dropped. The angle brackets denote the ensemble average of all possible (V_i, V_j) pairs satisfying the condition

$$\epsilon_i \le \|V_i - V_j\| \le \epsilon_i + \Delta\epsilon_i, \quad i = 1, 2, 3, \cdots, \qquad (13.6)$$

where ϵ_i and $\Delta\epsilon_i$ are prescribed small distances. Geometrically, a pair of ϵ_i and $\Delta\epsilon_i$ define a shell, with the former being the diameter of the shell and the latter the thickness of the shell. $\Delta\epsilon_i$ is not necessarily a constant. Since the computation is carried out for a sequence of shells, the effect of noise can be largely eliminated.

Gao and Zheng suggest that for a fixed small k, an optimal m is such that $\Lambda(m, L, k)$ no longer decreases much when further increasing m. After m is chosen, L can be selected by minimizing $\Lambda(m, L, k)$. This optimization procedure ensures that the motion in the reconstructed phase space is the *most uniform*, like that shown in Fig. 13.4(c). It is found that for model chaotic systems, the embedding parameters

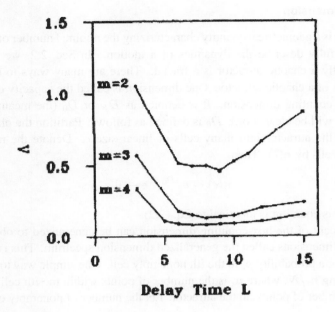

Figure 13.5. $\Lambda(L)$ vs. L for $k = 9\delta t$ and a few different m for the Rossler attractor, where $\delta t = \pi/25$ is the sampling time. 2000 points were used in the computation (adapted from Gao and Zheng [179]).

obtained by this method are the same as those obtained by the false nearest neighbor method. An example for the Rossler attractor,

$$\begin{aligned} dx/dt &= -(y + z), \\ dy/dt &= x + ay, \\ dz/dt &= b + z(x - c), \end{aligned} \tag{13.7}$$

where $a = 0.15$, $b = 0.2$, and $c = 10.0$, is shown in Fig. 13.5, where we observe that the optimal embedding parameters are $m = 3$, $L = 8$. It should be noted that the method works well on short, noisy time series because of the introduction of a series of shells instead of a small ball.

Finally, we note that $\Lambda(k)$ is related to the largest positive Lyapunov exponent. We shall have more to say on this later.

13.2 CHARACTERIZATION OF CHAOTIC ATTRACTORS

In Sec. 2.2, we briefly explained the essence of chaotic dynamics. In this section, we introduce three important measures used to characterize chaotic attractors. At this point, it would be useful to review the first few pages of Secs. 2.1 and 2.2.

13.2.1 Dimension

Dimension is a geometrical quantity characterizing the minimal number of variables needed to fully describe the dynamics of a motion. In Sec. 2.2, we explained that typically a chaotic attractor is a fractal. There are many ways to define the dimensions of a chaotic attractor. One dimension is called the capacity dimension or the box-counting dimension. It is denoted as D_F or D_0 (the meaning of the subscript 0 will be clear soon). D_0 is defined as follows: Partition the phase space containing the attractor into many cells of linear size ϵ. Denote the number of nonempty cells by $n(\epsilon)$. Then

$$n(\epsilon) \sim \epsilon^{-D_0}, \quad \epsilon \to 0.$$

Note that this equation is identical to Eq. (2.2).

The concept of the box-counting dimension can be generalized to obtain a sequence of dimensions called the generalized dimension spectrum. This is obtained by assigning a probability p_i to the ith nonempty cell. One simple way to calculate p_i is by using n_i/N, where n_i is the number of points within the ith cell and N is the total number of points on the attractor. Let the number of nonempty cells be n. Then

$$D_q = \frac{1}{q-1} \lim_{\epsilon \to 0} \left(\frac{\log \sum_{i=1}^{n} p_i^q}{\log \epsilon} \right), \tag{13.8}$$

where q is real. In general, D_q is a nonincreasing function of q. D_0 is simply the box-counting or capacity dimension, since $\sum_{i=1}^{n} p_i^q = n$. D_1 gives the information dimension D_I,

$$D_I = \lim_{\epsilon \to 0} \frac{\sum_{i=1}^{n} p_i \log p_i}{\log \epsilon}. \tag{13.9}$$

Typically, D_I is equal to the pointwise dimension α defined as

$$p(l) \sim l^{\alpha}, \quad l \to 0, \tag{13.10}$$

where $p(l)$ is the measure (i.e., probability) for a neighborhood of size l centered at a reference point. D_2 is called the correlation dimension.

The D_q spectrum is a multifractal characterization. An alternative multifractal formulation is to use singular measures. The basic idea is that a chaotic attractor is comprised of many interwoven fractals, each with a different fractal dimension. Let $f(\alpha)$ be the dimension of points with pointwise dimension α. Then we have

$$D_q = \frac{1}{q-1} [q\alpha(q) - f(\alpha(q))]. \tag{13.11}$$

Differentiating Eq. (13.11), we find

$$\alpha = \frac{d}{dq} [(q-1)D_q] \tag{13.12}$$

and then

$$f(\alpha) = (1 - q)D_q + q\alpha . \tag{13.13}$$

It is clear that the D_q and $f(\alpha)$ spectra give the same amount of information.

In practice, when the dimension of the phase space is high and the length of the data is not very great, calculating dimension by partitioning the phase space into small boxes is not an efficient method. The practical algorithm is the Grassberger-Procaccia algorithm. It involves computing the correlation integral

$$C(\epsilon) = \lim_{N \to \infty} \frac{1}{N^2} \sum_{i,j=1}^{N} H(\epsilon - \|V_i - V_j\|), \tag{13.14}$$

where V_i and V_j are points on the attractor, $H(y)$ is the Heaviside function (1 if $y \geq 0$ and 0 if $y < 0$), and N is the number of points randomly chosen from the entire dataset. The Heaviside function simply counts the number of points within the radius ϵ of the points denoted by V_i, and $C(\epsilon)$ gives the average fraction of points within a distance of ϵ. One then checks the following scaling behavior:

$$C(\epsilon) \sim \epsilon^{D_2}, \quad as \quad \epsilon \to 0. \tag{13.15}$$

Ding et al. [107] have shown that the dimensions calculated by the Grassberger-Procaccia algorithm and that defined by Eq. (13.8) (with $q = 2$) are equivalent.

When calculating the correlation integral, one may compute pairwise distances, excluding points V_i and V_j that are too close in time (i.e., i and j are too close). A rule of thumb suggested by Theiler [430] is that

$$|i - j| > w,$$

where w may be chosen as a decorrelation time. Gao and Zheng [179] pointed out that when V_i and V_j are close in time, they may be on the same orbit. The dimension corresponding to such tangential motion is 1, while the Lyapunov exponent is 0. This point will be made clearer later.

As an example, we compute the dimension for the Lorenz attractor:

$$\begin{aligned} dx/dt &= -16(x - y) + D\eta_1(t), \\ dy/dt &= -xz + 45.92x - y + D\eta_2(t), \\ dz/dt &= xy - 4z + D\eta_3(t). \end{aligned} \tag{13.16}$$

For later convenience, we have added independent Gaussian noise forcing terms $\eta_i(t)$, $i = 1, 2, 3$ with mean 0 and variance 1. D characterizes the strength of noise. When $D = 0$, the system is clean. The top plot of Fig. 13.6 shows the correlation integral for the $x(t)$ time series of the Lorenz attractor, where the curves from top to bottom correspond to $m = 4, 6, \cdots, 12$. The bottom plot of Fig. 13.6 is the so-called plateau plot of the correlation integrals, where we observe that a plateau common to different m occurs at 2.05. This is the correlation dimension for the Lorenz attractor.

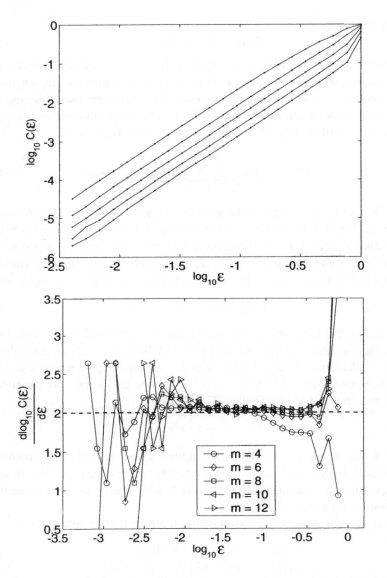

Figure 13.6.   The correlation integral (top) and the correlation dimension (bottom) for the Lorenz attractor. The sampling time is 0.06. 10000 points were used in the computation.

13.2.2 Lyapunov exponents

General considerations: Lyapunov exponents are dynamical quantities. In Sec. 2.2, we introduced the concept of Lyapunov exponents by considering the evolution of an infinitesimal line segment. We now discuss the entire spectrum of the Lyapunov exponents by considering the evolution of an infinitesimal ball of radius dr. Fig-

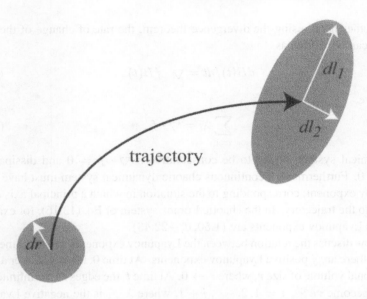

Figure 13.7.   Calculation of the Lyapunov exponents in the two-dimensional case: evolution of a small circle to an ellipsoid.

ure 13.7 schematically shows the evolution of an infinitesimal ball to an ellipsoid. Let $l_i(t)$ be the ith principal axis of the ellipsoid at time t. Generalizing Eq. (2.4), we can write

$$l_i(t) \sim dr \cdot e^{\lambda_i t}.$$

More formally, we have

$$\lambda_i = \lim_{dr \to 0, \, t \to \infty} \frac{1}{t} \ln \frac{l_i(t)}{dr}. \tag{13.17}$$

The Lyapunov exponents are conventionally listed in descending order: $\lambda_1 \geq \lambda_2 \geq \lambda_3 \cdots$. When λ_1 is positive, the dynamics of the system are said to be chaotic. A dynamical system may have multiple positive Lyapunov exponents.

We now examine the general properties of the Lyapunov exponents for an n-dimensional system described by

$$dV/dt = f(V),$$

where $V = (v_1, v_2, \cdots, v_n)$. Let the volume of a small ball at time $t = 0$ be denoted by $B(0)$. At time t, the volume becomes

$$B(t) = B(0)e^{(\sum_{i=1}^{n} \lambda_i)t}. \tag{13.18}$$

This is the solution to

$$dB(t)/dt = \left(\sum_{i=1}^{n} \lambda_i \right) B(t). \tag{13.19}$$

On the other hand, using the divergence theorem, the rate of change of the phase volume can be written as

$$dB(t)/dt = \nabla \cdot f B(t) . \qquad (13.20)$$

Therefore,

$$\sum_{i=1}^{n} \lambda_i = \nabla \cdot f . \qquad (13.21)$$

A dynamical system is said to be conservative if $\nabla \cdot f = 0$ and dissipative if $\nabla \cdot f < 0$. Furthermore, a continuous chaotic dynamical system must have a zero Lyapunov exponent, corresponding to the situation in which a principal axis $l_i(t)$ is parallel to the trajectory. In the chaotic Lorenz system of Eq. (13.16), for example, the three Lyapunov exponents are $(1.50, 0, -22.46)$.

We now discuss the relation between the Lyapunov exponents and the dimension. Assume there are j positive Lyapunov exponents. At time 0, let us consider a $j+1$-dimensional volume of size r, where $r \to 0$. At time t, the edges of the infinitesimal volume become $re^{\lambda_i t}$, $i = 1, 2, \cdots, j+1$, where λ_{j+1} is the negative Lyapunov exponent with the smallest magnitude. Now let us cover this volume by boxes of size $\epsilon = re^{\lambda_{j+1} t}$. It is easy to see that we need

$$N(\epsilon) = \Pi_{i=1}^{j}(e^{\lambda_i t}/e^{\lambda_{j+1} t}) = e^{\sum_{i=1}^{j}(\lambda_i - \lambda_{j+1})t} \sim \epsilon^{-D_L}$$

boxes to cover it, where the dimension is now indexed by L and is called the Lyapunov dimension or Kaplan-Yorke dimension. Therefore,

$$D_L = j - \sum_{i=1}^{j} \lambda_i/\lambda_{j+1} = j + \frac{\lambda_1 + \lambda_2 + \cdots + \lambda_j}{|\lambda_{j+1}|} . \qquad (13.22)$$

Note that $e^{\lambda_i t}$, $i > j+1$ converge to 0 more rapidly than $e^{\lambda_{j+1} t}$, since λ_i, $i > j+1$ are more negative than λ_{j+1}. Hence, those Lyapunov exponents do not have any effect on the dimension.

Numerical computations: There are a few straightforward ways to calculate the largest positive Lyapunov exponent from a time series. In order of increasing sophistication, we describe three of them here.

(1) Wolf et al.'s algorithm [478]: The basic idea is to select a reference trajectory and follow the divergence of a neighboring trajectory from it. Let the spacing between the two trajectories at time t_i and t_{i+1} be d'_i and d_{i+1}, respectively. The rate of divergence of the trajectory over a time interval of $t_{i+1} - t_i$ is then

$$\frac{\ln(d_{i+1}/d'_i)}{t_{i+1} - t_i} .$$

To ensure that the separation between the two trajectories is always small, when d_{i+1} exceeds certain threshold value, it has to be renormalized: a new point in the direction of the vector of d_{i+1} is picked up so that d'_{i+1} is very small compared to the size of the attractor. After n repetitions of stretching and renormalizing the spacing, one obtains the following formula:

$$\lambda_1 = \sum_{i=1}^{n-1} \left[\frac{t_{i+1} - t_i}{\sum_{i=1}^{n-1}(t_{i+1} - t_i)} \right] \left[\frac{\ln(d_{i+1}/d'_i)}{t_{i+1} - t_i} \right] = \frac{\sum_{i=1}^{n-1} \ln(d_{i+1}/d'_i)}{t_n - t_1} . \quad (13.23)$$

Note that this algorithm assumes but does not verify exponential divergence. In fact, the algorithm can yield a positive value of λ_1 for any type of noisy process so long as all the distances involved are small. As pointed out by Gao and Zheng, the reason for this is that when d'_i is small, evolution would move d'_i to the most probable spacing. Then, d_{i+1}, being in the middle step of this evolution, will be larger than d'_i; therefore, a quantity calculated based on Eq. (13.23) will be positive. This argument makes it clear that the algorithm cannot distinguish chaos from noise. We will come back to this issue later.

(2) Rosenstein et al.'s [372] and Kantz's [248] algorithm: In this method, one first chooses a reference point and finds its ϵ-neighbors V_j. One then follows the evolution of all these points and computes an average distance after a certain time. Finally, one chooses many reference points and takes another average. Following the notation of Eq. (13.5), these steps can be described by

$$\Lambda(k) = \left\langle \ln \left\langle \|V_{i+k} - V_{j+k}\| \right\rangle_{average\ over\ j} \right\rangle_{average\ over\ i}, \quad (13.24)$$

where V_i is a reference point and V_j are neighbors to V_i, satisfying the condition $\|V_i - V_j\| < \epsilon$. If $\Lambda(k) \sim k$ for a certain intermediate range of k, then the slope is the largest Lyapunov exponent (The reason for the intermediate range of k will be explained when we discuss the third method.)

While in principle this method can distinguish chaos from noise, with finite noisy data it may not function as desired. One of the major reasons is that in order for the *average over j* to be well defined, ϵ has to be small. In fact, sometimes the ϵ-neighborhood of V_i is replaced by the nearest neighbor of V_i. For this reason, the method cannot handle short, noisy time series well.

(3) Gao and Zheng's method [178–180]: This method contains three basic ingredients: Eq. (13.5), Eq. (13.6), and the condition

$$|i - j| > w. \quad (13.25)$$

When the Lyapunov exponent is calculated, it is usually assumed that the embedding parameters have been properly chosen. In the rest of the book, we shall simply write $\Lambda(k)$ instead of $\Lambda(m, L, k)$. Inequality (13.25) ensures that tangential motions, corresponding to the condition that V_i and V_j follow each other along the orbit, are removed. Tangential motions contribute a Lyapunov exponent of zero and, hence, severely underestimate the positive Lyapunov exponent. An example is shown

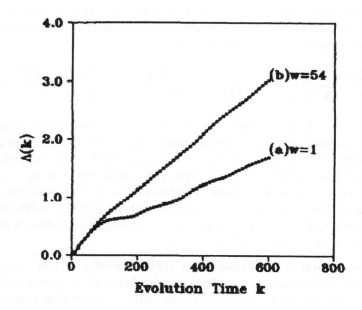

Figure 13.8. $\Lambda(k)$ vs. k curves for the Lorenz system. When $w = 1$, the slope of the curve severely underestimates the largest Lyapunov exponent. When w is increased to 54, the slope correctly estimates the largest Lyapunov exponent.

in Fig. 13.8, where we observe that $w = 1$ severely underestimates the largest Lyapunov exponent, while $w = 54$ solves the problem. In practice, w can be chosen to be larger than one orbital time, such as in the case of the Lorenz attractor and the Rossler attractor. If an orbital time cannot be defined, it can be more or less arbitrarily set to be a large integer if the dataset is not too small. Note that this parameter is closely related to the Theiler decorrelation time used to improve the calculation of correlation dimension, as explained earlier. However, relating w to autocorrelation time is a bit misleading, since the origin of this time scale is due to tangential motions along the orbit.

The condition described by Inequality (13.25) constitutes one of the major differences between this method and the method of Rosenstein et al. There are two more major differences between this method and the method of Rosenstein et al: (1) Gao and Zheng's method calculates $\Lambda(k)$ for a sequence of shells, not just the ϵ-neighbors (which is a ball). As will be explained in Sec. 13.3, this step is critical for the method to be a dynamical test for deterministic chaos. (2) Gao and Zheng's method takes the logarithm of $\|V_{i+k} - V_{j+k}\|$ before taking the average. Rosenstein et al.'s method, by contrast, takes the average over j first, then takes the logarithm, and then takes another average over i.

We will return to this method in Sec. 13.3 when we discuss tests for low-dimensional chaos.

13.2.3 Entropy

General considerations: Entropy characterizes the rate of creation of information in a system. Let us consider an imaginary dynamical system. To calculate the entropy of the motion, we can partition the phase space into small boxes of size ϵ, compute the probability p_i that box i is visited by the trajectory, and finally use the formula $I = -\sum p_i \ln p_i$. For many systems, information increases linearly with time:

$$I(\epsilon, t) = I_0 + Kt, \tag{13.26}$$

where I_0 is the initial entropy and K is the KS entropy (its more precise definition will follow shortly). Now suppose that the system is initially in a particular region of the phase space, and all initial probabilities are zero except the corresponding probability for that region, which is 1. Therefore, $I_0 = 0$.

We now consider three cases of the system: (1) deterministic, nonchaotic, (2) deterministic, chaotic, and (3) random. For case (1), during the time evolution of the system, phase trajectories remain close together. After a time T, nearby phase points are still close to each other and can be grouped in some other small region of the phase space. Therefore, there is no change in information. For case (2), due to exponential divergence, the number of phase space regions available to the system after a time T is $N \propto e^{(\sum \lambda^+)T}$, where λ^+ are positive Lyapunov exponents. Assuming that all of these regions are equally likely, the information function then becomes

$$I(T) = -\sum_{i=1}^{N} p_i(T) \ln p_i(T) = \left(\sum \lambda^+\right)T , \tag{13.27}$$

where it is assumed that $p_i(T) \sim 1/N$. Therefore, $K = \sum \lambda^+$. More generally, if these phase space regions are not visited with equal probability, then

$$K \leq \sum \lambda^+ . \tag{13.28}$$

Grassberger and Procaccia, however, suggest that equality usually holds. Finally, for case (3), we can easily imagine that after a short time, the entire phase space may be visited. Therefore, $I \sim \ln N$. When $N \to \infty$, we have $K = \infty$.

In summary, K is zero for regular motions, positive and finite for chaotic motions, and infinite for random motions.

Formal definitions: First, let us precisely define the KS entropy. Consider a dynamical system with F degrees of freedom. Suppose that the F-dimensional phase space is partitioned into boxes of size ϵ^F. Suppose that there is an attractor in phase space and consider a transient-free trajectory $\vec{x}(t)$. The state of the system is now measured at intervals of time τ. Let $p(i_1, i_2, \cdots, i_d)$ be the joint probability that $\vec{x}(t = \tau)$ is in box i_1, $\vec{x}(t = 2\tau)$ is in box i_2, $\cdots$, and $\vec{x}(t = d\tau)$ is in box i_d.

The KS entropy is then

$$K = -\lim_{\tau \to 0} \lim_{\epsilon \to 0} \lim_{d \to \infty} \frac{1}{d\tau} \sum_{i_1, \cdots, i_d} p(i_1, \cdots, i_d) \ln p(i_1, \cdots, i_d). \quad (13.29)$$

Alternatively, we may first introduce the block entropy:

$$H_d(\epsilon, \tau) = -\sum_{i_1, \cdots, i_d} p(i_1, \cdots, i_d) \ln p(i_1, \cdots, i_d). \quad (13.30)$$

It is on the order of $d\tau K$; then we take difference between $H_{d+1}(\epsilon, \tau)$ and $H_d(\epsilon, \tau)$ and normalize by τ:

$$h_d(\epsilon, \tau) = \frac{1}{\tau}[H_{d+1}(\epsilon, \tau) - H_d(\epsilon, \tau)]. \quad (13.31)$$

Let

$$h(\epsilon, \tau) = \lim_{d \to \infty} h_d(\epsilon, \tau). \quad (13.32)$$

It is clear that the KS entropy can also be obtained by taking proper limits in Eq. (13.32):

$$K = \lim_{\tau \to 0} \lim_{\epsilon \to 0} h(\epsilon, \tau) = \lim_{\tau \to 0} \lim_{\epsilon \to 0} \lim_{d \to \infty} \frac{1}{\tau}[H_{d+1}(\epsilon, \tau) - H_d(\epsilon, \tau)]. \quad (13.33)$$

The KS entropy can be easily extended to order-q Renyi entropies:

$$K_q = -\lim_{\tau \to 0} \lim_{\epsilon \to 0} \lim_{d \to \infty} \frac{1}{d\tau} \frac{1}{q-1} \ln \sum_{i_1, \cdots, i_d} p^q(i_1, \cdots, i_d). \quad (13.34)$$

When $q \to 1$, $K_q \to K$. As we have seen in Sec. 11.1, in the case of unequal probabilities, the Renyi entropy of order-q is a nonincreasing function of q.

Numerical calculations: There are three simple ways to calculate the KS entropy from a time series. One is to first estimate all the positive Lyapunov exponents, and then use the summation of these exponents to estimate the KS entropy. Alternatively, one can estimate K by approximating the probabilities $p(i_1, \cdots, i_m)$, as proposed by Cohen and Procaccia [80] and Eckmann and Ruelle [116]. Let the length of a time series be N, and let the m-dimensional embedding vectors be explicitly denoted by $V_i^{(m)}$, as we did in Sec. 13.1.3. For ease of interpretation, assume the delay time to be 1. Let $n_i^{(m)}(\epsilon)$ be the number of vectors $V_j^{(m)}$ satisfying $\|V_j^{(m)} - V_i^{(m)}\| \le \epsilon$. Cohen and Procaccia noted that

$$C_i^{(m)}(\epsilon = n_i^{(m)}/(N - m + 1)$$

approximates the probability $p(i_1, \cdots, i_m)$ for boxes of size 2ϵ. They then proposed to estimate $H_m(\epsilon)$ by

$$H_m(\epsilon) = -\frac{1}{N - m + 1} \sum_i \ln C_i^{(m)}. \quad (13.35)$$

Then

$$K = \lim_{\tau \to 0} \lim_{\epsilon \to 0} \lim_{m \to \infty} \frac{1}{\delta t} \Big[H_{m+1}(\epsilon) - H_m(\epsilon) \Big], \qquad (13.36)$$

where δt is the sampling time. Note that with the delay time $L = 1$, when the maximum norm,

$$\|V_j^{(m)} - V_i^{(m)}\| = \max_{0 \le k \le m-1} |x(i+k) - x(j+k)|,$$

is used, $H_{m+1}(\epsilon) - H_m(\epsilon)$ is the logarithm of the conditional probability

$$|x(i+m) - x(j+m)| \le \epsilon,$$

given that

$$|x(i+k) - x(j+k)| \le \epsilon \text{ for } k = 0, 1, \cdots, m-1$$

averaged over i.

The third method is to approximate K by estimating K_2 through the correlation integral (defined in Eq. (13.14)):

$$C_m(\epsilon) \sim \epsilon^{D_2} e^{-m\tau K_2}, \qquad (13.37)$$

where $\tau = L\delta t$ is the actual delay time. Note that $C_m(\epsilon)$ is the average of $C_i^{(m)}(\epsilon)$; therefore, it amounts to $\sum_{i_1, \cdots, i_m} p^2(i_1, \cdots, i_m)$. Equation (13.37) can also be expressed as

$$K_2 = \lim_{\tau \to 0} \lim_{\epsilon \to 0} \lim_{m \to \infty} \frac{1}{\tau} [\ln C_m(\epsilon) - \ln C_{m+1}(\epsilon)]. \qquad (13.38)$$

In actual computations, due to the finite size of the data, one cannot take the limits. Instead, one focuses on scaling behavior. For example, when one works with the correlation integral, for true low-dimensional chaotic dynamics, in a plot of $\ln C_m(\epsilon)$ vs. $\ln \epsilon$ with m as a parameter, one observes a series of parallel straight lines, with the slope being the correlation dimension D_2, and the spacing between the lines estimating K_2 (where lines for larger m lie below those for smaller m). See the upper plot of Fig. 13.6. Note that τ has to be small. Otherwise, K_2 may be underestimated.

As can be seen, noisy experimental data may not possess fractal scaling behavior (e.g., defined by Eq. (13.37)) on finite scales. And one cannot take the limit of $\lim_{\epsilon \to 0}$! Then what should one do? To answer this question, a number of derivatives of the KS entropy have been proposed. They include the approximate entropy, the (ϵ, τ)-entropy, and the sample entropy. The approximate entropy amounts to Eq. (13.36) without taking the limits of $\lim_{\epsilon \to 0}$ and $\lim_{m \to \infty}$. It is usually computed using the maximum norm. The (ϵ, τ)-entropy is simply defined by Eq. (13.32). The sample entropy amounts to Eq. (13.38) without taking the limits of $\lim_{\epsilon \to 0}$ and $\lim_{m \to \infty}$. It is clear that these three entropy measures are closely related. Gaspard and Wang [183] have carefully considered the dependence of the (ϵ, τ)-entropy on the scale ϵ. Unfortunately, it is hard to estimate these entropy measures accurately from finite data. This has motivated development of the concept of a finite-size Lyapunov exponent (FSLE). However, the standard method of calculating the FSLE still has severe limitations. To overcome these limitations, and to make the concept applicable to the study of continuous but nondifferential stochastic processes, in Chapter 15 we develop a new concept, the scale-dependent Lyapunov exponent (SDLE).

Finally, we note that recently, Costa et al. [83, 84] have developed a method, multiscale entropy analysis, by calculating the sample entropy on the original data as well as on the smoothed data defined by Eq. (8.3) and examining the dependence of the sample entropy on the block size used for smoothing.

13.3   TEST FOR LOW-DIMENSIONAL CHAOS

To determine whether a time series measured from a complex system is regular, deterministically chaotic, or random is a fundamental issue in science and engineering. Often it is assumed that a numerically estimated positive value for the Lyapunov exponent and a noninteger value for the fractal dimension are sufficient indicators of chaoticness in a time series. This assumption led to a great deal of research in many areas of the natural and social sciences as well as engineering, resulting in the conclusion that *chaos is everywhere*! Osborne and Provenzale [328, 357] observed that $1/f^{\beta}$ noise generates time series with a finite correlation dimension and converging K_2 entropy estimates. Since then, it has been realized that the many so-claimed chaotic motions are not chaotic but random. The difficulty of determining whether a complex time series is truly low-dimensional chaos or not has motivated some researchers to tackle related but less difficult problems, such as whether embedding of a time series defines directional vectors or whether the trajectory in the reconstructed phase space is continuous and smooth. Here we show that this issue can in fact be neatly handled using Gao and Zheng's method. In this section, we discuss the generic behaviors of the $\Lambda(k)$ curves for low-dimensional chaos and then consider two types of noise, iid random variables and surrogate data of low-dimensional chaotic signals. The latter are obtained by taking the Fourier transform of a chaotic signal, randomizing the phases of the Fourier transform, and then taking the inverse Fourier transform. Using the Wiener-Khintchine theorem, we see that surrogate data have the same autocorrelation function as the chaotic signal. $\Lambda(k)$ curves also have distinctive behaviors for random fractals, including $1/f^{\beta}$ processes. However, we shall postpone discussion of random fractals in Chapters 14 and 15.

To illustrate the generic behavior of the $\Lambda(k)$ curves for chaotic systems, we study the Lorenz system described by Eq. (13.16). Fig. 13.9(a) shows the $\Lambda(k)$ curves calculated from the x-component of the clean Lorenz system (i.e., $D = 0$). We observe that the curves have three characteristics:

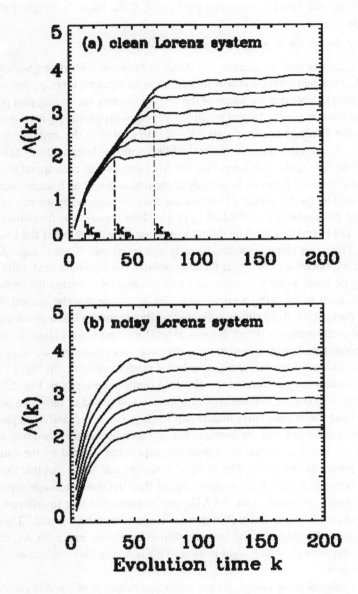

Figure 13.9.   Time-dependent exponent $\Lambda(k)$ vs. evolution time k curves for (a) clean and (b) noisy Lorenz systems. Six curves, from bottom to top, correspond to shells $(2^{-(i+1)/2}, 2^{-i/2})$ with $i = 8, 9, 10, 11, 12,$ and 13. The sampling time for the system is 0.03 s, and embedding parameters are $m = 4, L = 3$. 5000 points were used in the computation.

1. They are linearly increasing for $0 \leq k \leq k_a$.

2. They are still linearly increasing for $k_a \leq k \leq k_p$, but with a slightly different slope.

3. They are flat for $k \geq k_p$.

Feature 3 indicates that on average, the distance between embedding vectors V_{i+k} and V_{j+k} has reached the most probable separation on the attractor (e.g., the diameter of the attractor). Note that the slope of the second linearly increasing part provides an accurate estimate of the largest positive Lyapunov exponent. k_a is related to the time scale for a pair of nearby points (X_i, X_j) to evolve to the unstable manifold of V_i or V_j. It is on the order of the embedding window length, $(m - 1)L$. k_p is the prediction time scale. It is longer for the $\Lambda(k)$ curves that correspond to smaller shells. The difference between the slopes of the first and second linearly increasing parts is caused by the discrepancy between the direction defined by the pair of points (V_i, V_j) and the unstable manifold of V_i or V_j. This feature was first observed by Sano et al. [384] and was used by them to improve the estimation of the Lyapunov exponent. The larger slope in the first linearly increasing part is due to superposition of multiple exponential growths, as recently pointed out by Smith et al. [405]. This region may be made smaller or can even be eliminated by adjusting the embedding parameters, such as by using a larger value for m. Note that the second linearly increasing parts of the $\Lambda(k)$ curves collapse together to form a linear envelope. This feature makes the method a direct dynamical test for deterministic chaos, since such a feature cannot be found in any nonchaotic data. As examples, we have shown the $\Lambda(k)$ for independent, uniformly distributed random variables in Fig. 13.10 and the $\Lambda(k)$ for the surrogate data of the chaotic Lorenz time series in Fig. 13.11. In those figures, we observe that the $\Lambda(k)$ curves are composed of only two parts, an increasing (and sometimes fairly linear) part for $k \leq (m - 1)L$ and a flat part. The flat part again indicates that on average, the distance between embedding vectors V_{i+k} and V_{j+k} has reached the most probable separation defined by the data (now there is no longer an attractor). The first part is simply due to the fact that the initial separation between V_i and V_j is usually smaller than the most probable separation. More importantly, for noisy data, the $\Lambda(k)$ curves corresponding to different shells separate from each other, and a common envelope cannot be defined. Therefore, if one estimates the slopes as the largest positive Lyapunov exponent λ_1, then the value of λ_1 depends on which shell is used. This is a very clear indication that the data are random.

Next, we discuss noisy chaos. As expected, the behavior of the $\Lambda(k)$ curves for a noisy, chaotic system lies between that of the $\Lambda(k)$ curves for a clean, chaotic system and that of the $\Lambda(k)$ curves for white noise or for the surrogate data of a chaotic signal. This is indeed so, as can be easily observed from Fig. 13.9(b). Note that the separation is larger between the $\Lambda(k)$ curves corresponding to smaller shells. This indicates that the effect of noise on the small-scale dynamics is stronger than that on the large-scale dynamics. Also note that $k_a + k_p$ is now on the order of

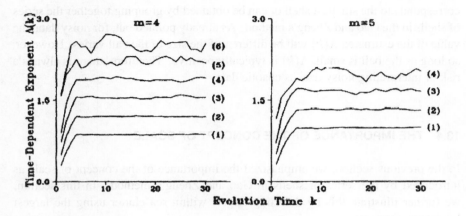

Figure 13.10. $\Lambda(k)$ curves for 6000 uniformly distributed random variables. Curves (1) to (6) correspond to shells $\left(2^{-(i+1)/2}, 2^{-i/2}\right)$ with $i = 4, 5, \cdots, 9$ (when $m = 5$, curves (5) and (6) are not resolved, since there are no points in the corresponding shells). Adapted from Gao and Zheng [179].

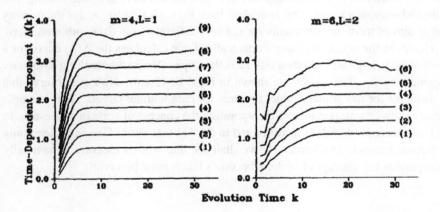

Figure 13.11. $\Lambda(k)$ curves for 6000 points of the surrogate data of the chaotic Lorenz signal. Curves (1)-(9) correspond to shells $\left(2^{-(i+1)/2}, 2^{-i/2}\right)$ with $i = 4, 5, \cdots, 12$. For the case of $m = 6, L = 2$, only curves (1) to (6) are resolved. Adapted from Gao and Zheng [179].

the embedding window length and is almost the same for all the $\Lambda(k)$ curves. With stronger noise $(D > 4)$, the $\Lambda(k)$ curves will be more like those for white noise.

At this point, it is pertinent to comment on why methods that do not employ shells cannot be used to reliably distinguish low-dimensional chaos from noise. The reason is as follows. Those methods amount to estimating the Lyapunov exponent by $\Lambda(k)/k$, where $\Lambda(k)$ is computed based on a single ball. The latter either corresponds to the smallest shell or can be obtained by grouping together the series of shells in the Gao and Zheng's method. As already pointed out, for noisy data, the value of the estimated $\Lambda(k)$ will be different if the size of the ball varies. However, so long as the ball is small, $\Lambda(k)$ is typically positive. Therefore, there is always a risk of interpreting noisy data as chaotic data.

13.4 THE IMPORTANCE OF THE CONCEPT OF SCALE

In the previous section, we emphasized the importance of the concept of scale as introduced by the series of shells in Gao and Zheng's method. In this section, we further illustrate this by detecting targets within sea clutter using the largest Lyapunov exponent.

Like the study of other complex time series, determining whether sea clutter is chaotic or random has been a hot topic. Recent works demonstrating that sea clutter may not be truly chaotic are mostly based on the observation that the Lyapunov exponent, λ_c, obtained using canonical methods is similar for sea clutter and some stochastic processes that share certain features with sea clutter. Extending this idea to target detection, one can readily find that λ_c is not effective in distinguishing data with and without a target. This is evident from Figs. 13.12(a,b): λ_c for the primary target bins of most measurements are not very different from data without targets.

However, the situation changes drastically if one calculates the $\Lambda(k)$ curve for a shell of fairly large size and then estimates the slope. Denote the resulting Lyapunov exponent by λ_ϵ. The results are shown in Figs. 13.12(c,d). Now we observe that λ_ϵ is larger for the primary target bin than for data without targets. This example clearly shows the importance of incorporating the concept of scale in a measure. In fact, the concept of scale is incorporated in the $\Lambda(k)$ curves of Gao and Zheng only in a static manner. In Chapter 15, we shall see that when a measure dynamically incorporates the concept of scale, it becomes much more powerful.

13.5 BIBLIOGRAPHIC NOTES

Much of the material covered in this chapter can also be found in [1, 25, 249]. For time delay embedding, we refer to [298, 334, 385, 421]. For optimal embedding, we refer to [8, 65, 140, 255, 285]. For characterization of chaotic time series, we refer to [80, 107, 108, 116, 178–180, 203, 204, 248, 344, 372, 373, 378, 384, 478]. For approximate entropy, sample entropy, (ϵ, τ)-entropy, and multiscale entropy analysis, we refer to [83, 84, 183, 345, 367]. For distinguishing chaos from noise, we refer to [28, 104, 105, 179, 180, 182, 197, 229, 251, 256, 339, 352, 358, 380, 416, 451, 469]. For chaos in fluid flows, we refer to [417, 418]. For surrogate data, we refer to [432]. Finally, for worm detection, we refer to [423] and references therein.

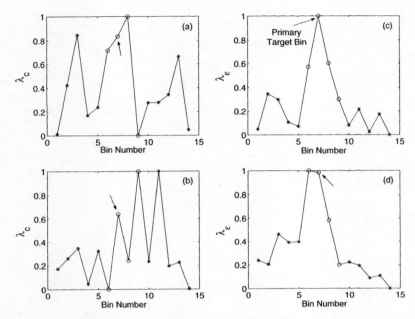

Figure 13.12.   (a,b) The variations of the largest Lyapunov exponent λ_c (estimated by a canonical method) for two measurements, each with 14 range bins. Open circles denote the range bins with a target, while asterisks denote the bins without a target. The primary target bin is indicated by an arrow. (c,d) The variations of the Lyapunov exponent, denoted as λ_ϵ, corresponding to large shells for the same two measurements.

13.6 EXERCISES

1. Generate a time series from $x(t) = \sin \omega_1 t + a \sin \omega_2 t$ and ω_1/ω_2 is an irrational number by choosing appropriate parameters where a is a constant. Embed the time series to a two-dimensional phase space. Do your plots look like a torus?

2. Write a program to compute the correlation integral from the chaotic Lorenz attractor data (see Sec. A.4 of Appendix A), then compute the correlation dimension and the correlation entropy. (Hint: If you have difficulty writing the program, you may compare your code with the one listed in the book's website explained in Sec. A.4 of Appendix A.)

3. Write a program to compute the $\Lambda(k)$ curves from various data downloadable at the book's website (see Sec. A.4 of Appendix A).

Figure 12.12

12.6 EXERCISES

1.

2.

3.

CHAPTER 14

POWER-LAW SENSITIVITY TO INITIAL CONDITIONS (PSIC) — AN INTERESTING CONNECTION BETWEEN CHAOS THEORY AND RANDOM FRACTAL THEORY

Chaos theory shows that apparently irregular behaviors in a complex system may be generated by nonlinear deterministic interactions of only a few numbers of degrees of freedom. Noise or intrinsic randomness does not play any role. Random fractal theory, on the other hand, assumes that the dynamics of the system are inherently random. Since the foundations of chaos theory and random fractal theory are entirely different, different conclusions may be drawn, depending upon which theory is used to analyze the data. In fact, as we explained in Chapter 13, much of the research in the past was devoted to determining whether a complex time series is generated by a chaotic or a random system. Such a categorical study, however, may discourage cross-talks between researchers specializing in chaos theory and random fractal theory, hindering the development of new methods for the multiscale analysis of complex data. Therefore, it would be very desirable to develop a more general framework to encompass chaos theory and random fractal theory as special cases. In this chapter, we show that power-law sensitivity to initial conditions (PSIC) provides such a framework.

14.1 EXTENDING EXPONENTIAL SENSITIVITY TO INITIAL CONDITIONS TO PSIC

As we have shown in Chapter 13, deterministic chaos is defined by exponential sensitivity to initial conditions (ESIC). In order to characterize a type of motion whose complexity is neither regular nor fully chaotic/random, recently the concept of ESIC has been generalized to PSIC by Tsallis and co-workers [450]. Mathematically, the formulation of PSIC closely parallels that of nonextensive entropy formalism, as briefly described in Sec. 11.1.4. In particular, there is also a parameter, the entropic index q, in PSIC. However, as will be explained momentarily, q in PSIC and q in Tsallis entropy are not the same.

To understand the essence of PSIC, let us focus on the one-dimensional case and consider

$$\xi(t) = \lim_{\Delta x(0) \to 0} \frac{\Delta x(t)}{\Delta x(0)},$$

where $\Delta x(0)$ is the infinitesimal discrepancy in the initial condition and $\Delta x(t)$ is the discrepancy at time $t > 0$. When the motion is chaotic, then

$$\xi(t) = e^{\lambda_1 t}.$$

$\xi(t)$ satisfies

$$\frac{d\xi(t)}{dt} = \lambda_1 \xi(t). \tag{14.1}$$

Tsallis and co-workers have generalized Eq. (14.1) to

$$\frac{d\xi(t)}{dt} = \lambda_q \xi(t)^q, \tag{14.2}$$

where q is called the entropic index, and λ_q is interpreted to be equal to K_q, the generalization of the KS entropy. Equation (14.2) defines the PSIC in the one-dimensional case. Obviously, PSIC reduces to ESIC when $q \to 1$. The solution to Eq. (14.2) is

$$\xi(t) = [1 + (1 - q)\lambda_q t]^{1/(1-q)}. \tag{14.3}$$

Notice the striking similarity between Eq. (14.3) and Eqs. (11.21) and (11.23), which give Tsallis and generalized Tsallis distributions. When t is large and $q \neq 1$, $\xi(t)$ increases with t as a power law,

$$\xi(t) \sim C t^{1/(1-q)}, \tag{14.4}$$

where $C = [(1 - q)\lambda_q]^{1/(1-q)}$. For Eq. (14.4) to define an unstable motion with $\lambda_q > 0$, we must have $q \leq 1$. In contrast, in Tsallis entropy, $-\infty < q < \infty$. Therefore, q in PSIC and q in Tsallis entropy are different. Below we shall map different types of motions to different ranges of q.

To apply PSIC to the analysis of time series data, one can first construct a phase space by constructing embedding vectors, $V_i = [x(i), x(i + L), \cdots, x(i + (m -$

1)$L)$], where $x(i), i = 1, 2, \cdots$ is the given scalar time series. Equation (14.3) can then be generalized to the high-dimensional case,

$$\xi(t) = \lim_{\|\Delta V(0)\| \to 0} \frac{\|\Delta V(t)\|}{\|\Delta V(0)\|} = \left[1 + (1 - q)\lambda_q^{(1)} t \right]^{1/(1-q)}, \qquad (14.5)$$

where $\|\Delta V(0)\|$ is the infinitesimal discrepancy between two orbits at time 0, $\|\Delta V(t)\|$ is the distance between the two orbits at time $t > 0$, q is the entropic index, and $\lambda_q^{(1)}$ is the first q-Lyapunov exponent, corresponding to the power-law increase of the first principal axis of an infinitesimal ball in the phase space. $\lambda_q^{(1)}$ may not be equal to K_q. This is understood by recalling that for chaotic systems, the KS entropy is the sum of all the positive Lyapunov exponents. We believe that a similar relation may hold between the q-Lyapunov exponents and K_q. When there are multiple unstable directions, then in general $\lambda_q^{(1)}$ may not be equal to K_q. When t is large and $q \neq 1$, Eq. (14.5) again gives a power-law increase of $\xi(t)$ with t.

We now consider the general computational framework for PSIC. Given a finite time series, the condition of $\|\Delta V(0)\| \to 0$ cannot be satisfied. In order to find the law governing the divergence of nearby orbits, one has to examine how the neighboring points, (V_i, V_j), in the phase space evolve with time by forming suitable ensemble averages. Notice that if (V_{i1}, V_{j1}) and (V_{i2}, V_{j2}) are two pairs of nearby points, when $\|V_{i1} - V_{j1}\| \ll \|V_{i2} - V_{j2}\| \ll 1$, the separations such as $\|V_{i1+t} - V_{j1+t}\|$ and $\|V_{i2+t} - V_{j2+t}\|$ cannot be simply averaged to provide estimates for q and $\lambda_q^{(1)}$. In fact, it would be most convenient to consider ensemble averages of pairs of points (V_i, V_j) that all fall within a very thin shell, $r_1 \leq \|V_i - V_j\| \leq r_2$, where r_1 and r_2 are close. These arguments suggest that the time-dependent exponent curves defined by Eq. (13.5) provide a natural framework to assess PSIC from a time series. This is indeed so. In fact, we have

$$\ln \xi(t) \approx \Lambda(t) = \left\langle \ln \left(\frac{\|X_{i+t} - X_{j+t}\|}{\|X_i - X_j\|} \right) \right\rangle. \qquad (14.6)$$

Now, by the discussions in Chapter 13, it is clear that PSIC is a generalization of ESIC: As long as the $\Lambda(t)$ curves from different shells form a linear envelope, $q = 1$ and the motion is chaotic. The next question is: Does PSIC also include random fractals as special cases? The answer is yes. This will be shown in the next section. There we will also gain a better understanding of the meaning of $\lambda_q^{(1)}$.

14.2 CHARACTERIZING RANDOM FRACTALS BY PSIC

As we have discussed in Chapters 7 and 8, Levy processes and $1/f^\beta$ processes with long-range correlations constitute two major types of random fractals. Below we show, both analytically and through numerical simulations, that both types of processes can be readily characterized by PSIC.

14.2.1 Characterizing $1/f^\beta$ processes by PSIC

In Chapters 6 and 8, we demonstrated the ubiquity of $1/f^\beta$ processes in science and engineering. Two important prototypical models for such processes are the fractional Brownian motion (fBm) processes and ON/OFF intermittency with power-law distributed ON and OFF periods, discussed in Chapters 6 and 8, respectively. To put $1/f^\beta$ processes in the framework of PSIC, as before, we introduce a parameter H through the equation

$$\beta = 2H + 1, \quad 0 < H < 1.$$

The defining property for a $1/f^\beta$ process is that its variance increases with t as t^{2H}. Irrespective of which embedding dimension is chosen, this property can be translated into

$$\xi(t) = t^H . \tag{14.7}$$

Therefore,

$$\frac{d\xi(t)}{dt} = Ht^{H-1} . \tag{14.8}$$

Expressing t in terms of ξ, we have

$$\frac{d\xi(t)}{dt} = H\xi^{1-\frac{1}{H}} . \tag{14.9}$$

Comparing with the defining Eq. (14.2) of PSIC, we find that

$$q = 1 - \frac{1}{H} = 1 - \frac{2}{\beta - 1}, \tag{14.10}$$

$$\lambda_q^{(1)} = H = \frac{\beta - 1}{2} . \tag{14.11}$$

Noting that $0 < H < 1$, from Eq. (14.10) we have $-\infty < q < 0$.

Computationally, the key is to demonstrate the behaviors described by Eq. (14.7). This can be readily done by calculating the time-dependent exponent curves defined by Eq. (14.6) (see also Eq. (13.5)). Let us first examine the fBm processes. Figure 14.1 shows three examples for the fBm processes with $H = 0.25, 0.5$, and 0.75. Clearly, the slopes of the straight lines correctly estimate the H parameter used in the simulations.

Next, we examine ON/OFF processes with Pareto-distributed ON and OFF periods as that described by Eq. (3.24). Figure 14.2 shows three examples for $\mu = 1.2, 1.6$, and 2.0. Noting that $H = (3 - \mu)/2$, we see that the slopes of the straight lines again correctly estimate the H parameter. Also note that when $0 < \mu < 1, 1 < H < 1.5$. Therefore, Eqs. (14.7) and (14.10) are still correct when $1 < H < 1.5$. The entire range of $H = (3 - \mu)/2$, with $0 < \mu \leq 2$, determines that $-1 \leq q < 1/3$ for Pareto-distributed ON/OFF processes.

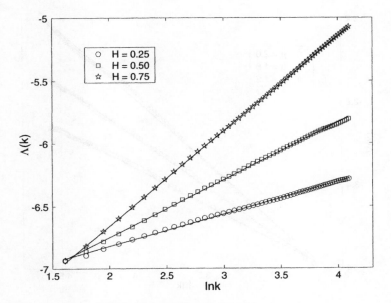

Figure 14.1. $\Lambda(k)$ vs. $\ln k$ curves for fBm.

14.2.2 Characterizing Levy processes by PSIC

In Chapter 7, we discussed α-stable distributions and Levy processes. To clarify the following discussions, we briefly review the major elements in Chapter 7.

A (standard) symmetric α-stable Levy process $\{L_\alpha(t), t \geq 0\}$ is a stochastic process that is almost surely zero at $t = 0$, has independent increments, and $L_\alpha(t) - L_\alpha(s)$ follows an α-stable distribution with characteristic function $e^{-(t-s)|u|^\alpha}$, where $0 \leq s < t < \infty$. Equation (7.7) shows that a random variable Y is called (strictly) stable if the distribution for $\sum_{i=1}^{n} Y_i$ is the same as that for $n^{1/\alpha}Y$,

$$\sum_{i=1}^{n} Y_i \overset{d}{=} n^{1/\alpha}Y,$$

where $Y_1, Y_2, \cdots$ are independent random variables, each having the same distribution as Y. This means that $n\mathrm{Var}Y = n^{2/\alpha}\mathrm{Var}Y$. For the distribution to be valid, $0 < \alpha \leq 2$. When $\alpha = 2$, the distribution is Gaussian, and hence, the corresponding Levy process is just the standard Brownian motion (Bm). When $0 < \alpha < 2$, the distribution is heavy-tailed, $P[X \geq x] \sim x^{-\alpha}$, $x \to \infty$, and has infinite variance. Furthermore, when $0 < \alpha \leq 1$, the mean is also infinite.

The symmetric α-stable Levy motion is $1/\alpha$ self-similar. That is, for $c > 0$, the processes $\{L_\alpha(ct), t \geq 0\}$ and $\{c^{1/\alpha}L_\alpha(t), t \geq 0\}$ have the same finite-dimensional distributions. By this argument as well as Eq. (7.7), it is clear that the length of the motion in a time span of Δt, $\Delta L(\Delta t)$ is given by the following scaling:

$$\Delta L(\Delta t) \propto \Delta t^{1/\alpha}. \tag{14.12}$$

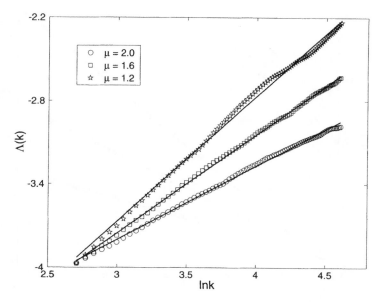

Figure 14.2. $\Lambda(k)$ vs. $\ln k$ curves for ON/OFF model.

Comparing to $1/f^{2H+1}$ processes, we identify that $1/\alpha$ plays the role of H. Therefore,

$$q = 1 - \alpha, \qquad (14.13)$$
$$\lambda_q^{(1)} = 1/\alpha. \qquad (14.14)$$

Noting that $0 < \alpha \leq 2$, from Eq. (14.13) we have $-1 \leq q < 1$.

We have simulated a number of Levy processes with different α. Examples for $\alpha = 1$ and 1.5 are shown in Fig. 14.3. The slopes of the straight lines correctly estimate the values of α used in the simulations.

14.3 CHARACTERIZING THE EDGE OF CHAOS BY PSIC

In nature and in engineering, many types of motion are neither regular nor fully chaotic/random. An example is motions around the edge of chaos. The ubiquity of such motions is perhaps the reason that truly chaotic dynamics have rarely been observed in time series data. To qualify as a unified theory, PSIC must be able to characterize such types of motion. Pleasingly, this can be done, as illustrated by the following discussions on the deterministic and noisy logistic map:

$$x_{n+1} = ax_n(1 - x_n) + \sigma\eta_n, \qquad (14.15)$$

where a is the bifurcation parameter and η_n is a white Gaussian noise with mean 0 and variance 1. The parameter σ characterizes the strength of noise. For the

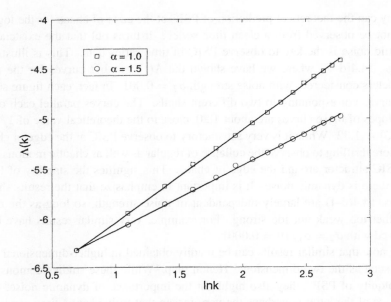

Figure 14.3. $\Lambda(k)$ vs. $\ln k$ curves for Levy flights.

clean system ($\sigma = 0$), the edge of chaos occurs at the accumulation point, $a_\infty = 3.569945672\cdots$. We shall study three parameter values: $a_1 = a_\infty - 0.001$, a_∞, and $a_2 = a_\infty + 0.001$. When noise is absent, a_1 corresponds to a periodic motion with period 2^5, while a_2 corresponds to a truly chaotic motion. We shall only study transient-free time series. In Figs. 14.4(a–c), we have plotted the $\Lambda(t)$ vs. t curves for parameter values a_1, a_∞, and a_2, respectively. We observe in Fig. 14.4(a) that the variation of $\Lambda(t)$ with t is periodic (with period 16, which is half of the period of the motion) when the motion is periodic. This is a generic feature of the $\Lambda(t)$ curves for discrete periodic attractors when the radius of the shell is larger than the smallest distance between two points on the attractor (when a periodic attractor is continuous, $\Lambda(t)$ can be arbitrarily close to zero).

Tsallis and co-workers found that at the edge of chaos for the logistic map, $\xi(t)$ is given by Eq. (14.3) with $q \approx 0.2445$. Surprisingly, we do not observe such a divergence in Fig. 14.4(b). In fact, if one plots $\Lambda(t)$ vs. $\ln t$, one only observes a curve that increases very slowly. The more interesting pattern is the one shown in Fig. 14.4(c), where we observe a linearly increasingly $\Lambda(t)$ vs. t curve. In fact, Fig. 14.4(c) shows two such curves corresponding to two different shells. Interestingly, the two curves collapse to form a common envelope in the linearly increasing part of the curve. The slope of the envelope gives a good estimate of the largest positive Lyapunov exponent. This is a generic feature of chaos, as explained in Chapter 13. Since the chaos studied here is close to the edge of chaos, the curves shown in Fig. 14.4(c) are less smooth than those discussed earlier.

Why can the theoretical prediction of PSIC at the edge of chaos for the logistic map not be observed from a clean time series? It turns out that the existence of dynamic noise is the key to observe PSIC in time series data. This is illustrated in Figs. 14.4(d–f), where we have shown the $\Lambda(t)$ vs. $\ln t$ curves for the three parameters considered, with noise strength $\sigma_1 = 0.001$. In fact, each figure shows two curves corresponding to two different shells. The curves parallel each other. The slopes of those curves are about 1.20, close to the theoretical value of $1/(1 - 0.2445) \approx 1.32$. While it is very satisfactory to observe PSIC at the edge of chaos, it is more thrilling to observe the collapse of regular as well as chaotic motions onto the PSIC attractor around the edge of chaos. This signifies the stability of PSIC when there is dynamic noise. It is important to emphasize that the results shown in Figs. 14.4(d–f) are largely independent of noise strength, so long as the noise is neither too weak nor too strong. For example, very similar results have been observed with $\sigma_2 = \sigma_1/10 = 0.0001$.

We note that similar results can be readily obtained in higher-dimensional systems, such as the two-dimensional Henon map. While these studies demonstrate the ubiquity of PSIC, they also highlight the importance of dynamic noise. The existence of the latter is perhaps the very reason that truly chaotic time series can seldom be observed.

In closing this section, we note that the edge of chaos is characterized by $q \approx 0.2445$, which is larger than 0 but smaller than 1.

14.4 BIBLIOGRAPHIC NOTES

Most of the materials discussed here have not appeared elsewhere. They have been developed based on [176, 289, 450]. See also [56, 85, 276, 436–441] on applications of PSIC in model dynamical systems. Note that the formalism of PSIC closely parallels that of nonextensive statistical mechanics (NESM) of Tsallis [448], which has found numerous applications to the study of systems with long-range interactions [12, 277, 278, 349], multifractal behavior [13, 289], and fully developed turbulence [13, 37, 38], among many others.

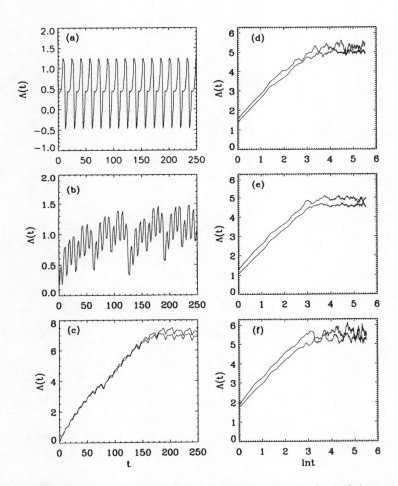

Figure 14.4. $\Lambda(t)$ vs. t curves for time series generated from the noise-free logistic map with (a) $a_1 = a_\infty - 0.001$, where the motion is periodic with period 2^5, (b) $a_\infty = 3.569945672\cdots$, and (c) $a_2 = a_\infty + 0.001$, where the motion is chaotic. Plotted in (d–f) are $\Lambda(t)$ vs. $\ln t$ curves for the noisy logistic map with $\sigma = 0.001$. Very similar results were obtained when $\sigma = 0.0001$. Shown in (c–f) are two curves corresponding to two different shells. A total of 10^4 points were used in the computation, with embedding parameters $m = 4$, $L = 1$. However, so long as $m > 1$, the results are largely independent of embedding. When $m = 1$, the $\Lambda(t)$ curves are not smooth, and the estimated $1/(1 - q)$ is much smaller than the theoretical value.

Figure 14.4. A(t) ... time series generated from the noise-free logistic map with (a) $a_1 = a_{12} = 0.000$ ...

CHAPTER 15

MULTISCALE ANALYSIS BY THE SCALE-DEPENDENT LYAPUNOV EXPONENT (SDLE)

In Chapter 1, we emphasized the importance of developing scale-dependent measures to simultaneously characterize behaviors of complex multiscaled signals on a wide range of scales. This has helped us uncover nonlinear structures in sea clutter, as demonstrated in Chapter 13. We now develop an effective algorithm to compute an excellent measure, the scale-dependent Lyapunov exponent (SDLE), and show that the SDLE can readily classify known types of complex motions, neatly solve the classic problem of distinguishing chaos from noise, effectively deal with non-stationarity, and aptly characterize complex real-world signals, including financial time series, EEG, heart rate variability (HRV) data, and sea clutter data.

15.1 BASIC THEORY

The SDLE is a variant of the finite-size Lyapunov exponent (FSLE) (see Sec. C.1 of Appendix C). The latter is closely related to another scale-dependent measure, the ϵ-entropy, which has been discussed in Sec. 13.2.3. As described in Sec. C.1, the algorithm for calculating the FSLE is very similar to Wolf et al.'s algorithm. It computes the average r-fold time by monitoring the divergence between a reference trajectory and a perturbed trajectory. To do so, it needs to define the nearest neighbors, as well as to perform a renormalization whenever the distance between the reference and the perturbed trajectory becomes too large. Such a procedure requires very long time series and therefore is not practical. To facilitate the derivation of a fast algorithm that works on short data, as well as to ease discussion of continuous but nondifferentiable stochastic processes, we define the SDLE as follows.

Consider an ensemble of trajectories. Denote the initial separation between two nearby trajectories by ϵ_0 and their *average separation* at times t and $t + \Delta t$ by ϵ_t and $\epsilon_{t+\Delta t}$, respectively. Being defined in an average sense, ϵ_t and $\epsilon_{t+\Delta t}$ can be readily computed from any processes, even if they are nondifferentiable. Next, we examine the relation between ϵ_t and $\epsilon_{t+\Delta t}$ where Δt is small. When $\Delta t \to 0$, we have

$$\epsilon_{t+\Delta t} = \epsilon_t e^{\lambda(\epsilon_t)\Delta t} \tag{15.1}$$

or

$$\lambda(\epsilon_t) = \frac{\ln \epsilon_{t+\Delta t} - \ln \epsilon_t}{\Delta t}, \tag{15.2}$$

where $\lambda(\epsilon_t)$ is the SDLE. Equivalently, we have a differential equation for ϵ_t,

$$\frac{d\epsilon_t}{dt} = \lambda(\epsilon_t)\epsilon_t. \tag{15.3}$$

Given time series data, the smallest Δt possible is the sampling time τ.

The definition of the SDLE suggests a simple ensemble average-based scheme to compute it. A straightforward way would be to find all the pairs of vectors in the phase space whose distance is approximately ϵ and then calculate their average distance after a time Δt. The first half of this description amounts to introducing a shell (indexed as k),

$$\epsilon_k \leq \|V_i - V_j\| \leq \epsilon_k + \Delta\epsilon_k,$$

where V_i, V_j are reconstructed vectors and ϵ_k (the radius of the shell) and $\Delta\epsilon_k$ (the width of the shell) are arbitrarily chosen small distances. Such a shell may be considered as a differential element that would facilitate computation of the conditional probability. To expedite the computation, it is advantageous to introduce a sequence of shells, $k = 1, 2, 3, \cdots$. We thus arrive at the same computational procedure for computing the time-dependent exponent (TDE) curves discussed in Chapter 13. Specifically, with all these shells, we can then monitor the evolution of all of the pairs of vectors (V_i, V_j) within a shell and take the average. When each shell is very thin, by assuming that the order of averaging and taking the logarithm in Eq. (15.2) can be interchanged, we have

$$\lambda(\epsilon_t) = \frac{\left\langle \ln \|V_{i+t+\Delta t} - V_{j+t+\Delta t}\| - \ln \|V_{i+t} - V_{j+t}\| \right\rangle}{\Delta t}, \tag{15.4}$$

where t and Δt are integers in a unit of the sampling time, and the angle brackets denote the average within a shell. Note that contributions to the SDLE at a specific scale from different shells can be combined, with the weight for each shell being determined by the number of pairs of vectors (V_i, V_j) in that shell. In the following, to see more clearly how each shell characterizes the dynamics of the data on different scales, we shall plot the $\lambda(\epsilon)$ curves for different shells separately.

In the above formulation, it is assumed that the initial separation, $\|V_i - V_j\|$, aligns with the most unstable direction instantly. For high-dimensional systems this is not true, especially when the growth rate is nonuniform and/or the eigenvectors of the Jacobian are nonnormal. Fortunately, the problem is not as serious as one might think, since our shells are not infinitesimal. When computing the TDE, we have found that when difficulties arise, it is often sufficient to introduce an additional condition,

$$|j - i| \geq (m - 1)L, \tag{15.5}$$

when finding pairs of vectors within each shell. This amounts to removing the time scale k_a shown in Fig. 13.9(a). Such a scheme also works well when computing the SDLE. This means that after taking a time comparable to the embedding window $(m - 1)L$, it would be safe to assume that the initial separation has evolved to the most unstable direction of the motion. In turn, the time index t in Eq. (15.4) also has to satisfy a similar inequality, $t \geq (m - 1)L$.

What is the relation between the SDLE and other complexity measures such as the largest positive Lyapunov exponent λ_1? To find the answer, we recall that Wolf et al.'s algorithm for computing λ_1 involves monitoring the exponential divergence between a reference and a perturbed trajectory. It involves a scale parameter ϵ^* such that whenever the divergence exceeds this chosen scale, a renormalization procedure is performed. Therefore, λ_1 estimated by Wolf et al.'s algorithm can be computed from the SDLE by the following relation

$$\lambda_1 = \int_0^{\epsilon^*} \lambda(\epsilon)p(\epsilon)d\epsilon, \tag{15.6}$$

where $p(\epsilon)$ is the probability density function for the scale ϵ. Since our algorithm involves shells described by condition (15.4), $p(\epsilon)$ is proportional to the derivative of the Grassberger-Procaccia's correlation integral,

$$p(\epsilon) = Z\frac{dC(\epsilon)}{d\epsilon}, \tag{15.7}$$

where Z is a normalization constant satisfying

$$\int_0^{\epsilon^*} p(\epsilon)d\epsilon = 1 \tag{15.8}$$

and $C(\epsilon)$ is the correlation integral given by Eq. (13.14).

Before proceeding, we wish to emphasize the major difference between our algorithm and the standard method for calculating the FSLE. As we have pointed out, to compute the FSLE, two trajectories, one as reference, another as perturbed, have to be defined. This requires huge amounts of data. Our algorithm avoids this by employing two critical operations to fully utilize information about the time evolution of the data: (1) The reference and perturbed trajectories are replaced by time evolution of all pairs of vectors satisfying Inequality (15.5) and falling within a shell, and (2) introduction of a sequence of shells ensures that the number of pairs of vectors within the shells is large and the ensemble average within each shell is well defined. Let the number of points needed to compute the FSLE by standard methods be N. These two operations imply that the method described here requires only about $\sqrt{N}$ points to compute the SDLE. In the following, we shall illustrate the effectiveness of our algorithm by examining various types of complex motions.

15.2 CLASSIFICATION OF COMPLEX MOTIONS

To understand the SDLE as well as appreciate its power, we apply it to classify seven major types of complex motions.

15.2.1 Chaos, noisy chaos, and noise-induced chaos

Obviously, for truly low-dimensional chaos, $\lambda(\epsilon)$ equals the largest positive Lyapunov exponent and, hence, must be independent of ϵ over a wide range of scales. For noisy chaos, we expect $\lambda(\epsilon)$ to depend on small ϵ. To illustrate both features, we consider the chaotic Lorenz system with stochastic forcing described by Eq. (13.16). Figure 15.1(a) shows five curves for the cases of $D = 0, 1, 2, 3, 4$. The computations are done with 10,000 points and $m = 4, L = 2$. We observe a few interesting features:

- For the clean, chaotic signal, $\lambda(\epsilon)$ fluctuates slightly around a constant (which numerically equals the largest positive Lyapunov exponent) when ϵ is smaller than a threshold value that is determined by the size of the chaotic attractor. The reason for the small fluctuations in $\lambda(\epsilon)$ is that the divergence rate varies from one region of the attractor to another.

- When there is stochastic forcing, $\lambda(\epsilon)$ is no longer a constant when ϵ is small, but increases as $-\gamma \ln \epsilon$ when the scale ϵ is decreased. The coefficient γ does not seem to depend on the strength of the noise. This feature suggests that entropy generation is infinite when the scale ϵ approaches zero. Note that the relation of $\lambda(\epsilon) \sim -\gamma \ln \epsilon$ has also been observed for the FSLE and the ϵ-entropy. In fact, such a relation can be readily proven for the ϵ-entropy.

- When the noise is increased, the part of the curve with $\lambda(\epsilon) \sim -\gamma \ln \epsilon$ shifts to the right. In fact, little chaotic signature can be identified when D is increased beyond 3. When noise is not too strong, this feature can be readily used to quantify the strength of noise.

Next, we consider noise-induced chaos. To illustrate the idea, we study the noisy logistic map

$$x_{n+1} = \mu x_n (1 - x_n) + P_n, \quad 0 < x_n < 1, \tag{15.9}$$

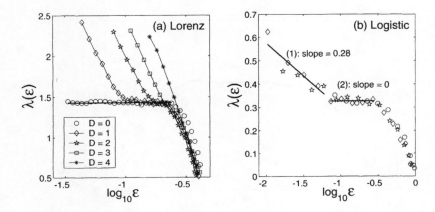

Figure 15.1. $\lambda(\epsilon)$ curves for (a) the clean and noisy Lorenz systems and (b) the noise-induced chaos in the logistic map. Curves from different shells are designated by different symbols.

where μ is the bifurcation parameter and P_n is a Gaussian random variable with zero mean and standard deviation σ. It has been found that at $\mu = 3.74$ and $\sigma = 0.002$, noise-induced chaos occurs, and it may be difficult to distinguish noise-induced chaos from clean chaos. In Fig. 15.1(b), we have plotted the $\lambda(\epsilon_t)$ for this particular noise-induced chaos. The computation was done with $m = 4$, $L = 1$ and 10,000 points. We observe that Fig. 15.1(b) is very similar to the curves of noisy chaos plotted in Fig. 15.1(a). Hence, noise-induced chaos is similar to noisy chaos but different from clean chaos.

At this point, two comments are noteworthy: (1) On very small scales, the effect of measurement noise is similar to that of dynamic noise. (2) The $\lambda(\epsilon)$ curves shown in Fig. 15.1 are based on a fairly small shell. The curves computed based on larger shells collapse on the right part of the curves shown in Fig. 15.1. For this reason, for chaotic systems, one or a few small shells are sufficient. If one wishes to know the behavior of λ on ever smaller scales, one has to use longer and longer time series.

Finally, we consider the Mackey-Glass delay differential system,

$$dx/dt = \frac{ax(t + \Gamma)}{1 + x(t + \Gamma)^c} - bx(t).$$

When $a = 0.2, b = 0.1, c = 10, \Gamma = 30$, it has two positive Lyapunov exponents, with the largest Lyapunov exponent close to 0.007. Having two positive Lyapunov exponents while the value of the largest Lyapunov exponent of the system is not much greater than 0, one might be concerned that it may be difficult to compute the SDLE of the system. This is not the case. In fact, this system can be analyzed as straightforwardly as other dynamical systems, including the Henon map and the Rossler system. An example of the $\lambda(\epsilon)$ curve is shown in Fig. 15.2, where we have used $m = 5$, $L = 1$, and 5000 points sampled with a time interval of 6. We observe a well-defined plateau, with its value close to 0.007. This example illustrates that

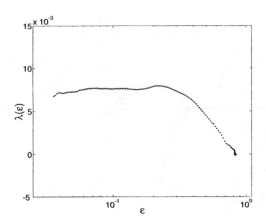

Figure 15.2.   The $\lambda(\epsilon)$ curve for the Mackey-Glass system. The computation was done with $m = 5, L = 1$, and 5000 points were sampled with a time interval of 6.

when computing the SDLE, one does not need to be concerned about a nonuniform growth rate in high-dimensional systems.

15.2.2 $1/f^\beta$ processes

As we have discussed in depth in Chapters 6 and 8, a $1/f^\beta$ process is one of the most important classes of random fractals. Two important prototypical models for such processes are the fractional Brownian motion (fBm) process and the ON/OFF intermittency with power-law distributed ON and OFF periods. Depending on whether $H = (\beta - 1)/2$ is smaller than, equal to, or larger than $1/2$, the process is said to have antipersistent correlation, short-range correlation, or persistent long-range correlation.

It is well known that the variance of such stochastic processes increases with t as t^{2H}. Translating into the average distance between nearby trajectories, we can write

$$\epsilon_t = \epsilon_0 t^H. \tag{15.10}$$

Using Eq. (15.2), we readily find $\lambda(\epsilon_t) \sim H/t$. Expressing t by ϵ_t, we finally find

$$\lambda(\epsilon_t) \sim H \epsilon_t^{-1/H}. \tag{15.11}$$

Note that the same functional relation can be derived for ϵ-entropy. However, using ϵ-entropy, it is not easy to estimate H through analysis of short time series. In contrast, Eq. (15.11) can be conveniently used to estimate H. To illustrate this, we first study fBm processes $B_H(t)$. Figure 15.3(a) shows three curves for $H = 0.33, 0.5$ and 0.7, where the calculation is done with 2^{14} points and $m = 2, L = 1$. We observe that the estimated $1/H$ clearly match those used in simulating these processes.

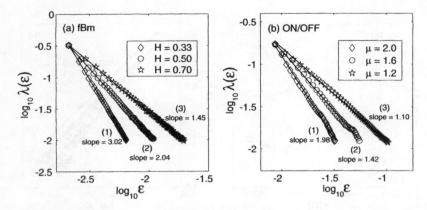

Figure 15.3. $\lambda(\epsilon)$ curves for (a) fBm processes with $H = 0.33, 0.50$, and 0.70, and (b) ON/OFF models with $\mu = 1.2, 1.6$, and 2.0.

Next, we study ON/OFF models with power-law distributed ON and OFF periods,

$$P[X \geq x] = \left(\frac{b}{x}\right)^\mu, \quad x \geq b > 0, \quad 0 < \mu \leq 2.$$

Figure 15.3(b) shows three curves for $\mu = 1.2, 1.6, 2.0$, which correspond to $H = 0.9, 0.7$, and 0.5, respectively. Recalling that $H = (3 - \mu)/2$, again we find that the estimated H accurately reflect the values of μ used in the simulations.

What is the meaning of small ϵ in the power-law distributed ON/OFF model? As we have pointed out, when $\mu < 2$, ON/OFF processes have infinite variance. To effectively characterize ON/OFF processes, the absolute value of ϵ also has to be large. Therefore, ϵ being small is only relative to the largest scale resolved by the dataset. This note is also pertinent to the Levy flights to be studied next. In fact, the example of a Levy flight to be discussed has not only infinite variance but also infinite mean. Interestingly, the defining parameter for such a Levy flight can still be readily estimated by our method.

Finally, we discuss in which scale range Eq. (15.11) may be valid. Let Eq. (15.10) be valid between two time scales, t_1 and t_2, where $t_1 < t_2$. Using Eq. (15.10), we have $\epsilon_2/\epsilon_1 = (t_2/t_1)^H$. This is the scale range for Eq. (15.11) to be valid. Let t_2/t_1 be fixed, it becomes smaller when H is decreased.

15.2.3   Levy flights

We now consider Levy processes, another important type of random fractal models that have found numerous applications.

There are two types of Levy processes. One is Levy flights, which are random processes consisting of many independent steps, each characterized by a stable law and consuming a unit time regardless of its length. The other is Levy walks, where each step takes time proportional to its length. A Levy walk can be viewed

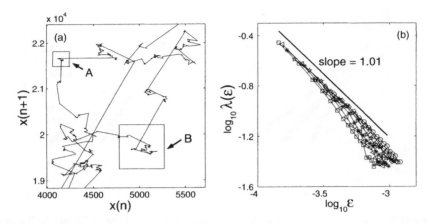

Figure 15.4.   (a) Two-dimensional Levy flights with $\alpha = 1.0$, (b) $\lambda(\epsilon)$ curves for the Levy process (curves from different shells are designated by different symbols).

as sampled from a Levy flight with a uniform speed. The increment process of a Levy walk, obtained by differencing the Levy walk, is very similar to an ON/OFF train with power-law distributed ON and OFF periods. Therefore, in the following, we shall not be concerned about it. We shall focus on Levy flights. As we have seen in Chapter 8, Levy flights, having independent steps, are memoryless processes characterized by $H = 1/2$, irrespective of the value of the exponent α characterizing the stable laws. In other words, methods such as DFA cannot be used to estimate the α parameter.

In order to derive an analytic expression for $\lambda(\epsilon)$ to estimate α, it is sufficient to note Eq. (14.12) derived in Chapter 14:

$$\Delta L(\Delta t) \propto \Delta t^{1/\alpha}.$$

Comparing this to $1/f^{2H+1}$ processes, we find that $1/\alpha$ plays the role of H. Therefore, the scaling for the SDLE is

$$\lambda(\epsilon_t) \sim \frac{1}{\alpha}\epsilon_t^{-\alpha}. \tag{15.12}$$

We have simulated a number of Levy flights with different α. One realization for the flights with $\alpha = 1$ is shown in Fig. 15.4(a). The computed $\lambda(\epsilon_t)$ curves (based on 2^{15} points and $m = 2, L = 1$) are shown in Fig. 15.4(b). In fact, the $\lambda(\epsilon_t)$ curves from a number of different shells are plotted together. We observe that the slope of the *envelope* is -1, consistent with the value of α chosen in simulating the flights.

To understand why the envelope of $\lambda(\epsilon_t)$ gives a better estimate of α, we resort to the two small boxes, denoted A and B, in Fig. 15.4(a). When those boxes are enlarged, they show patterns similar to those in Fig. 15.4(a). Obviously, the scales involved in A and B are different. Those different scales are captured by shells of different sizes. Since the mean and variance of the flights are both infinite, while the number of points in each small box is not large, $\lambda(\epsilon_t)$ from each shell cannot have a long scaling region. However, when $\lambda(\epsilon_t)$ from different shells are plotted together, they form an envelope with a long scaling region. Using this argument, it is clear that when computing $\lambda(\epsilon_t)$ for a Levy flight, a number of shells are more advantageous than a single shell, especially when the time series is not too long.

15.2.4 SDLE for processes defined by PSIC

In Chapter 14, we saw that PSIC provides a common foundation for chaos and random fractals. Now that we have characterized chaos, $1/f^\beta$ noise, and Levy processes by the SDLE, we derive a simple equation relating the λ_q and q of PSIC to the SDLE.

First, we recall that PSIC is defined by

$$\xi_t = \lim_{\|\Delta V(0)\| \to 0} \frac{\|\Delta V(t)\|}{\|\Delta V(0)\|} = \left[1 + (1-q)\lambda_q^{(1)}t\right]^{1/(1-q)}$$

Since $\epsilon_t = \|\Delta V(0)\|\xi_t$, it is now more convenient to express the SDLE as a function of ξ_t. Using Eq. (15.2), we find that

$$\lambda(\xi_t) = \frac{\ln \xi_{t+\Delta t} - \ln \xi_t}{\Delta t}. \tag{15.13}$$

When $\Delta t \to 0$, $1 + (1-q)\lambda_q^{(1)}t \gg (1-q)\lambda_q^{(1)}\Delta t$. Simplifying Eq. (15.13), we obtain

$$\lambda(\xi_t) = \lambda_q^{(1)}\xi_t^{q-1}. \tag{15.14}$$

We now consider three cases:

- For chaotic motions, $q = 1$; therefore, $\lambda(\xi_t) = \lambda_q = $ constant.
- For $1/f^\beta$ noise, using Eqs. (14.10) and (14.11), we have $q = 1 - \frac{1}{H}$ and $\lambda_q^{(1)} = H$. Therefore,

$$\lambda(\xi_t) = H\xi_t^{-\frac{1}{H}}.$$

This is equivalent to Eq. (15.11).

- For Levy flights, using Eqs. (14.13) and (14.14), we have $q = 1 - \alpha$ and $\lambda_q^{(1)} = 1/\alpha$. Therefore,

$$\lambda(\xi_t) = \frac{1}{\alpha}\xi_t^{-\alpha}.$$

This is equivalent to Eq. (15.12).

15.2.5 Stochastic oscillations

Stochastic oscillation is an important type of complex motion that has been observed in many different disciplines of science and engineering. A stochastic oscillator, having structures due to oscillatory motions but not having closed orbits in the phase space, may be interpreted as chaos. Which functional form of $\lambda(\epsilon_t)$ characterizes such motions? To gain insights into this problem, we study a stochastically driven Van der Pol's oscillator:

$$\begin{aligned} dx/dt &= y + D_1\eta_1(t), \\ dy/dt &= -(x^2 - 1)y - x + D_2\eta_2(t), \end{aligned} \tag{15.15}$$

where $< \eta_i(t) >= 0$, $< \eta_i(t)\eta_j(t') >= \delta_{ij}\delta(t - t')$, $i, j = 1, 2$, and the parameters D_i, $i = 1, 2$ characterize the strength of noise. Figure 15.5(a) shows short segment of the $y(t)$ time series for $D_1 = D_2 = 0.02$. Since the noise is small, the reconstructed phase diagram is only slightly diffused, as shown in Fig. 15.5(b). Interestingly, we have found two generic functional forms of $\lambda(\epsilon_t)$ for the $y(t)$ (and $x(t)$) time series of the stochastic Van der Pol's oscillator as well as other stochastic oscillators, independent of the noise level (so long as the noise is not extremely small). One functional form for $\lambda(\epsilon_t)$ is $\lambda(\epsilon) \sim -\ln\epsilon$, observed when the embedding dimension m and delay time L are both small. This is shown in Fig. 15.5(c). Another type of behavior is $\lambda(\epsilon) \sim \epsilon^{-1/H}$, where $H \approx 1/2$, observed when $(m - 1)L$ is comparable to half of the period of the oscillation. An example of the latter is shown in Fig. 15.5(d).

The relation $\lambda(\epsilon) \sim -\ln\epsilon$ for a small embedding window suggests that locally, the dynamics of a stochastic oscillator are just like other noisy dynamics, such as those shown in Fig. 15.1. The relation $\lambda(\epsilon) \sim \epsilon^{-2}$ for a large embedding window suggests that the motion is like a Brownian motion (Bm). This reflects the variation in amplitude of the oscillation on longer time scales. Such variations have also been observed in experimental data, such as pathologic tremor data and velocity fluctuations in the near wake of a circular cylinder. In fact, pathologic tremor data are often characterized by more general relations $\lambda(\epsilon) \sim \epsilon^{-1/H}$, with $H \neq 1/2$.

Up to now, we have focused on the positive portion of $\lambda(\epsilon_t)$. It turns out that when t is large, $\lambda(\epsilon)$ becomes oscillatory, with a mean about 0. Such oscillatory behavior can be in fact identified from Fig. 15.5(c). Denote the corresponding scales by ϵ_∞ and call them the limiting or characteristic scales. They are the stationary portion of ϵ_t, and hence, they may still be a function of time. When we take the Fourier transform of the limiting scales, we find that its power spectral density (PSD), shown in Fig. 15.5(f), is very similar to that of the original $y(t)$ time series shown in Fig. 15.5(e). This simple discussion suggests that the limiting scales capture the structured component of the data. This feature will be further illustrated in Secs. 15.4 and 15.7.2 when we discuss identification of hidden frequencies and analysis of HRV data. Therefore,the positive portion of $\lambda(\epsilon_t)$ and the concept of limiting scale provide a comprehensive characterization of the signals.

15.2.6   Complex motions with multiple scaling behaviors

Some dynamical systems may exhibit multiple scaling behaviors, such as chaotic

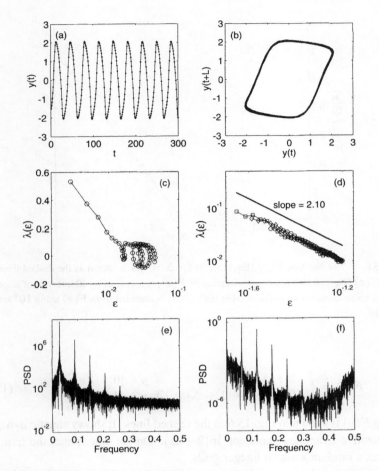

Figure 15.5. (a) $y(t)$ time series of the Van der Pol's oscillator with $D_1 = D_2 = 0.02$. (b) the reconstructed 2-D phase diagram. (c,d) $\lambda(\epsilon)$ curves for the $y(t)$ time series, where embedding parameters are $m = 4$, $L = 1$ for (c) and $m = 8$, $L = 2$ for (d). In (d), three curves from three different shells are designated by different symbols. (e,f) are, respectively, the PSD for $y(t)$ data and the stationary portion of ϵ_t (which corresponds to $\lambda_\epsilon \approx 0$). The data were sampled with a time interval of 0.2. The period of the oscillation is about 30 sample points. The embedding parameters of $m = 8$, $L = 2$ yield an embedding window $(m - 1)L$ of about half of the period.

behavior on small scales but diffusive behavior on large scales. To see how the SDLE can characterize such systems, we study the following map:

$$x_{n+1} = [x_n] + F(x_n - [x_n]) + \sigma \eta_t, \tag{15.16}$$

where $[x_n]$ denotes the integer part of x_n, η_t is a noise uniformly distributed in the interval $[-1, 1]$, σ is a parameter quantifying the strength of noise, and $F(y)$ is

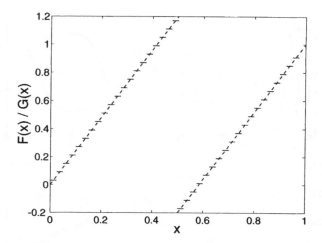

Figure 15.6. The function $F(x)$ (Eq. (15.17)) for $\Delta = 0.4$ is shown as the dashed lines. The function $G(x)$ (Eq. (15.18)) is an approximation of $F(x)$, obtained using 40 intervals of slope 0. In the case of noise-induced chaos discussed in [64], $G(x)$ is obtained from $F(x)$ using 10^4 intervals of slope 0.9.

given by

$$F(y) = \begin{cases} (2 + \Delta)y & \text{if } y \in [0, 1/2) \\ (2 + \Delta)y - (1 + \Delta) & \text{if } y \in (1/2, 1] \end{cases} . \tag{15.17}$$

The map $F(y)$ is shown in Fig. 15.6 as the dashed lines. It shows chaotic dynamics with a positive Lyapunov exponent $\ln(2 + \Delta)$. On the other hand, the term $[x_n]$ introduces a random walk on integer grids.

It turns out that this system is very easy to analyze. When $\Delta = 0.4$, with only 5000 points and $m = 2, L = 1$, we can resolve both the chaotic behavior on very small scales, and the normal diffusive behavior (with slope -2) on large scales. See Fig. 15.7(a).

We now ask a question: Given a small dataset, which type of behavior, chaotic or diffusive, is resolved first? To answer it, we have tried to compute the SDLE with only 500 points. The result is shown in Fig. 15.7(b). It is interesting to observe that chaotic behavior can be well resolved by only a few hundred points. However, diffusive behavior needs much more data to resolve. Intuitively, this makes sense, since diffusive behavior amounts to a Bm on the integer grids and is of much higher dimension than the small-scale chaotic behavior. Therefore, more data are needed to resolve it. As will be shown in the next section, this important result will allow us to determine whether a deterministic Bm is of low or high dimension.

We have also studied the noisy map. The resulting SDLE for $\sigma = 0.001$ is shown in Fig. 15.7(a) as squares. We have again used 5000 points. While the behavior of the SDLE suggests noisy dynamics, with 5000 points we are not able to resolve

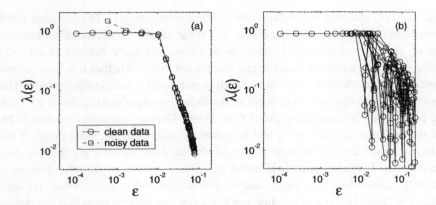

Figure 15.7. $\lambda(\epsilon)$ for the model described by Eq. (15.16): (a) 5000 points were used; for the noisy case, $\sigma = 0.001$; (b) 500 points were used.

well the relation $\lambda(\epsilon) \sim -\ln \epsilon$. This indicates that for the noisy map, on very small scales, the dimension is very high.

Map (15.16) can be modified to produce an interesting system with noise-induced chaos. This can be done by replacing the function $F(y)$ in map (15.16) by $G(y)$ to obtain the following map:

$$x_{t+1} = [x_t] + G(x_t - [x_t]) + \sigma\eta_t, \tag{15.18}$$

where η_t is a noise uniformly distributed in the interval $[-1, 1]$, σ is a parameter quantifying the strength of noise, and $G(y)$ is a piecewise linear function that approximates $F(y)$ in Eq. (15.17). An example of $G(y)$ is shown in Fig. 15.6. In our numerical simulations, we have followed Cencini et al. and used 10^4 intervals of slope 0.9 to obtain $G(y)$. With such a choice of $G(y)$, in the absence of noise the time evolution described by map (15.18) is nonchaotic, since the largest Lyapunov exponent $\ln(0.9)$ is negative. With an appropriate noise level (e.g., $\sigma = 10^{-4}$ or 10^{-3}), the SDLE for the system becomes indistinguishable from the noisy SDLE shown in Fig. 15.7 for map (15.16). Having a diffusive regime on large scales, this is a more complicated noise-induced chaos than the one we have found from the logistic map.

15.3 DISTINGUISHING CHAOS FROM NOISE

15.3.1 General considerations

Distinction between chaos and noise is a classic problem arising from the life sciences, finance, ecology, physics, fluid mechanics, and geophysics. Although tremendous efforts have been made to solve this problem, it remains wide open. Two major difficulties in solving this problem are that (1) chaos can be induced by noise and (2) standard Bm's may have a deterministic origin. To overcome these two difficulties, we ask two questions: (1) What are the fundamental differences among clean low-dimensional chaos, noisy chaos, and noise-induced chaos? (2) When a Bm has a deterministic origin, can we determine whether it is from a low-dimensional deterministic system or a high-dimensional deterministic system? The above questions motivate us to propose the following algorithmic scheme to solve the problem of distinguishing chaos from noise. Denote two scales resolvable by the resolution of the data by ϵ_1 and ϵ_2, where $\epsilon_1 < \epsilon_2$. On this scale range, if the behavior of the data is the same as that of chaotic data, then we say that the data are chaotic. If the behavior is like that of a Bm, then we say that the data are a Bm. Of course, there may exist scale ranges disjoint with (ϵ_1, ϵ_2), where the data behave neither like a chaotic motion nor like a Bm, but have features that can define other types of motion. The data on those scales will be classified accordingly. The feasibility of such a scheme depends critically on whether, by analyzing short noisy time series, one can (1) classify different types of motions and (2) characterize the time series by automatically identifying different scale ranges where different types of motion are manifested. Note that when dealing with complex multiscaled data, the choice of a scale-dependent classification scheme is most natural. However, when a Bm is generated by a very high-dimensional deterministic system, we shall simply treat it as a type of stochastic process, so long as it has the defining properties for a Bm.

At this point, we must comment on the concept of a resolvable scale. Denote a time series under investigation by $x(1), x(2), \cdots, x(n)$. Using time delay embedding, one obtains vectors of the form $V_i = [x(i), x(i + L), \cdots, x(i + (m - 1)L)]$, where the embedding dimension m and the delay time L are chosen according to certain optimization criteria. In the reconstructed phase space, the dataset determines two scales: the maximum and the minimum of the distances between two vectors, $||V_i - V_j||$, where $i \neq j$. Denote them by $\epsilon_{\max}$ and $\epsilon_{\min}$, respectively. This is the resolvable scale range. Analysis of data has to be confined within these two scales. Of course, with more data available, the resolvable scale range can be enlarged. Note that one can treat $\epsilon_{\max}$ as one unit. Alternatively, one may normalize the time series $x(1), x(2), \cdots, x(n)$ into the unit interval $[0, 1]$ before further analysis.

15.3.2   A practical solution

Now that we understand that the key to distinguish chaos from noise is to identify different scale ranges where different types of motion are manifested, we are ready to determine how chaos can be distinguished from noise. In order for our discussion to be useful for practical applications, we assume that our dataset is not only finite, but of small or medium size. Note that if the dataset can be extremely large, then on scales smaller than those where the behavior of $\lambda(\epsilon) \sim -\gamma \ln \epsilon$ is observed, one can observe another plateau, indicating the existence of a high-entropic chaotic state.

Nonlinear maps used as pseudorandom number generators belong in this category. But we shall not be concerned about this here.

We first ask a question. When all three distinctive behaviors of the SDLE,

(1) $\lambda(\epsilon) \sim -\ln \epsilon$,

(2) $\lambda(\epsilon) \sim$ constant,

(3) $\lambda(\epsilon) \sim \epsilon^{-1/H}$,

coexist, is it typically true that the scales where they occur are in ascending order? The answer is yes, as can be readily understood by the following argument. Behavior (1) amounts to stochastic forcing. Its effect is to kick the dynamics to larger scales. For the chaotic motion to be resolvable, its effect has to be limited to scales that are smaller than the scales showing chaotic motion. To understand why diffusive behavior (3) has to occur on scales larger than those showing chaotic motions, it suffices to note that diffusive motion is a nonstationary process. It needs a huge or even an unbounded region of phase space to play out. It is unimaginable to have chaotic motion beyond those scales still resolvable by finite data. Therefore, the scales for the three behaviors to occur have to be in increasing order.

The above discussion suggests that if a Bm is generated by a deterministic system, then the dimension of the system can be readily determined. This is because, for deterministic low-dimensional systems such as map (15.16) to generate large-scale diffusive motions, its local dynamics have to be unstable. Low-dimensional local unstable dynamics amount to low-dimensional chaotic motion. Therefore, we must have a constant $\lambda(\epsilon)$ on small scales. In other words, coexistence of $\lambda(\epsilon) \sim$ constant and $\lambda(\epsilon) \sim \epsilon^{-2}$ would indicate that the Bm is from a low-dimensional deterministic system. We emphasize that $\lambda(\epsilon) \sim$ constant should occur on the scales smaller than those for diffusion to occur and that the behavior of $\lambda(\epsilon) \sim -\ln \epsilon$ should not occur, since the system is low-dimensionally deterministic.

We are now ready to determine to what extent chaos can be distinguished from noise. First, in order to say that a time series under study has a chaotic scaling regime, $\lambda(\epsilon)$ has to show a plateau (i.e., almost constant) for a scale range (ϵ_1, ϵ_2). Since many arbitrary local functions could be treated as a constant, if $\lambda(\epsilon)$ is constant only for ϵ_2 very close to ϵ_1, then it should not be interpreted as a plateau. In other words, ϵ_2 has to be considerably larger than ϵ_1. Unfortunately, it is not easy to state exactly how much larger ϵ_2 should be than ϵ_1, since this depends on the dimension of the data. We believe that $\log_{10}(\epsilon_2/\epsilon_1)$ has to be at least around $1/2$.

Once we know about the existence of a chaotic scaling regime in the data, it is natural to ask if the dynamics are deterministically chaotic or not. This can be readily inferred by checking the behavior of $\lambda(\epsilon)$ on scales smaller than ϵ_1. If, on those scales, $\lambda(\epsilon)$ is above the plateau, then we have good reason to believe that the data are noisy. If, on those scales, $\lambda(\epsilon) \sim -\ln \epsilon$, then we can be sure that the data are noisy chaos or noise-induced chaos. However, if all we know is the time series, then we will not be able to distinguish further between noisy chaos and noise-induced chaos.

In most situations, when the data are found to be noisy, it is important to find out which type of noise process the data represent. The power-law behavior of the SDLE is especially interesting for this purpose. For example, we can combine analysis using the SDLE with spectral analysis and distributional analysis to determine whether the data are a type of $1/f^\beta$ process, or Levy processes, or stochastic oscillations. More precisely, power-law behavior of the SDLE together with sharp spectral peaks revealed by Fourier analysis would indicate a stochastic oscillation, whereas power-law behavior of the SDLE together with a Gaussian-like distribution would indicate fBm-like data. If, instead, heavy-tailed distributions are observed, then the data are an ON/OFF intermittency type or are similar to Levy processes. While many excellent methods for estimating the Hurst parameter have been proposed (see the recent comparative study by Gao et al. [157]), the SDLE offers another effective approach. In fact, as we pointed out in Sec. 15.2, the SDLE provides better characterization of Levy flights and stochastic oscillations, since most other methods only give $H = 0.5$ for Levy flights, irrespective of the values of the defining parameter α, and fail to characterize the power-law behavior of the SDLE for stochastic oscillations.

15.4 CHARACTERIZING HIDDEN FREQUENCIES

Defining and characterizing large-scale, orderly motions is a significant problem in many scientific disciplines. One of the most important types of large-scale, orderly motions is oscillatory motions. An interesting type of oscillatory motions is associated with the so-called hidden frequency phenomenon. That is, when the dynamics of a complicated system are monitored through the temporal evolution of a variable x, Fourier analysis of $x(t)$ may not suggest any oscillatory motions. However, if the dynamics of the system are monitored through the evolution of another variable, say, z, then the Fourier transform of $z(t)$ may contain a well-defined spectral peak indicating oscillatory motions. An example is the chaotic Lorenz system described in Eq. (13.16). Figures 15.8(a–c) show the PSD of the x, y, z components of the system. We observe that the PSD of $x(t)$ and $y(t)$ are simply broad. However, the PSD of $z(t)$ shows a very sharp spectral peak. Recall that geometrically the Lorenz attractor consists of two scrolls (see Fig. 15.9). The sharp spectral peak in the PSD of $z(t)$ of the Lorenz system is due to the circular motions along either of the scrolls.

The above example illustrates that if the dynamics of a system contain a hidden frequency that cannot be revealed by the Fourier transform of a measured variable (say, $x(t)$), then in order to reveal the hidden frequency, one has to embed $x(t)$ in a suitable phase space. This idea has led to the development of two interesting methods for identifying hidden frequencies. One method, proposed by Ortega [326, 327], involves computing the temporal evolution of density measures in the reconstructed phase space. Another, proposed by Chern et al. [74], involves

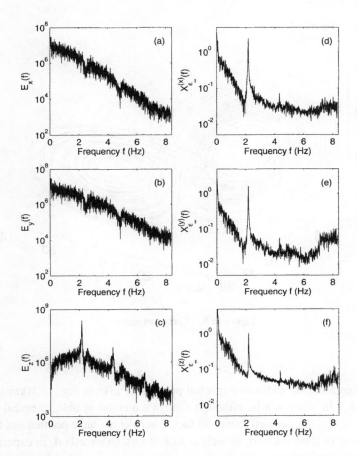

Figure 15.8. Power-spectral density (PSD) for (a) $x(t)$, (b) $y(t)$, (c) $z(t)$, and the magnitude of the Fourier transform for (d) $\epsilon_\infty^{(x)}(t)$, (e) $\epsilon_\infty^{(y)}(t)$, and (f) $\epsilon_\infty^{(z)}(t)$. The SDLE was computed with a shell of size $(2^{-9/2}, 2^{-5})$, 5000 points sampled at $\tau = 0.06$, and embedding parameters $m = 4, L = 2$.

taking singular value decomposition (see Appendix B) of local neighbors. Ortega's method has been applied to an experimental time series recorded from a far-infrared laser in a chaotic state (see Sec. A.4 in Appendix A). The laser dataset, shown in Fig. 15.10(a), contains 10,000 points, sampled with a time interval of 80 ns. The PSD of the data is shown in Fig. 15.10(b), where one observes a sharp peak around 1.7 MHz. Figure 15.10(c) shows the PSD of the density time series, where one notes an additional spectral peak around 37 kHz. This peak is due to the envelope of chaotic pulsations, which is discernible from Fig. 15.10(a).

To appreciate the strength of the additional spectral peak in Fig. 15.10(c), we have to note that the units of the PSD in Figs. 15.10(b,c) are arbitrary; the largest peak is treated as 1 unit, as was done by Ortega. In fact, the actual energy of the original laser intensity time series is much greater than that of the density time

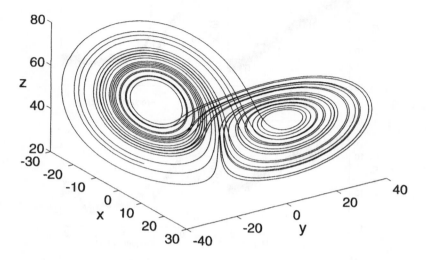

Figure 15.9. Lorenz attractor.

series. Therefore, the additional spectral peak of 37 kHz in Fig. 15.10(c) is really quite weak. In other words, although Ortega's method is able to reveal hidden frequencies, it is not very effective. In fact, Chern et al. have pointed out that the effectiveness of their method, as well as that of Ortega's method, in experimental data analysis remains uncertain.

One of the major difficulties with the above two methods is conceptual; both methods are based on very small-scale neighbors, while the phenomenon itself is large-scale. Recognizing this, one can readily understand that the concept of limiting scale, developed in Sec. 15.2.5, provides an excellent solution to the problem. In other words, one can simply take the Fourier transform of the temporal evolution of the limiting scale and expect additional well-defined spectral peaks if the dynamics contain hidden frequencies. To illustrate the idea, we have shown in Figs. 15.8(d–f) the magnitude of the Fourier transform of the limiting scales of $x(t)$, $y(t)$, and $z(t)$ of the Lorenz system. We observe very well-defined spectral peaks in all three cases. In fact, now the z component no longer plays a more special role than the x and y components. The method is also amazingly effective in identifying the hidden frequency from the laser data, as shown in Fig. 15.10(d). We now not only observe an additional spectral peak around 37 KHz in (d), but also that this peak is even more dominant than the peak around 1.7 MHz. The reason for the exchange of this dominance is the fairly large sampling time; although 80 ns is small, it is only able to sample about eight points in each oscillation. When the embedding window,

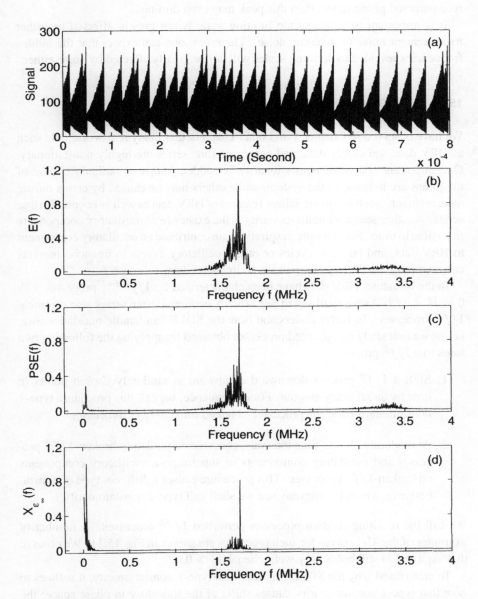

Figure 15.10. (a) Laser intensity data, (b) the PSD of the original laser intensity time series, (c) the PSD of the density measure of the phase space reconstructed from the laser data, and (d) the magnitude of the Fourier transform of the limiting scale of the laser data. In (c) and (d), the embedding parameters are $m = 4, L = 1$.

$(m - 1)L$, becomes greater than 4, the Nyquist sampling theorem is violated in the reconstructed phase space; then this peak may even diminish.

It is important to note that the limiting scale is not greatly affected by either measurement noise or dynamic noise. Therefore, one can expect that the hidden frequencies revealed by limiting scales will not be greatly affected by noise either.

15.5   COPING WITH NONSTATIONARITY

We have observed in Chapter 1 and other chapters that many real-world data such as HRV data, sea clutter data, and economic time series are highly nonstationary. One important source of nonstationarity is sudden jumps or outliers. Some of the jumps are intrinsic to the system, while others may be caused by errors during measurement. Such jumps are salient features of HRV data as well as economic time series. Another source of nonstationarity is the existence of oscillatory components from time to time. For example, respiration can contribute an oscillatory component to HRV data, and business cycles or other oscillatory events in financial markets can contribute oscillatory components to financial data.

In the literature, HRV data have been characterized by $1/f^{2H+1}$ processes with $0 < H < 1/2$. As we shall see in Sec. 15.7.3, economic time series also resemble $1/f^\beta$ processes. To better understand how the SDLE can handle nonstationarity, below we first study complicated processes obtained by applying the following two steps to a $1/f^\beta$ process:

1. Shift a $1/f^\beta$ process downward or upward at randomly chosen points in time by an arbitrary amount. For convenience, we call this procedure type-1 nonstationarity and the processes obtained broken-$1/f^\beta$ processes.

2. At randomly chosen time intervals, concatenate randomly broken-$1/f^\beta$ processes and oscillatory components or superimpose oscillatory components on broken-$1/f^\beta$ processes. This procedure causes a different type of nonstationarity, which for convenience we shall call type-2 nonstationarity.

We call the resulting random processes perturbed $1/f^\beta$ processes. A number of examples of the $\lambda(\epsilon)$ curves for such processes are shown in Fig. 15.11. We observe that Eq. (15.11) still holds very well when $\lambda(\epsilon) > 0.02$.

To understand why the SDLE can deal with type-1 nonstationarity, it suffices to note that type-1 nonstationarity causes shifts of the trajectory in phase space; the greater the nonstationarity, the larger the shifts. The SDLE, however, cannot be affected much by shifts, especially large ones, since it is based on the coevolution of pairs of vectors within chosen small shells. In fact, the effect of shifts is to exclude a few pairs of vectors that were originally counted in the ensemble average. Therefore, so long as the shifts are not frequent, the effect of shifts can be neglected, since ensemble averaging within a shell involves a large number of pairs of vectors.

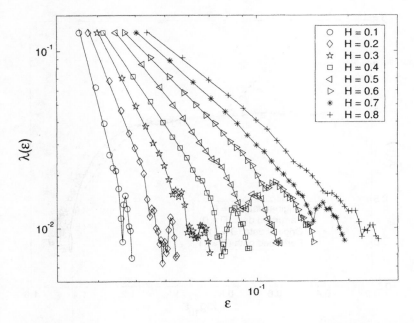

Figure 15.11. $\lambda(\epsilon)$ vs. ϵ curves for perturbed $1/f$ processes. Eight different H are considered. To put all the curves on one plot, the curves for different H (except the smallest one considered here) are arbitrarily shifted rightward.

Let us now turn to type-2 nonstationarity, which involves oscillatory components. Being regular, these components can only affect $\lambda(\epsilon)$ where it is close to 0. Therefore, type-2 nonstationarity cannot affect the positive portion of $\lambda(\epsilon)$ either.

Finally, we examine how the two perturbations may affect deterministically chaotic or noisy chaotic signals. We find that the characteristic features of the SDLE for chaotic signals are not affected much either, as shown in Fig. 15.12.

15.6 RELATION BETWEEN SDLE AND OTHER COMPLEXITY MEASURES

In Chapter 1, we mentioned that in order to understand brain dynamics and diagnose brain pathologies, a number of complexity measures from information theory, chaos theory, and random fractal theory have been used to analyze EEG data. Since these theories have different foundations, it has been difficult to compare studies based on different complexity measures. We shall now show that this problem can be solved by relating those complexity measures to the value of the SDLE at specific scales. For coherence of the presentation, we shall directly explain the results, postponing descriptions of the complexity measures to Appendix C, if they have not been already discussed.

To compare the SDLE with other complexity measures, *seven patients' multiple*-channel intracranially measured EEG data, each with a duration of a few hours, with

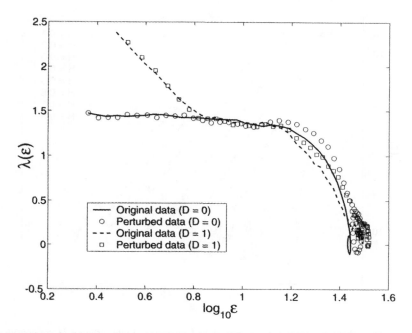

Figure 15.12. $\lambda(\epsilon)$ vs. ϵ curves for the perturbed Lorenz data. It may be useful to compare this figure with Fig. 15.1(a).

a sampling frequency of 200 Hz, have been analyzed. As is typically done, the long EEG signals are partitioned into short windows of length W points, and the measures of interest for each window are calculated. While the following results are based on $W = 2048$ points, it has been found that the variations in these measures with time are largely independent of the window size W. The relations among the measures studied here are the same for all patients' EEG data, so we illustrate the results based on only one patient's EEG signals.

We first consider the Lempel-Ziv (LZ) complexity, denoted as C_{LZ}. As is explained in Sec. C.2 of Appendix C, the LZ complexity is a popular measure used to characterize the randomness of a signal. It is closely related to the Kolmogorov complexity and the Shannon entropy. Figure 15.13(a) shows the variation in C_{LZ} with time. The vertical red dashed lines indicate seizure occurrence times. We note that at the onset of seizure, C_{LZ} increases sharply; then it decreases sharply, followed by a gradual increase. While the sudden sharp increase in C_{LZ} is not observed in all seven patients' data, the sharp decrease and then the gradual increase in C_{LZ} are typical features. This indicates that the dynamics of the brain first become more regular right after the seizure; then their irregularity increases as they approach the normal state.

Next, we examine the permutation entropy (PE), denoted as E_p. As is explained in Sec. C.3 of Appendix C, it is a measure from chaos theory. It measures the departure of the time series under study from a completely random one: the smaller the value of E_p, the more regular the time series is. Figure 15.13(b) shows the $E_p(t)$ for the same patient's EEG data. We observe that the variation of E_p with t is similar to that of C_{LZ} with t.

Third, we consider the correlation entropy, K_2. As is discussed in Chapter 13, K_2 is a tight lower bound of the Kolmogorov entropy. It is zero, positive, and infinite for regular, chaotic, and random data, respectively. Its variation with time is shown in Fig. 15.13(c). Its pattern is similar to that of Figs. 15.13(a,b). This is as expected, since all three measures are entropy measures.

As the fourth measure, we examine the correlation dimension, D_2. First, we recall from Chapter 13 that D_2 is a tight lower bound of the box-counting dimension and measures the minimal number of variables needed to fully describe the dynamics of the data. It is independent of the correlation entropy. However, from Fig. 15.13(d), we find that its variation with time is similar to $K_2(t)$ of Fig. 15.13(c) (as well as Figs. 15.13(a,b)). This is a puzzling observation.

It turns out that the relation between the largest Lyapunov exponent λ_1 and the correlation entropy is reciprocal, as can be seen by comparing Figs. 15.13(e) and 15.13(c). This is an even bigger puzzle for the following reason. The summation of the positive Lyapunov exponents is typically equal to the Kolmogorov entropy. As a major component contributing to the Kolmogorov entropy, $\lambda_1(t)$ is expected to be similar to $K_2(t)$. However, we have observed a reciprocal relation between $\lambda_1(t)$ and $K_2(t)$.

Finally, we consider the calculation of the Hurst parameter H using DFA, one of the most reliable methods for estimating H. The variation of H with time is shown in Fig. 15.13(f). We observe that $H(t)$ and $\lambda_1(t)$ are similar. This is another surprising relation, since the Hurst parameter is a measure from the random fractal theory, while $\lambda_1(t)$ is a measure from chaos theory.

We now calculate the SDLE for each segment of the EEG data. Two representative examples for seizure and nonseizure segments are shown in Fig. 15.14. We observe that on a specific scale ϵ^*, the two curves cross. Loosely, we may term any $\epsilon < \epsilon^*$ small scale, while any $\epsilon > \epsilon^*$ is large scale. Therefore, on small scales, $\lambda(\epsilon)$ is smaller for seizure than for nonseizure EEGs, while on large scales the opposite is true. The variations of $\lambda_{\text{small}-\epsilon}$ and $\lambda_{\text{large}-\epsilon}$ with time in this patient's data, where small $-\epsilon$ and large $-\epsilon$ stand for (more or less arbitrarily) chosen fixed small and large scales, are shown in Figs. 15.15(a) and 15.15(b), respectively. We observe that the pattern of variation of $\lambda_{\text{small}-\epsilon}(t)$ is the reciprocal of that of $\lambda_{\text{large}-\epsilon}(t)$. This result can be expected from Fig. 15.14. More importantly, the patterns of $\lambda_{\text{small}-\epsilon}(t)$ and $\lambda_{\text{large}-\epsilon}(t)$ encompass the two patterns for the six measures presented in Fig. 15.13.

We are now ready to resolve all of the curious relations observed among the six measures presented in Fig. 15.13.

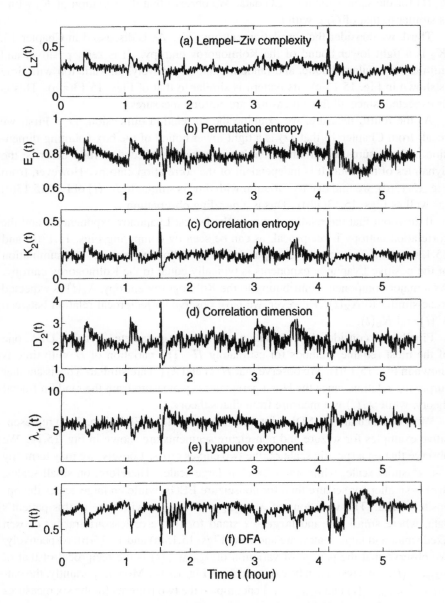

Figure 15.13. Complexity measures for EEG.

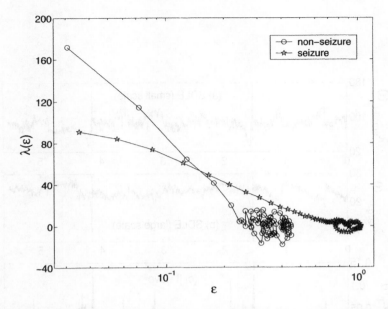

Figure 15.14.   Representative $\lambda(\epsilon)$ (per second) vs. ϵ for seizure and nonseizure EEG segments.

1. Generally, entropy measures the randomness of a dataset. This pertains to small scale. Therefore, $C_{LZ}(t)$, $E_P(t)$, and $K_2(t)$ should be similar to $\lambda_{small-\epsilon}(t)$. This is indeed the case. Therefore, we can conclude that the variation of entropy is represented by $\lambda_{small-\epsilon}(t)$, regardless of how entropy is defined.

2. To understand why $\lambda_1(t)$ calculated by the algorithm of Wolf et al. [478] corresponds to $\lambda_{large-\epsilon}(t)$, it is sufficient to note that when the algorithm of Wolf et al. is applied to a time series with only a few thousand points, in order to obtain a well-defined Lyapunov exponent, a fairly large scale parameter has to be chosen. Typically, the probability in Eq. (15.6) is larger on large scales than on small scales, therefore, the Lyapunov exponent and $\lambda_{large-\epsilon}$ are similar. In fact, the scale we have chosen to calculate $\lambda_1(t)$ is even larger than that for calculating $\lambda_{large-\epsilon}(t)$. This is the reason that the value of $\lambda_1(t)$ shown in Fig. 15.13(e) is smaller than that of $\lambda_{large-\epsilon}(t)$ shown in Fig. 15.15(b).

3. It is easy to see that if one fits the $\lambda(\epsilon)$ curves shown in Fig. 15.14 by a straight line, then the variation of the slope with time should be similar to $\lambda_{small-\epsilon}(t)$ but the reciprocal of $\lambda_{large-\epsilon}(t)$. Such a pattern will be preserved even if one takes the logarithm of $\lambda(\epsilon)$ first and then does the fitting. Note that the slope plays the role of the Hurst parameter. Therefore, even if the EEG is not ideally of the $1/f^{2H+1}$ type, qualitatively the relation $\lambda(\epsilon) \sim \epsilon^{-1/H}$ holds. This, in turn, implies that $D_2 \sim 1/H$.

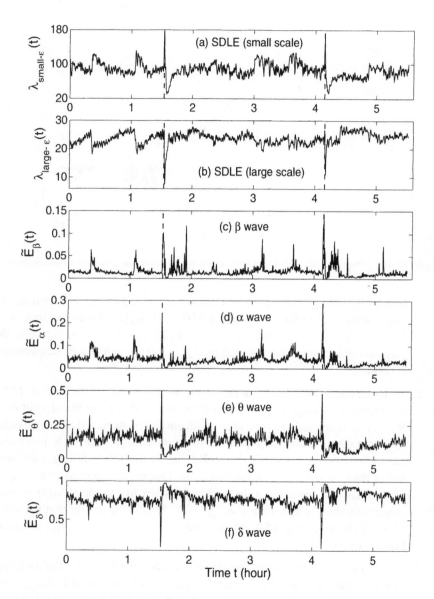

Figure 15.15.   The SDLE and the four waves for the EEG signals.

With the above arguments, it is clear that the seemingly puzzling relations among the measures considered here can be readily understood by the $\lambda(\epsilon)$ curves. Most importantly, we have established that commonly used complexity measures can be related to the values of the SDLE at specific scales.

Before ending this section, we make two comments. (1) To comprehensively characterize the complexity of complicated data such as EEG data, a wide range of scales has to be considered, since the complexity may be different on different scales. For this purpose, the entire $\lambda(\epsilon)$ curve, where ϵ is such that $\lambda(\epsilon)$ is positive, provides a good solution. This point is particularly important when one wishes to compare the complexity between two signals; the complexity of one signal may be higher on some scales but lower on other scales. The situation shown in Fig. 15.14 may be considered one of the simplest. (2) For detecting important events such as epileptic seizures, $\lambda_{\text{small}-\epsilon}$ and $\lambda_{\text{large}-\epsilon}$ appear to provide better-defined features than other commonly used complexity measures. This may be due to the fact that $\lambda_{\text{small}-\epsilon}$ and $\lambda_{\text{large}-\epsilon}$ are evaluated at fixed scales, while other measures are not. In other words, scale mixing may blur the features of events being detected, such as seizures.

15.7 BROAD APPLICATIONS

In Sec. 15.2.5, we have introduced the notion of limiting scales, ϵ_∞, and pointed out that they contain a lot of useful information about the structured components (i.e., regularly oscillating components) of the data. More generally, we can say that they are closely related to the total variation or the energy of the signal. For example, for a chaotic system, ϵ_∞ defines the size of the chaotic attractor. If one starts from $\epsilon_0 \ll \epsilon_\infty$, then, regardless of whether the data are deterministically chaotic or simply random, ϵ_t will initially increase with time and gradually settle at ϵ_∞. Consequently, $\lambda(\epsilon_t)$ will be positive before ϵ_t reaches ϵ_∞. On the other hand, if one starts from $\epsilon_0 \gg \epsilon_\infty$, then ϵ_t will simply decrease, yielding negative $\lambda(\epsilon_t)$, again regardless of whether the data are chaotic or random. When $\epsilon_0 \sim \epsilon_\infty$, $\lambda(\epsilon_t)$ will remain around 0. With these comments, we are now ready to tackle various types of complex data.

15.7.1 EEG analysis

In Sec. 15.6, we analyzed EEG by computing a number of different complexity measures. We complete our study of EEG by trying to understand the four waves — β, α, θ and δ of the EEG.

In Sec. 4.1, we pointed out that these waves are not characterized by sharp spectral peaks in the frequency domain. Therefore, it would make more sense to find the total energy of each type of wave. Denote them by $E_\beta, E_\alpha, E_\theta$, and E_δ, respectively. They can be obtained by integrating the PSD over their defining frequency bands. Furthermore, we normalize them by the total energy in the frequency band $(0, 40)$

Hz. Denote the normalized energy, or fraction of energy, by $\tilde{E}_\beta, \tilde{E}_\alpha, \tilde{E}_\theta$, and $\tilde{E}_\delta$. Note that normalization does not change the functional form for temporal variations of the energy of each type of wave. The variations of these normalized energies with time in the EEG data of the patient discussed in Sec. 15.6 are shown in Figs. 15.15(c–f). Comparing with Figs. 15.15(a,b), we observe that, overall, the variations of $\tilde{E}_\beta(t), \tilde{E}_\alpha(t)$, and $\tilde{E}_\theta(t)$ are similar to that of $\lambda_{\text{small}-\epsilon}(t)$, while the variation of $\tilde{E}_\delta(t)$ is similar to that of $\lambda_{\text{large}-\epsilon}(t)$. This suggests that small $-\epsilon$ corresponds to high frequency, or small temporal scales, while large $-\epsilon$ corresponds to low frequency, or large temporal scales. It should be emphasized, however, that the features uncovered by spectral analysis are not as good as those uncovered by the complexity measures discussed here. Furthermore, when the high-frequency bands, such as the β wave band, are used for seizure detection, the variation of the corresponding energy with time shows activities that are not related to seizures but are related to noise such as motion artifacts. Therefore, for the purpose of epileptic seizure detection/prediction, spectral analysis (or other methods, such as wavelet-based methods, for extracting equivalent features) may not be as effective as the complexity measures discussed here, especially the SDLE.

Let us now quantitatively understand the relation between $\tilde{E}_\beta(t), \tilde{E}_\alpha(t), \tilde{E}_\theta(t)$, and $\lambda_{\text{small}-\epsilon}(t)$. In Sec. 15.6, we pointed out that EEG signals may be approximated as a type of $1/f^{2H+1}$ process. Integrating $1/f^{2H+1}$ over a frequency band of $[f_1, f_2]$, we obtain the energy

$$E(f_1 \to f_2) = \int_{f_1}^{f_2} \frac{1}{f^{2H+1}} df = \frac{1}{2H}\left[f_1^{-(2H+1)} - f_2^{-(2H+1)}\right].$$

This suggests that the variation of $E(f_1 \to f_2)$ is similar to $1/H(t)$. Since the variation of H with t is reciprocal to that of $\lambda_{\text{small}-\epsilon}$, as can be seen in Figs. 15.13(f) and 15.15(a), we thus see that β, α, and θ waves can be approximated by $1/f^{2H+1}$-type processes. The δ wave, however, is not of the $1/f^{2H+1}$ type.

15.7.2 HRV analysis

We now examine HRV data and show how the SDLE can readily characterize the hidden differences in HRV under healthy and disease conditions.

We examine two types of HRV data, one for healthy subjects, another for subjects with congestive heart failure (CHF), a life-threatening disease. The data were downloaded from PhysioNet. There are 18 healthy subjects and 15 subjects with CHF. Parts of these datasets were analyzed by other methods. In particular, 12 of the 15 CHF datasets were analyzed by wavelet-based multifractal analysis and point process adaptive filtering for the purpose of distinguishing healthy subjects from CHF patients. For ease of comparison, we take the first 3×10^4 points of both groups of HRV data for analysis. Figures 15.16(a,b) show two typical $\lambda(\epsilon)$ vs. ϵ curves, one for a healthy subject and another for a patient with CHF. We observe that for the healthy subject, $\lambda(\epsilon)$ linearly decreases with $\ln \epsilon$ before λ approximately

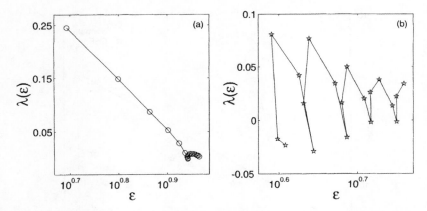

Figure 15.16. $\lambda(\epsilon)$ (per beat) vs. ϵ (on a semilog scale) for HRV data of (a) a healthy subject and (b) a subject with CHF.

reaches 0 or before ϵ settles around the characteristic scale, ϵ_∞. Recall that this is a characteristic of noisy dynamics (Fig. 15.1). For the CHF case plotted in Fig. 15.16(b), we observe that $\lambda(\epsilon)$ is oscillatory, with its value always close to 0; hence, the only scale resolvable is around ϵ_∞. Since the length of the time series used in our analysis for the healthy and CHF subjects is the same, the inability to resolve the $\lambda(\epsilon)$ behavior on scales much smaller than ϵ_∞ for patients with CHF strongly suggests that the dimension of the dynamics of the cardiovascular system for CHF patients is considerably higher than that of healthy subjects.

We now discuss how to distinguish between healthy subjects and patients with CHF from HRV analysis. We have devised two simple measures, or features. One is to characterize how well the linear relation between $\lambda(\epsilon)$ and $\ln \epsilon$ can be defined. We have quantified this by calculating the error between a fitted straight line and the actual $\lambda(\epsilon)$ vs. $\ln \epsilon$ plots of Figs. 15.16(a,b). The second feature is to characterize how well the characteristic scale ϵ_∞ is defined. This is quantified by the ratio between two scale ranges, one from the 2nd to the 6th point of the $\lambda(\epsilon)$ curves and the other from the 7th to the 11th point of the $\lambda(\epsilon)$ curves. Now each subject's data can be represented as a point in the feature plane, as shown in Fig. 15.17. We observe that for healthy subjects, feature 1 is generally very small but feature 2 is large, indicating that the dynamics of the cardiovascular system are like those of a nonlinear system with stochasticity, with resolvable small-scale behaviors and a well-defined characteristic scale ϵ_∞. The opposite is true for the patients with CHF: feature 1 is large but feature 2 is small, indicating not only that small-scale behaviors of the $\lambda(\epsilon)$ curves cannot be resolved but also that the characteristic scale ϵ_∞ is not well defined. Very interestingly, these two simple features completely separate the normal subjects from patients with CHF. In fact, *each feature alone can* almost perfectly separate the two groups of subjects studied here.

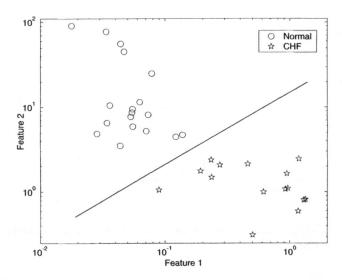

Figure 15.17.   Feature plane separating normal subjects from subjects with CHF.

In recent years, many efforts have been made to search for cardiac chaos. Due to the inability to unambiguously distinguish deterministic chaos from noise by calculating the largest positive Lyapunov exponent and the correlation dimension, it is still unclear whether the control mechanism of the cardiovascular system is truly chaotic or not. Our analysis here strongly suggests that if cardiac chaos does exist, it is more likely to be identified in healthy subjects than in pathologic groups. This is because the dimension of the dynamics of the cardiovascular system appear to be lower for healthy than for pathologic subjects. Intuitively, such an implication makes sense, because a healthy cardiovascular system is a tightly coupled system with coherent functions, while components in a malfunctioning cardiovascular system are somewhat loosely coupled and function incoherently.

15.7.3   Economic time series analysis

In Chapter 1, we pointed out that although economic time series are very complicated, recent studies have reported negative largest Lyapunov exponents for various types of economic time series. We now show that the reported negative largest Lyapunov exponent may correspond to the value of the SDLE at fairly large scales and therefore does not necessarily imply regular economic dynamics.

15.7.3.1   Analysis of a chaotic asset pricing model   We first analyze a chaotic asset pricing model with heterogeneous beliefs driven by dynamic noise proposed by Brock and Hommes [58]. The model is a nonlinear map of the form

$$x_t = F(x_{t-1}, x_{t-2}, x_{t-3}) + \sigma\eta_t. \tag{15.19}$$

Specifically, the model is described by the following equations:

$$x_t = \frac{1}{R} \sum_{h=1}^{4} n_{h,\,t}(g_h x_{t-1} + b_h) + \sigma \eta_t \tag{15.20}$$

$$n_{h,\,t} = \frac{e^{\beta U_{h,\,t-1}}}{\sum_{j=1}^{4} e^{\beta U_{j,\,t-1}}} \tag{15.21}$$

$$U_{h,\,t-1} = (x_{t-1} - R x_{t-2})(g_h x_{t-3} + b_h - R x_{t-2}). \tag{15.22}$$

Here x_t denotes the deviation of the price of a risky asset from its benchmark fundamental value (the discounted sum of expected future dividends), $R > 1$ is the constant risk-free rate, $n_{h,\,t}$ represents the discrete choice of agents using belief type h, $U_{h,\,t-1}$ is the profit generated by strategy h in the previous period, g_h and b_h characterize the linear belief with one time lag of strategy h, and the noise term $\sigma \eta_t$ is standard normally distributed noise with variance σ^2. For suitable choice of the parameter values, the model exhibits chaotic dynamics. In particular, Brock and Hommes have studied the model at parameter values $\beta = 90$, $R = 1.1$, $g_1 = b_1 = 0$, $g_2 = 1.1$, $b_2 = 0.2$, $g_3 = 0.9$, $b_3 = -0.2$, $g_4 = 1.21$, $b_4 = 0$. They found that the motion is chaotic with the largest Lyapunov exponent 0.1. More recently, Hommes and Manzan [222] studied the model at parameter values $\beta = 90$, $R = 1.01$, $g_1 = b_1 = 0$, $g_2 = 0.9$, $b_2 = 0.2$, $g_3 = 0.9$, $b_3 = -0.2$, $g_4 = 1.01$, $b_4 = 0$. When $\sigma = 0$, they found that the motion is chaotic with the largest Lyapunov exponent around 0.135. When σ was increased beyond 0.1, they found that the largest Lyapunov exponent becomes negative. Here we shall focus on these two parameter sets. For convenience, we call them the BH and HM parameter sets, respectively. We have calculated the $\lambda(\epsilon)$ curves for the model with the BH and HM parameter sets for various levels of noise. The results are shown in Figs. 15.18(a,b). We observe that they are very similar to those in Figs. 15.1. Specifically, we note the following:

1. The value of about 0.1 for the largest positive Lyapunov exponent shown in Fig. 15.18(b) is consistent with the Lyapunov exponent obtained by Brock and Hommes, but the value of about 0.3 for the largest positive Lyapunov exponent shown in Fig. 15.18(a) is larger than the value of the Lyapunov exponent obtained by Hommes and Manzan using the neural network–based Lyapunov exponent estimator (as well as the algorithm of Wolf et al.). We shall explain why this discrepancy exists shortly.

2. The scale range where $\lambda(\epsilon)$ is almost constant is much wider for the HM parameter set than for the BH parameter set. In fact, only 10,000 points were used to calculate Fig. 15.18(a), while 30,000 points were used for Fig. 15.18(b). The reason is that the correlation dimension for the chaotic attractor of the HM parameter set is only about 0.4, while that of the BH parameter set is about 2.0. This means that in order to resolve the behavior of $\lambda(\epsilon)$ on ever smaller scales, longer and longer time series have to be used.

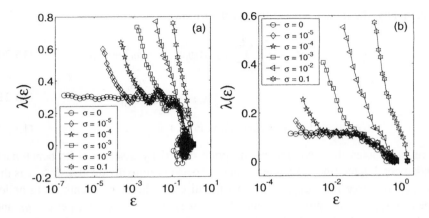

Figure 15.18. $\lambda(\epsilon)$ curves for the clean and the noisy asset pricing model with (a) the HM parameter set and (b) the BH parameter set.

More precisely, for a given dataset, if the smallest resolvable scale is ϵ_0, in order to resolve a smaller scale ϵ_0/r, where $r > 1$, a larger dataset has to be used. The larger the dimension of the attractor, the longer the time series has to be.

We can now understand why the Lyapunov exponent calculated by Hommes and Manzan using the neural network–based Lyapunov exponent estimator and the algorithm of Wolf et. al. is smaller than our estimate in the noise-free case. The answer is given by Eq. (15.6), noting the fact that the probability in Eq. (15.6) is typically larger on large scales than on small scales. As we pointed out earlier, when the algorithm of Wolf et al. is applied to a time series of finite length, in order to obtain a well-defined Lyapunov exponent, a fairly large-scale parameter has to be chosen. The discrepancy between our estimate and that of Hommes and Manzan suggests that the scale parameter used by them, using the algorithm of Wolf et al., is larger than the scales where $\lambda(\epsilon)$ is almost constant.

We now discuss the neural network–based Lyapunov exponent estimator. Here the Lyapunov exponent is estimated by the Jacobian matrix method. The main function of the neural network is to estimate a trajectory in the phase space so that the Jacobian matrices can be computed. When estimating the trajectory in the phase space, global optimization is employed such that the error between the given and estimated time series is minimized. Global optimization involves a scale parameter. Hommes and Manzan's analysis suggests that this scale parameter is comparable to the one used by the algorithm of Wolf et al.

Finally, we discuss why the estimated Lyapunov exponent can be negative in the strongly noise-driven case. The reason is as follows. In the strongly noise-driven case, the scales that can be resolved by a finite dataset are always close to the characteristic scale ϵ_∞. In order for an algorithm to numerically return a non-null result, the chosen scale parameter has to be larger than ϵ_∞. This then produces a negative Lyapunov exponent.

Our discussion up to now has made it clear that an estimated negative Lyapunov exponent may not necessarily imply absence of chaos and, therefore, regular dynamics. On the contrary, the dynamics are simply stochastic. Entropy generation is high even on fairly large scales and is much higher than the case of noise-free, low-dimensional chaos.

15.7.3.2 Analysis of U.S. foreign exchange rate data
Now that we understand the generic behavior of the $\lambda(\epsilon)$ curves for clean and noisy chaotic systems, we analyze some of the U.S. foreign exchange rate data. Altogether, we have analyzed about 20 datasets with a sampling time of 1 day. Three examples are shown in Figs. 15.19(b,d,f), for U.S.-Canada, U.S.-Mexico, and U.S.-Korea exchange rate data. We conclude that foreign exchange rate data, when sampled at a 1 day resolution, are like random walk–type processes and are well characterized by Eq. (15.11), with the Hurst parameter being able to take all types of values (close to, smaller than, and larger than 1/2 in Figs. 15.19(b,d,f), respectively). We note that among the 20 datasets, $H \approx 1/2$ is most prevalent. This indicates that a market is often efficient. However, deviations from the efficient market assumption do exist, as clearly shown in Figs. 15.19(d,f). Such deviations might be due to political factors. If so, then such deviations might be used to study economic/political ties between two nations.

15.7.4 Sea clutter modeling

In Sec. 8.9.1, we discussed target detection within sea clutter using the Hurst parameter. While the detection accuracy is high, we have emphasized that the scaling behavior is not well defined for sea clutter data. It turns out that sea clutter data are more complicated than other data examined in this book in the sense that the $\lambda(\epsilon)$ curves corresponding to different shells do not collapse together. Interestingly, each $\lambda(\epsilon)$ curve behaves very well. Figure 15.20(a) shows one example for a fairly large shell. When the slope of the $\lambda(\epsilon)$ curve is used to detect targets within sea clutter, we obtain figures like that shown in Fig. 15.20(b). Interestingly, if we form histograms of the slopes for data with the primary target and without the target, the histograms completely separate, as shown in the right subplot of Fig. 8.14. More importantly, the time scale for $\lambda(\epsilon) = -\gamma \ln \epsilon$ to be valid is up to 0.01 s, smaller than the time scale of $0.04 - 1$ s identified when we performed fractal analysis in Sec. 8.9.1. Furthermore, the pattern of the $\lambda(\epsilon)$ curves where they are about 0 is quite different for sea clutter data with and without a target. All this information can be combined to improve the accuracy of target detection.

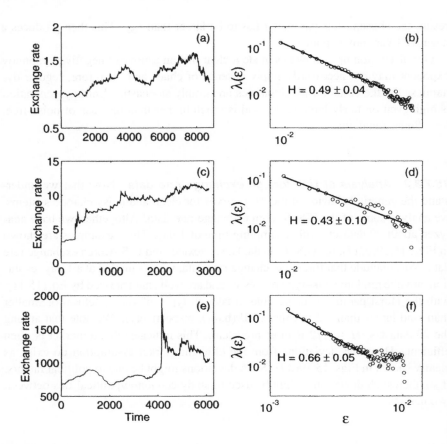

Figure 15.19.   Foreign exchange rate data between the United States and (a) Canada, (c) Mexico, and (e) Korea; (b,d,f) are the corresponding $\lambda(\epsilon)$ curves.

15.8 BIBLIOGRAPHIC NOTES

Most of the materials in this chapter are new. They have been developed based on the concepts of the SDLE [156] and time-dependent exponent curves [178–180]. The former is closely related to the finite-size Lyapunov exponent [16,53,64, 262] and (ϵ, τ)-entropy [183]. See also approximate entropy and sample entropy in Sec. 13.2.3. For distinguishing chaos from noise, we refer to [28, 104, 105, 179, 180, 182, 197, 229, 251, 256, 339, 352, 358, 380, 416, 451, 469]. In particular, [99, 135, 241, 354, 371, 410, 433] are about whether EEG signals are chaotic or not, while [11, 14, 220, 242, 275, 300, 356] discuss the appropriateness of using nonlinear measures for epileptic seizure forewarning. For complexity of biomedical signals, we refer to [2, 10, 208, 272, 317, 325, 332, 360, 365, 367, 412, 419, 442– 445, 468, 484, 485, 489–494]. For HRV analysis, we refer to [6, 15, 17, 29, 49, 52, 123, 127, 137, 194, 195, 221, 223, 232, 235, 236, 252, 261, 264, 292, 303, 342,

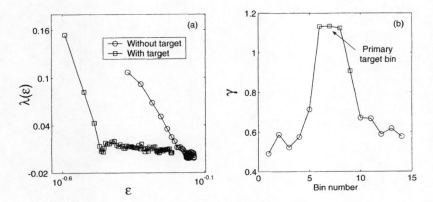

Figure 15.20.   (a) $\lambda(\epsilon)$ curves for sea clutter. The embedding parameters are $m = 4, L = 1$. (b) Target detection using γ parameter.

343, 346, 386, 427, 453, 486]. For chaos in cardiovascular systems, we refer to [17, 138, 181, 194, 195, 244, 252, 303]. For chaos in economic time series, we refer to [21, 34, 57–59, 97, 112, 200, 201, 222, 273, 389, 395, 401, 402]. For memory and multifractal features of economic time series, we refer to [329, 498]. For analysis of nonstationary signals, we refer to [115, 146, 147, 149, 152, 177, 446, 470]. For characterizing hidden frequencies in time series, we refer to [74, 326, 327]. For nonstationarity, we refer to [149, 230, 254, 299, 392].

APPENDIX A
DESCRIPTION OF DATA

A.1 NETWORK TRAFFIC DATA

The traffic data studied in this book include three sources. One is Ethernet traffic. There are four datasets, denoted as pAug.TL, pOct.TL, OctExt.TL, and OctExt4.TL, which can be obtained at ftp.bellcore.com under the directory /pub/world/wel/lan_traffic. Each dataset contains 1 million points representing data of arrival time stamps and packet sizes. The first two datasets were measured on the "purple cable." The last two sets were collected on Bellcore's link to the outside world. Originally all four datasets were simply called LAN traffic. Later, the last two datasets were reclassified as WAN traffic [425]. This classification is more appropriate, since the data were collected on Bellcore's link to the outside world.

Another data source, denoted as MPEG.data, is Bellcore's VBR video traffic data, available at ftp.telcordia.com under the directory /pub/vbr.video.trace. It consists of 174,136 integers representing the number of bits per video frame (at 24 frames/second for approximately 2 hrs).

The third source of data is vBNS (very-high-speed Backbone Network Service) traffic, collected by Dr. Ronn Ritke and his co-workers of the National Laboratory for Applied Network Research (NLANR) measurement and analysis group at the San Diego Supercomputer Center (SDSC) at a number of high-performance-connection (HPC) sites. The measurement durations of these traffic traces range from 0.5 min to several minutes, with half a million to several million arrivals. They are now available at our dedicated website (see Sec. A.4; for readers' convenience, the aforementioned Ethernet and VBR video traffic traces also appear there).

A.2   SEA CLUTTER DATA

Fourteen sea clutter datasets were obtained from a website maintained by Professor Simon Haykin: http://soma.ece.mcmaster.ca/ipix/dartmouth/datasets.html.
The measurement was made using the McMaster IPIX radar at Dartmouth, Nova Scotia, Canada. The radar was mounted in a fixed position on land 25-30 m above sea level, with an operating (carrier) frequency of 9.39 GHz (and hence a wavelength of about 3 cm). It was operated at low grazing angles, with the antenna dwelling in a fixed direction, illuminating a patch of ocean surface. The measurements were performed with the wave height in the ocean varying from 0.8 to 3.8 m (with peak heights up to 5.5 m) and the wind conditions varying from still to 60 km/hr (with gusts up to 90 km/hr). For each measurement, 14 areas, called antenna footprints or range bins, were scanned. Their centers are depicted as $B_1, B_2, \cdots, B_{14}$ in Fig. A.1. The distance between two adjacent range bins was 15 m. One or a few range bins (say, B_{i-1}, B_i and B_{i+1}) hit a target, which was a spherical block of styrofoam of diameter 1 m wrapped with wire mesh. The locations of the three targets were specified by their azimuthal angle and distance to the radar. They were $(128^0, 2660 \text{ m})$, $(130^0, 5525 \text{ m})$, and $(170^0, 2655 \text{ m})$, respectively. The range bin where the target is strongest is labeled as the primary target bin. Due to drift of the target, bins adjacent to the primary target bin may also have hit the target. They are called secondary target bins. For each range bin, there were 2^{17} complex numbers, sampled with a frequency of 1000 Hz.

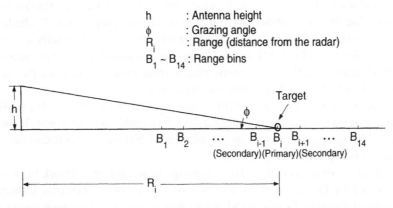

Figure A.1.   A schematic showing how the sea clutter data were collected.

A.3 NEURONAL FIRING DATA

Two sources of neuronal firing data were analyzed in this book. One source consists of firings of 104 cells, collected synchronously at Duke University when an owl monkey performed a three-dimensional reaching task involving a right-handed reach to food and subsequent placing of the food in the mouth [471]. The total observation time was about 36 min. While the details of the behavioral paradigm and the surgical procedure for chronic microwire recordings can be found in the literature [471], it is important to mention the components of the paradigm that are important for the discussions presented in Sec. 8.9.2. Microwire electrodes were implanted in four cortical regions with known motor associations [319]. The monkey's hand position, which was considered the desired signal by adaptive models, was also recorded (with a time-shared clock) and digitized with a 200 Hz sampling rate. On average, the time interval between two successive reaching tasks is about 8 s. From the neuronal firing data, spike detection was performed. In our analysis, both interspike interval data and spike-counting data (equivalent to firing rate) have been analyzed. Note that some neurons fired more than 10^4 times during about 36 min, while a few neurons only fired a few tens of times during this entire time period. This indicates the tremendous differences among the neurons.

The second source of data was three patients with Parkinson's disease. The data were collected in two areas, the globus pallidus externa (GPe) and the globus pallidus interna (GPi) of the brain. The sampling time is 20 kHz. These data are analyzed in Sec. 9.5.2.

A.4 OTHER DATA AND PROGRAM LISTINGS

The book has discussed many other types of data, some of which are also used in the exercises. A dedicated book website has been created, which contains two subdirectories, one for data, another for programs. For more details, please see the readme files at http://www.gao.ece.ufl.edu/GCTH_Wileybook.

APPENDIX B

PRINCIPAL COMPONENT ANALYSIS (PCA), SINGULAR VALUE DECOMPOSITION (SVD), AND KARHUNEN-LOÈVE (KL) EXPANSION

PCA is the eigenanalysis of the autocorrelation (or autocovariance) matrix R:

$$R = \begin{pmatrix} R_{11} & R_{12} & \cdots & R_{1n} \\ R_{21} & R_{22} & \cdots & R_{2n} \\ \vdots & \vdots & \vdots & \vdots \\ R_{n1} & R_{n2} & \cdots & R_{nn} \end{pmatrix}.$$

Let λ_i and ϕ_i be the ith eigenvalue and corresponding eigenvector; then we have

$$R\phi_i = \lambda_i\phi_i, \quad i = 1, 2, \cdots, n.$$

Since the matrix R is symmetric and positive-definite, the eigenvalues are all positive, and the eigenvectors corresponding to different eigenvalues are orthogonal.

The SVD of a matrix $A_{n \times m}$ is

$$A = U \Sigma V^T,$$

where $U_{n \times n}$ and $V_{m \times m}$ are orthogonal matrices, $\Sigma_{n \times m}$ is a specific matrix that will be specified shortly, and T denotes the transpose of a matrix. The computation can be carried out by first forming AA^T or $A^T A$, and then doing eigenanalysis by finding all the positive eigenvalues and eigenvectors. The elements of Σ are all zero except that $\Sigma_{ii} = \sigma_i$ for $i = 1, 2, \cdots, r$, where r is the number of positive eigenvalues from either AA^T or $A^T A$ and σ_i^2 is the ith eigenvalue of AA^T (or $A^T A$). The relation between PCA and SVD is clear if one forms the matrix A by taking delayed coordinates as its row vectors.

Under the KL expansion, a signal $x(t)$ is expanded as

$$x(t) = \sum_{n=1}^{\infty} c_n \psi_n(t) \quad 0 < t < T,$$

where $\psi(t)$ is a set of orthonormal functions in the interval $(0, T)$

$$\int_0^T \psi_n(t) \psi_m^*(t) dt = \delta[n - m]$$

and the coefficients c_n are random variables given by

$$c_n = \int_0^T x(t) \psi_n^*(t) dt,$$

where $*$ denotes a complex conjugate. The basis functions $\psi(t)$ are the solutions to the following integral equation:

$$\int_0^T R(t_1, t_2) \psi(t_2) dt_2 = \lambda \psi(t_1) \quad 0 < t_1 < T,$$

where $R(t_1, t_2)$ is the autocorrelation function of the process $x(t)$. Note that the KL expansion does not require the process $x(t)$ to be stationary.

APPENDIX C

COMPLEXITY MEASURES

The term complexity has many different meanings. Deterministic complexity, for example, measures the randomness of the data. It is a nondecreasing function of the entropy rate h_μ (e.g., Shannon entropy or Kolmogorov-Sinai (KS) entropy). See Fig. C.1 (left). The famous Kolmogorov-Chaitin complexity and the Lempel-Ziv (LZ) complexity both belong to this category. The structural complexity, on the other hand, is maximized for neither high nor low randomness (Fig. C.1 (right)). Complexity is a vast field, with many excellent schemes as well as numerous misleading results published. For details, we refer readers to books by Badii and Politi [19], Cover and Thomas (chapter 7) [86], Chaitin [67], Rissanen [369], Shaw [397], Watanabe [467], Bar-Yam [27], and a number of classic papers [18, 20, 39, 40, 66, 90–96, 102, 132, 133, 142, 185, 202, 212, 231, 267–270, 283, 286, 287, 301, 377, 408, 420, 464, 465, 477, 480–482].

In Chapters 11, 13, and 15, we discussed a number of complexity measures. In this appendix, we describe three more: the finite-size Lyapunov exponent (FSLE), the LZ complexity, and the permutation entropy (PE). The latter two were used in Chapter 15 to characterize EEG data.

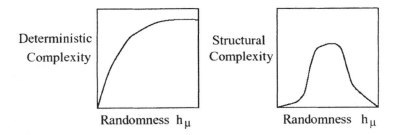

Figure C.1.   Deterministic vs. structural complexity.

C.1   FSLE

There exist a number of variants of the FSLE [16]. We describe the most popular one here.

Recall that Wolf et al.'s algorithm for calculating the largest Lyapunov exponent is used to monitor the exponential growth of a distance between a reference and a perturbed trajectory. For a given dataset, the growth rate, however, may not be truly exponential. When this is the case, the growth rate then depends on the actual distance between the reference and the perturbed trajectory. The FSLE is a simple way to quantify this possible dependence. Quantitatively, it is defined through the following steps:

1. One first constructs a suitable phase space from the time series data and then chooses a norm to define distance in the phase space.

2. One then introduces a series of scales, $\epsilon_n = r^n \epsilon_0$, $n = 1, \cdots, P$, where $1 < r \leq 2$.

3. One monitors, on average, how soon a small perturbation of size ϵ_i $i = 0, \cdots, P - 1$, grows to the size ϵ_{i+1}. Denote this average time by $T_r(\epsilon_i)$. When $r = 2$, $T_r(\epsilon_i)$ is the doubling time. When $r \neq 2$, $T_r(\epsilon_i)$ is called the r-fold time.

4. Assuming that

$$\epsilon_{i+1} = r\epsilon_i = \epsilon_i e^{T_r(\epsilon_i)\lambda(\epsilon_i)}, \tag{C.1}$$

one obtains the FSLE at scale ϵ_i, denoted by $\lambda(\epsilon_i)$, to be

$$\lambda(\epsilon_i) = \frac{\ln r}{T_r(\epsilon_i)}. \tag{C.2}$$

We now make a few comments:

- Equation (C.1) may be considered as a fitting using exponential function. This fitting is often acceptable, so long as $1 < r \leq 2$, even though the actual growth from scale ϵ_i to scale $\epsilon_{i+1} = r\epsilon_i$ may not be truly exponential.

- One major purpose of defining the FSLE is to identify interesting scales. The algorithm for computing the FSLE involves $P + 1$ scale parameters — the set of scales ϵ_i, $i = 0, \cdots, P - 1$, and a scale parameter needed to perform renormalization in order to use Wolf et al.'s algorithm. Determining these scales a priori through a process of trial and error is often a challenge.

- In order to calculate the r-fold time $T_r(\epsilon_i)$, a large dataset is usually needed to suitably define a reference and a perturbed trajectory and take averages.

- The r-fold time $T_r(\epsilon_i)$ is often connected with the prediction time scale. This interpretation is not suitable for stochastic systems. In fact, for $1/f^\beta$ and Levy processes, using Eqs. (15.11), (15.12), and (C.1), one has

$$T_r(\epsilon_i) = \frac{\ln r}{\mu} \, \epsilon_i^{\frac{1}{\mu}} \, , \tag{C.3}$$

where $\mu = H$ for $1/f^\beta$ processes and $\mu = 1/\alpha$ for Levy flights. In such cases, $T_r(\epsilon_i)$ may not be interpreted as a prediction time in the usual sense.

C.2 LZ COMPLEXITY

The LZ complexity and its derivatives, being easily implementable, very fast, and closely related to the Kolmogorov complexity, have found numerous applications in characterizing the randomness of complex data.

To compute the LZ complexity, a numerical sequence first has to be transformed into a symbolic sequence. The most popular approach is to convert the signal into a $0 - 1$ sequence by comparing the signal with a threshold value S_d [491]. That is, whenever the signal is larger than S_d, one maps the signal to 1; otherwise, one maps it to 0. One good choice of S_d is the median of the signal [317]. When multiple threshold values are used, one may map the numerical sequence to a multisymbol sequence. Note that if the original numerical sequence is a nonstationary random walk–type process, one should analyze the stationary differenced data instead of the original nonstationary data.

After the symbolic sequence is obtained, it can then be parsed to obtain distinct words, and the words can be encoded. Let $L(n)$ denote the length of the encoded sequence for those words. The LZ complexity can be defined as

$$C_{LZ} = \frac{L(n)}{n}. \tag{C.4}$$

Note this is very much in the spirit of the Kolmogorov complexity [267, 268].

There exist many different methods to perform parsing. One popular scheme was proposed by the original authors of the LZ complexity [282, 497]. For convenience, we call this Scheme 1. Another attractive method is described by Cover and Thomas [86], which we shall call Scheme 2. For convenience, we describe them in the context of binary sequences.

• **Scheme 1:** Let $S = s_1 s_2 \cdots s_n$ denote a finite-length $0-1$ symbolic sequence; $S(i, j)$ denote a substring of S that starts at position i and ends at position j, that is, when $i \leq j$, $S(i, j) = s_i s_{i+1} \cdots s_j$ and when $i > j$, $S(i, j) = \{\}$, the null set; $V(S)$ denote the vocabulary of a sequence S. It is the set of all substrings, or words, $S(i, j)$ of S, (i.e., $S(i, j)$ for $i = 1, 2, \cdots, n; j \geq i$). For example, let $S = 001$; we then have $V(S) = \{0, 1, 00, 01, 001\}$. The parsing procedure involves a left-to-right scan of the sequence S. A substring $S(i, j)$ is compared to the vocabulary that is comprised of all substrings of S up to $j - 1$, that is, $V(S(1, j - 1))$. If $S(i, j)$ is present in $V(S(1, j - 1))$, then update $S(i, j)$ and $V(S(1, j - 1))$ to $S(i, j + 1)$ and $V(S(1, j))$, respectively, and the process repeats. If the substring is not present, then place a dot after $S(j)$ to indicate the end of a new component, update $S(i, j)$ and $V(S(1, j - 1))$ to $S(j + 1, j + 1)$ (the single symbol in the $j + 1$ position) and $V(S(1, j))$, respectively, and the process continues. This parsing operation begins with $S(1, 1)$ and continues until $j = n$, where n is the length of the symbolic sequence. For example, the sequence 1011010100010 is parsed as $1 \cdot 0 \cdot 11 \cdot 010 \cdot 100 \cdot 010\cdot$. By convention, a dot is placed after the last element of the symbolic sequence. In this example, the number of distinct words is six.

• **Scheme 2:** The sequence $S = s_1 s_2 \cdots$ is sequentially scanned and rewritten as a concatenation $w_1 w_2 \cdots$ of words w_k chosen in such a way that $w_1 = s_1$ and w_{k+1} is the shortest word that has not appeared previously. In other words, w_{k+1} is the extension of some word w_j in the list, $w_{k+1} = w_j s$, where $0 \leq j \leq k$, and s is either 0 or 1. The sequence 1011010100010 in Scheme 1 is parsed as $1 \cdot 0 \cdot 11 \cdot 01 \cdot 010 \cdot 00 \cdot 10\cdot$. Therefore, a total of seven distinct words are obtained. This number is larger than the six of Scheme 1 by one.

The words obtained by Scheme 2 can be readily encoded. One simple way is as follows [86]. Let $c(n)$ denote the number of words in the parsing of the source sequence. For each word, we use $\log_2 c(n)$ bits to describe the location of the prefix to the word and one bit to describe the last bit. For our example, let 000 describe an empty prefix; then the sequence can be described as $(000, 1)(000, 0)(001, 1)(010, 1)(100, 0)(010, 0)(001, 0)$. The total length of the encoded sequence is $L(n) = c(n)[\log_2 c(n) + 1]$. Equation (C.4) then becomes

$$C_{LZ} = c(n)[\log_2 c(n) + 1]/n. \tag{C.5}$$

When n is very large, $c(n) \leq n/\log_2 n$ [86, 282]. Replaing $c(n)$ in Eq. (C.5) by $n/\log_2 n$, one obtains

$$C_{LZ} = \frac{c(n)}{n/\log_2 n}. \tag{C.6}$$

The commonly used definition of C_{LZ} takes the same functional form as Eq. (C.6), except that $c(n)$ is obtained by Scheme 1. Typically, $c(n)$ obtained by Scheme 1 is smaller than that obtained by Scheme 2. However, encoding the words obtained by Scheme 1 requires more bits than that obtained by Scheme 2. We surmise that the complexity defined by Eq. (C.4) is similar for both schemes. Indeed, numerically, we have observed that the functional dependence of C_{LZ} on n (based on Eqs. (C.5) and (C.6)) is similar for both schemes.

Rapp et al. [364] have considered the issue of normalizing the LZ complexity to make it independent of the sequence length. Recently, this issue has been reconsidered by Hu et al. [226], using an analytic approach. Specifically, they derived formulas for the LZ complexity for random equiprobable sequences as well as periodic sequences with an arbitrary period m and proposed a simple formula to perform normalization. It should be emphasized, however, that there may not exist universal normalization schemes that work for complex data with vastly different time scales.

Finally, we note that in the literature, the LZ complexity is sometimes interpreted as characterizing the structure of the data, possibly due to the perception that structure is a more appealing word than randomness. This is inappropriate, since asymptotically, the LZ complexity is equal to the Shannon entropy. This very characteristic of the LZ complexity — characterizing the randomness of the data — also dictates that the LZ complexity alone should not be used for the purpose of distinguishing deterministic chaos from noise.

C.3   PE

PE is introduced in [26] as a convenient means of analyzing a time series. It may be considered as a measure from chaos theory, since embedding vectors are used in the analysis. Using the notations of [61], it can be described as follows. For a given but otherwise arbitrary i, the m number of real values of $X_i = [x(i), x(i+L), \cdots, x(i+(m-1)L)]$ are sorted in $[x(i + (j_1 - 1)L) \leq x(i + (j_2 - 1)L) \leq \cdots \leq x(i + (j_m - 1)L]$. When an equality occurs, e.g., $x(i + (j_{i1} - 1)L) = x(i + (j_{i2} - 1)L)$, the quantities x are ordered according to the values of their corresponding j's; that is, if $j_{i1} < j_{i2}$, then we write $x(i + (j_{i1} - 1)L) \leq x(i + (j_{i2} - 1)L)$. Therefore, the vector X_i is mapped onto $(j_1, j_2, \cdots, j_m)$, which is one of the $m!$ permutations of m distinct symbols $(1, 2, \cdots, m)$. When each such permutation is considered as a symbol, the reconstructed trajectory in the m-dimensional space is represented by a symbol sequence. Let the probability for the $K \leq m!$ distinct symbols be $P_1, P_2, \cdots, P_K$. Then PE, denoted by E_p, for the time series $\{x(i), i = 1, 2, \cdots\}$ is defined as

$$E_p(m) = -\sum_{j=1}^{K} P_j \ln P_j. \tag{C.7}$$

The maximum of $E_P(m)$ is $\ln(m!)$ when $P_j = 1/(m!)$. It is convenient to work with

$$0 \leq E_p = E_p(m)/\ln(m!) \leq 1. \tag{C.8}$$

Thus E_p gives a measure of the departure of the time series under study from a completely random one: the smaller the value of E_p, the more regular the time series is.

REFERENCES

1. Abarbanel, H.D.I. (1996) *Analysis of Observed Chaotic Data.* Springer.

2. Abásolo, D., Hornero, R., Gómez, C., García, M., and López, M. (2005) Analysis of EEG background activity in Alzheimer's disease patients with Lempel-Ziv complexity and central tendency measure. *Med. Eng. Phys.* **28**, 315–322.

3. Abry, P. and Sellan, F. (1996) The wavelet-based synthesis for fractional Brownian motion — Proposed by Sellan, F. and Meyer, Y.: Remarks and fast implementation. *Applied and Comput. Harmonic Analysis* **3**, 377–383.

4. Abry, P. and Veitch, D. (1998) Wavelet analysis of long-range-dependent traffic. *IEEE Trans. on Info. Theory* **44**, 2–15.

5. Addie, R.G., Zukerman, M., and Neame, T. (1995) Fractal traffic: Measurements, modeling and performance evaluation. In *Proc. IEEE InfoCom*, Boston, MA, pp. 985–992.

6. Akselrod, S., Gordon, D., Ubel, F.A., Shannon, D.C., Barger, M.A., and Cohen, R.J. (1981) Power spectrum analysis of heart rate fluctuation: A quantitative probe of beat-to-beat cardiovascular control. *Science* **213**, 220–222.

7. Albert, R. and Barabasi, A.L. (2002) Statistical mechanics of complex networks. *Rev. Modern Phys.* **74**, 47–97.

8. Aleksic, Z. (1991) Estimating the embedding dimension. *Physica D* **52**, 362–368.

9. Alizadeh, A.A. et al. (2000) Distinct types of diffuse large B-cell lymphoma identified by gene expression profiling. *Nature* **403**, 503–511.

10. Amigó, J.M., Szczepaski, J., Wajnryb, E., and Sanchez-Vives, M.V. (2004) Estimating the entropy rate of spike trains via Lempel-Ziv complexity. *Neural Comput.* **16**, 717–736.

11. Andrzejak, R.G., Lehnertz, K., Mormann, F., Rieke, C., David, P., and Elger, C.E. (2001) Indications of nonlinear deterministic and finite-dimensional structures in time series of brain electrical activity: Dependence on recording region and brain state. *Phys. Rev. E* **64**, 061907.

12. Antoni, M. and Ruffo, S. (1995) Clustering and relaxation in Hamiltonian long-range dynamics. *Phys. Rev. E* **52**, 2361–2374.

13. Arimitsu, T. and Arimitsu, N. (2000) Tsallis statistics and fully developed turbulence. *J. Phys. A* **33**, L235–L241.

14. Aschenbrenner-Scheibe, R., Maiwald, T., Winterhalder, M., Voss, H.U., Timmer, J., and Schulze-Bonhage, A. (2003) How well can epileptic seizures be predicted? An evaluation of a nonlinear method. *Brain* **126**, 2616–2626.

15. Ashkenazy, Y., Ivanov, P.C., Havlin, S., Peng, C.K., Goldberger, A.L., and Stanley, H.E. (2001) Magnitude and sign correlations in heartbeat fluctuations. *Phys. Rev. Lett.* **86**, 1900–1903.

16. Aurell, E., Boffetta, G., Crisanti, A., Paladin, G., and Vulpiani, A. (1997) Predictability in the large: An extension of the concept of Lyapunov exponent. *J. Physics A* **30**, 1–26.

17. Babyloyantz, A. and Destexhe, A. (1988) Is the normal heart a periodic oscillator? *Biol. Cybern.* **58**, 203–211.

18. Bachas, C.P. and Huberman, B.A. (1986) Complexity and the relaxation of hierarchical structures. *Phys. Rev. Lett.* **57**, 1965–1969.

19. Badii, R. and Politi, A. (1997) *Complexity: Hierarchical Structures and Scaling in Physics*. Cambridge University Press.

20. Badii, R. and Politi, A. (1997) Thermodynamics and complexity of cellular automata. *Phys. Rev. Lett.* **78**, 444–447.

21. Bailey, B.A. (1996) Local Lyapunov exponents: Predictability depends on where you are. In *Nonlinear Dynamics and Economics*, edited by Barnett, W.A., Kirman, A.P., and Salmon, M., Cambridge University Press, pp. 345–359.

22. Bak, P. (1996) *How Nature Works: The Science of Self-Organized Criticality*. Copernicus.

23. Bak, P., Tang, C., and Wiesenfeld, K. (1987) Self-organized criticality — An explanation of $1/f$ noise. *Phys. Rev. Lett.* **59**, 381–384.

24. Bak, P., Tang, C., and Wiesenfeld, K. (1988) Self-organized criticality. *Phys. Rev. B* **38**, 364–374.

25. Baker, G.L. and Gollub, J.P. (1996) *Chaotic Dynamics: An Introduction*. Cambridge University Press.

26. Bandt, C. and Pompe, B. (2002) Permutation entropy: A natural complexity measure for time series. *Phys. Rev. Lett.* **88**, 174102.

27. Bar-Yam, Y. (1997) *Dynamics of Complex Systems*. Addison-Wesley.

28. Barahona, M. and Poon, C.S. (1996) Detection of nonlinear dynamics in short, noisy time series. *Nature* **381**, 215–217.

29. Barbieri, R. and Brown, E.N. (2006) Analysis of heartbeat dynamics by point process adaptive filtering. *IEEE Trans. Biomed. Eng.* **53**, 4–12.

30. Barcelo, F. and Jordan, J. (1999) Channel holding time distribution in public cellular telephony. In *Teletraffic Engineering in a Competitive World. Proceedings of the International Teletraffic Congress*, ITC-16, Vol. 3a, edited by Key, P. and Smith, D., Elsevier Science, pp. 107–116.

31. Barenblatt, G.I. (2003) *Scaling*. Cambridge University Press.

32. Barenblatt, G.I. and Chorin, A.J. (1998) New perspectives in turbulence: Scaling laws, asymptotics and intermittency. *SIAM Rev.* **40**, 265–291.

33. Barnsley, M. (1993) *Fractals Everywhere*, 2nd ed. Morgan Kaufmann.

34. Barnett, W.A. and Serletis, A. (2000) Martingales, nonlinearity and chaos. *J. Econ. Dyn. Control* **24**, 703–724.

35. Bartumeus, F., Peters, F., Pueyo, S., Marrase, C., and Catalan, J. (2003) Helical Levy walks: Adjusting searching statistics to resource availability in microzooplankton. *Proc. Natl. Acad. Sci. USA* **100**, 12771–12775.

36. Bassingthwaighte, J.B., Liebovitch, L.S., and West, B.J. (1994) *Fractal Physiology*. Oxford University Press.

37. Beck, C. (2000) Application of generalized thermostatistics to fully developed turbulence. *Physica A* **277**, 115–123.

38. Beck, C., Lewis, G.S., and Swinney, H.L. (2001) Measuring nonextensivity parameters in a turbulent Couette-Taylor flow. *Phys. Rev. E* **63**, 035303.

39. Bennett, C.H. (1986) On the nature and origin of complexity in discrete, homogeneous, locally-interacting systems. *Found. Phys.* **16**, 585–592.

40. Bennett, C.H. (1990) How to define complexity in physics, and why? In *Complexity, Entropy and the Physics of Information*, edited by Zurek, W.H. Addison-Wesley, pp. 137–148.

41. Bennett, J.C.R., Partridge, C., and Shectman, N. (1999) Packet reordering is not pathological network behavior. *IEEE-ACM Trans. on Networking* **7**, 789–798.

42. Benzi, R., Ciliberto, S., Baudet, C., and Chavarria, G.R. (1995) On the scaling of 3-dimensional homogeneous and isotropic turbulence. *Physica D* **80**, 385–398.

43. Benzi, R., Ciliberto, S., Tripiccione, R., Baudet, C., Massaioli, F., and Succi, S. (1993) Extended self-similarity in turbulent flows. *Phys. Rev. E* **48**, R29–R32.

44. Beran, J., Sherman, R., Taqqu, M.S., and Willinger, W. (1995) Long-range dependence in variable-bit-rate video traffic. *IEEE Trans. on Commun.* **43**, 1566–1579.

45. Berge, P., Pomeau, Y., and Vidal, C. (1984) *Order within Chaos: Towards a Deterministic Approach to Turbulence*. Hermann.

46. Bernamont, J. (1937) Fluctuations de potentiel aux bornes d'un conducteur metallique de faible volume parcouru par un courant. *Ann. Phys.* (Leipzig) **7**, 71–140.

47. Berry, E.X. (1967) Cloud droplet growth by collection. *J. Atmos. Sci.* **24**, 688–701.

48. Berry, M.V. and Lewis, Z.V. (1980) On the Weierstrass-Mandelbrot fractal function. *Proc. R. Soc. Lond. A* **370**, 459–484.

49. Bigger, J.T., Jr., Steinman, R.C., Rolnitzky, L.M., Fleiss, J.L., Albrecht, P., and Cohen, R,J. (1996) Power law behavior of RR-interval variability in healthy middle-aged persons, patients with recent acute myocardial infarction, and patients with heart transplants. *Circulation* **93**, 2142–2151.

50. Billock, V.A. (2000) Neural acclimation to $1/f$ spatial frequency spectra in natural images transduced by the human visual system. *Physica D* **137**, 379–391.

51. Billock, V.A., de Guzman, G.C., and Kelso, J.A.S. (2001) Fractal time and $1/f$ spectra in dynamic images and human vision. *Physica D* **148**, 136–146.

52. Bloomfield, P. (1976) Fourier analysis of time series: An introduction. In *Fourier Analysis of Time Series: An Introduction*. Wiley Series in Probability and Mathematical Statistics, edited by Bradley, A.B., Hunter, J.S., Kendall, D.G., and Watson, S.G. Wiley, pp. 1–258.

53. Boffetta, G., Cencini, M., Falcioni, M., and Vulpiani, A. (2002) Predictability: A way to characterize complexity. *Phys. Rep.* **356**, 367–474.

54. Bolotin, V.A., Levy, Y., and Liu, D. (1999) Characterizing data connection and messages by mixtures of distributions on logarithmic scale. In *Teletraffic Engineering in a Competitive World. Proceedings of the International Teletraffic Congress*, ITC-16, Vol. 3a, edited by Key, P. and Smith, D., Elsevier Science, pp. 887–894.

55. Borgas, M.S. (1992) A comparison of intermittency models in turbulence. *Phys. of Fluids* **4**, 2055–2061.

56. Borges, E.P., Tsallis, C., Ananos, G.F.J., and de Oliveira, P.M.C. (2002) Nonequilibrium probabilistic dynamics of the Logistic map at the edge of chaos. *Phys. Rev. Lett.* **89**, 254103.

57. Brock, W.A., Dechert, W.D., Scheinkman, J.A., and LeBaron, B. (1996) A test for independence based on the correlation dimension. *Econ. Rev.* **15**, 197–235.

58. Brock, W.A. and Hommes, C.H. (1998) Heterogeneous beliefs and routes to chaos in a simple asset pricing model. *J. Econ. Dyn. Control* **22**, 1235–1274.

59. Brock, W.A. and Sayers, C.L. (1988) Is the business cycle characterized by deterministic chaos? *J. Monetary Econ.* **22**, 71–90.

60. Butz, A.R. (1972) A theory of $1/f$ noise. *J. Stat. Phys.* **4**, 199–216.

61. Cao, Y.H., Tung, W.W., Gao, J.B., Protopopescu, V.A., and Hively, L.M. (2004) Detecting dynamical changes in time series using the permutation entropy. *Phys. Rev. E* **70**, 046217.

62. Carlson, J.M. and Doyle, J. (1999) Highly optimized tolerance: A mechanism for power laws in designed systems. *Phys. Rev. E* **60**, 1412–1427.

63. Casilari, E., Reyes, A., Diaz-Estrella, A., and Sandoval, F. (1999) Classification and comparison of modelling strategies for VBR video traffic. In *Teletraffic Engineering in a Competitive World. Proceedings of the International Teletraffic Congress*, ITC-16, Vol. 3b, edited by Key, P. and Smith, D., Elsevier Science, pp. 817–826.

64. Cencini, M., Falcioni, M., Olbrich, E., Kantz, H., and Vulpiani, A. (2000) Chaos or noise: Difficulties of a distinction. *Phys. Rev. E* **62**, 427–437.

65. Cenys, A. and Pyragas, K. (1988) Estimation of the number of degrees of freedom from chaotic time-series. *Phys. Lett. A* **129**, 227–230.

66. Chaitin, G.J. (1966) On length of programs for computing finite binary sequences. *J. Assoc. Comp. Mach.* **13**, 547–569.

67. Chaitin, G.J. (1987) *Information, Randomness and Incompleteness*, World Scientific.

68. Chambers, J.M., Mallows, C.L., and Stuck, B.W. (1976) Method for simulating stable random variables. *J. Am. Statistical Assn.* **71**, 340–344.

69. Chan, H.C. (1990) Radar sea-clutter at low grazing angles. *Proc. Inst. Elect. Eng.* **F137**, 102–112.

70. Chapin, J.K., Moxon, K.A., Markowitz, R.S., and Nicolelis, M.A.L. (1999) Real-time control of a robot arm using simultaneously recorded neurons in the motor cortex. *Nature Neurosci.* **2**, 664–670.

71. Chen, Y., Ding, M., and Kelso, J.A.S. (1997) Long memory processes ($1/f$ alpha type) in human coordination. *Phys. Rev. Lett.* **79**, 4501–4504.

72. Chen, Z., Hu, K., Carpena, P., Bernaola-Galvan, P., Stanley, H.E., and Ivanov, P.C. (2005) Effect of nonlinear filters on detrended fluctuation analysis. *Phys. Rev. E* **71**, 011104.

73. Chen, Z., Ivanov, P.C., Hu, K., and Stanley, H.E. (2002) Effect of nonstationarities on detrended fluctuation analysis. *Phys. Rev. E* **65**, 041107.

74. Chern, J.L., Ko, J.Y., Lih, J.S., Su, H.T., and Hsu, R.R. (1998) Recognizing hidden frequencies in a chaotic time series. *Phys. Lett. A* **238**, 134–140.

75. Chorin, A.J. (1993) *Vorticity and Turbulence*, Springer-Verlag.

76. Chorin, A.J. and Hald, O.H. (2006) *Stochastic Tools in Mathematics and Science*, Springer-Verlag.

77. Chorin, A.J., Hald, O.H., and Kupferman, R. (2002) Optimal prediction with memory. *Physica D* **166**, 239–257.

78. Chu, S., DeRisi, J., Eisen, M., Mulholland, J., Botstein, D., Brown, P.O., and Herskowitz, I. (1998) The transcriptional program of sporulation in budding yeast. *Science* **282**, 699–705.

79. Chung, F., Lu, L.Y., and Vu, V. (2003) The spectra of random graphs with given expected degrees. *Proc. Natl. Acad. Sci. USA* **100**, 6313–6318.

80. Cohen, A. and Procaccia, I. (1985) Computing the Kolmogorov entropy from time series of dissipative and conservative dynamical systems, *Phys. Rev. A* **31**, 1872–1882.

81. Collins, J.J. and DeLuca, C.J. (1994) Random walking during quiet standing. *Phys. Rev. Lett.* **73**, 764–767.

82. Collins, P.G., Fuhrer, M.S., and Zettl, A. (2000) $1/f$ noise in carbon nanotubes. *Appl. Phys. Lett.* **76**, 894–896.

83. Costa, M., Goldberger, A.L., and Peng, C.K. (2002) Multiscale entropy analysis of complex physiologic time series. *Phys. Rev. Lett.* **89**, 068102.

84. Costa, M., Goldberger, A.L., and Peng, C.K. (2005) Multiscale entropy analysis of biological signals. *Phys. Rev. E* **71**, 021906.

85. Costa, U.M.S., Lyra, M.L., Plastino, A.R., and Tsallis, C. (1997) Power-law sensitivity to initial conditions within a logisticlike family of maps: Fractality and nonextensivity. *Phys. Rev. E* **56**, 245–250.

86. Cover, T.M. and Thomas, J.A. (1991) *Elements of Information Theory*. Wiley.

87. Cox, D.R. (1984) Long-range dependence: A review. In *Statistics: An Appraisal*, edited by David, H.A. and Davis, H.T. The Iowa State University Press, pp. 55–74.

88. Crick, F. and Koch, C. (2003) A framework for consciousness. *Nature Neurosci.* **6**, 119–126.

89. Crovella, M.E. and Bestavros, A. (1997) Self-similarity in World Wide Web traffic: Evidence and possible causes. *IEEE/ACM Trans. on Networking* **5**, 835–846.

90. Crutchfield, J.P. (1994) The Calcuci of emergence — computation, dynamics and induction. *Physica D* **75**, 11–54.

91. Crutchfield, J.P. and McNamara, B.S. (1987) Equations of motion from a data series. *Complex Systems*, **1**, 417–452.

92. Crutchfield, J.P. and Packard, N.H. (1982) Symbolic dynamics of one-dimensional maps — entropies, finite precision, and noise. *J. Theo. Phys.* **21**, 433–466.

93. Crutchfield, J.P. and Packard, N.H. (1983) Symbolic dynamics of noisy chaos. *Physica D* **7**, 201–223.

94. Crutchfield, J.P. and Shalizi, C.R. (1999) Thermodynamic depth of causal states: Objective complexity via minimal representations. *Phys. Rev. E* **59**, 275–283.

95. Crutchfield, J.P. and Young, K. (1989) Inferring statistical complexity. *Phys. Rev. Lett.* **63**, 105–108.

96. Csordas, A. and Szepfalusy, P. (1989) Singularities in Renyi information as phase transitions in chaotic states. *Phys. Rev. A* **39**, 4767–4777.

97. Dacorogna, M.M., Gencay, R., Muller, U., Olsen, R.B., and Pictet, O.V. (2001) *An Introduction to High-Frequency Finance*. Academic Press.

98. Dalton, F. and Corcoran, D. (2001) Self-organized criticality in a sheared granular stick-slip system. *Phys. Rev. E* **63**, 061312.

99. daSilva, F.H.L., Pijn, J.P., Velis, D., and Nijssen, P.C.G. (1997) Alpha rhythms: Noise, dynamics and models. *Int. J. Psychophysiol.* **26**, 237–249.

100. Davis, A., Marshak, A., Wiscombe, W., and Cahalan, R. (1994) Multifractal characterizations of nonstationarity and intermittency in geophysical fields — observed, retrieved, or simulated. *J. Geophys. Res.* **99**, 8055–8072.

101. Davis, A., Marshak, A., Wiscombe, W., and Cahalan, R. (1996) Multifractal characterizations of intermittency in nonstationary geophysical signals and fields. In *Current Trends in Nonstationary Analysis*, edited by Trevilo, G., Hardin, J., Douglas, B., and Andreas, E. World Scientific, pp. 97–158.

102. Delgado, J. and Sole, R.V. (1997) Collective-induced computation. *Phys. Rev. E* **55**, 2338–2344.

103. DeLong, M.R. (1971) Activity of pallidal neurons during movement. *J. Neurophysiol.* **34**, 414–427.

104. Dettmann, C.P. and Cohen, E.G.D. (2000) Microscopic chaos and diffusion. *J. Stat. Phys.* **101**, 775–817.

105. Dettmann, C.P. and Cohen, E.G.D. (2001) Note on chaos and diffusion. *J. Stat. Phys.* **103**, 589–599.

106. Devaney, R. (2003) *Introduction to Chaotic Dynamical Systems*, 2nd ed. Westview.

107. Ding, M.Z., Grebogi, C., Ott, E., Sauer, T., and Yorke, J.A. (1993) Estimating correlation dimension from a chaotic time series: When does plateau onset occur? *Physica D* **69**, 404–424.

108. Ding, M.Z., Grebogi, C., Ott, E., Sauer, T., and Yorke, J.A. (1993) Plateau onset for correlation dimension: when does it occur? *Phys. Rev. Lett.* **70**, 3872–3875.

109. Ding, M.Z. and Yang, W.M. (1995) Distribution of the first return time in fractional brownian motion and its application to the study of on-off intermittency. *Phys. Rev. E* **52**, 207–213.

110. Dorogovtsev, S.N., Goltsev, A.V., Mendes, J.F.F., et al. (2003) Spectra of complex networks. *Phys. Rev. E* **68**, 046109.

111. Doyle, J. and Carlson, J.M. (2000) Power laws, highly optimized tolerance, and generalized source coding. *Phys. Rev. Lett.* **84**, 5656–5659.

112. Drozdz, S., Kwapien, J., Grummer, F., Ruf, F., and Speth, J. (2003) Are the contemporary financial fluctuations sooner converging to normal? *Acta Phys. Pol. B* **34**, 4293–4306.

113. Dubrulle, B. (1994) Intermittency in fully-developed turbulence — log-poisson statistics and generalized scale covariance. *Phys. Rev. Lett.* **73**, 959–962.

114. Duda, R.O., Hart, P.E., and Stork, D.G. (2001) *Pattern Classification*, 2nd ed. Wiley.

115. Eckmann, J.P., Kamphorst, S.O., and Ruelle, D. (1987) Recurrence plots of dynamic systems. *Europhys. Lett.* **4**, 973–977.

116. Eckmann, J.P. and Ruelle, D. (1985) Ergodic theory of chaos and strange attractors. *Rev. Modern Phys.* **57**, 617–656.

117. Edelman, G.M. and Tononi, G. (2000) *A Universe of Consciousness*. Basic Books.

118. Elliott, F.W., Jr., Majda, A.J., Horntrop, D.J., and McLaughlin, R.M. (1995) Hierarchical Monte Carlo methods for fractal random fields. *J. Stat. Phys.* **81**, 717–736.

119. Erramilli, A., Narayan, O., and Willinger, W. (1996) Experimental queueing analysis with long-range dependent packet traffic. *IEEE/ACM Trans. on Networking* **4**, 209–223.

120. Erramilli, A., Singh, P.R., and Pruthi, P. (1995) An application of deterministic chaotic maps to model packet traffic. *Queueing Systems* **20**, 171–206.

121. Evans, W. and Pippenger, N. (1998) On the maximum tolerable noise for reliable computation by formulas. *IEEE Tran. Inform. Theory* **44**, 1299–1305.

122. Evans, W. and Schulman, L.J. (1999) Signal propagation and noisy circuits. *IEEE Tran. Inform. Theory* **45**, 2367–2373.

123. Ewing, D.J. (1984) Cardiac autonomic neuropathy. In *Diabetes and Heart Disease*, edited by Jarrett, R.J. Elsevier, pp. 122–132.

124. Falconer, K.J. (1990) *Fractal Geometry : Mathematical Foundations and Applications*. Wiley.

125. Faloutsos, M., Faloutsos, P., and Faloutsos, C (1999) On power-law relationships of the Internet topology. *SIGCOMM'99*, 251–262.

126. Farkas, I.J., Derenyi, I., Barabasi, A.L., et al. (2001) Spectra of "real-world" graphs: Beyond the semicircle law. *Phys. Rev. E* **64**, 026704.

127. Farrell, T.G., Bashir, Y., Cripps, T., Malik, M., Poloniecki, J., Bennett, E.D., Ward, D.E., and Camm, A.J. (1991) Risk stratification for arrhythmic events in postinfarction patients based on heart rate variability, ambulatory electrocardiographic variables and the signal-averaged electrocardiogram. *J. Am. Coll. Cardiol.* **18**, 687–697.

128. Fay, F.A., Clarke, J., and Peters, R.S. (1977) Weibull distribution applied to sea clutter. *Proc. IEE Conf. Radar'77*, London, pp. 101–103.

129. Feder, H.J.S. and Feder, J. (1991) Self-organized criticality in a stick-slip process. *Phys. Rev. Lett.* **66**, 2669–2672; erratum in *Phys. Rev. Lett.* **67**, 283.

130. Feder, J. (1988) *Fractals*. Plenum Press.

131. Feigenbaum, M.J. (1983) Universal behavior in nonlinear systems. *Physica D* **7**, 16–39.

132. Feldman, D.P. and Crutchfield, J.P. (1998) Measures of statistical complexity: Why? *Phys. Lett. A* **238**, 244–252.

133. Feldman, D.P. and Crutchfield, J.P. (2003) Structural information in two-dimensional patterns: Entropy convergence and excess entropy. *Phys. Rev. E* **67**, 051104.

134. Feldmann, A., Gilbert, A.C., and Willinger, W. (1998) Data networks as cascades: Investigating the multifractal nature of Internet WAN traffic. *ACM SIGCOMM'98 Conference*, Vancouver, Canada.

135. Fell, J., Roschke, J., and Schaffner, C. (1996) Surrogate data analysis of sleep electroencephalograms reveals evidence for nonlinearity. *Biol. Cybern.* **75**, 85–92.

136. Feller, W. (1971) *Probability Theory and Its Applications*, 2nd ed. Vols I and II. Wiley.

137. Fleisher, L.A., Pincus, S.M., and Rosenbaum, S.H. (1993) Approximate entropy of heart rate as a correlate of postoperative ventricular dysfunction. *Anesthesiology* **78**, 683–692.

138. Fortrat, J.O., Yamamoto, Y., and Hughson, R.L. (1997) Respiratory influences on non-linear dynamics of heart rate variability in humans. *Biol. Cybern.* **77**, 1–10.

139. Franceschetti, G., Iodice, A., Migliaccio, M., and Riccio, D. (1999) Scattering from natural rough surfaces modeled by fractional Brownian motion two-dimensional processes. *IEEE Trans Antennas Propagation* **47**, 1405–1415.

140. Fraser, A.M. and Swinney, H.L. (1986) Independent coordinates for strange attractors from mutual information. *Phys. Rev. A* **33**, 1134–1140.

141. Frederiksen, R.D., Dahm, W.J.A., and Dowling, D.R. (1997) Experimental assessment of fractal scale similarity in turbulent flows — Multifractal scaling. *J. Fluid Mech.* **338**, 127–155.

142. Freund, J., Ebeling, W., and Rateitschak, K. (1996) Self-similar sequences and universal scaling of dynamical entropies. *Phys. Rev. E* **54**, 5561–5566.

143. Frisch, U. (1995) *Turbulence—The Legacy of A.N. Kolmogorov*. Cambridge University Press.

144. Fritz, H., Said, S., and Weimerskirch, H. (2003) Scale-dependent hierarchical adjustments of movement patterns in a long-range foraging seabird. *Proc. R. Soc. Lond. Ser. B — Biol. Sci.* **270**, 1143–1148.

145. Gao, J.B. (1997) Recognizing randomness in a time series. *Physica D* **106**, 49–56.

146. Gao, J.B. (1999) Recurrence time statistics for chaotic systems and their applications. *Phys. Rev. Lett.* **83**, 3178–3181.

147. Gao, J.B. (2000) On the structures and quantification of recurrence plots. *Phys. Lett. A* **270**, 75–87.

148. Gao, J.B. (2000) Multiplicative multifractal modeling of long-range-dependent (LRD) traffic in computer communications networks. Ph.D dissertation, EE Dept, UCLA.

149. Gao, J.B. (2001) Detecting nonstationarity and state transitions in a time series. *Phys. Rev. E* **63**, 066202.

150. Gao, J.B. (2004) Analysis of amplitude and frequency variations of essential and parkinsonian tremors. *Med. Biol. Eng. Comput.* **52**, 345–349.

151. Gao, J.B., Billock, V.A., Merk, I., Tung, W.W., White, K.D., Harris, J.G., and Roychowdhury, V.P. (2006) Inertia and memory in ambiguous visual perception. *Cogn. Process* **7**, 105–112.

152. Gao, J.B., Cao, Y.H., Gu, L.Y., Harris, J.G., and Principe, J.C. (2003) Detection of weak transitions in signal dynamics using recurrence time statistics. *Phys. Lett. A* **317**, 64–72.

153. Gao, J.B., Cao, Y.H., and Lee, J.M. (2003) Principal component analysis of $1/f$ noise. *Phys. Lett. A* **314**, 392–400.

154. Gao, J.B., Cao, Y.H., Qi, Y., and Hu, J. (2005) Building innovative representations of DNA sequences to facilitate gene finding. *IEEE Intelligent Systems Nov/Dec*, special issue on Data Mining for Bioinformatics, 34–39.

155. Gao, J.B., Chen, C.C., Hwang, S.K, and Liu, J.M. (1999) Noise-induced chaos. *Int. J. Mod. Phys. B* **13**, 3283–3305.

156. Gao, J.B., Hu, J., Tung, W.W., and Cao, Y.H. (2006) Distinguishing chaos from noise by scale-dependent Lyapunov exponent. *Phys. Rev. E* **74**, 066204.

157. Gao, J.B., Hu, J., Tung, W.W., Cao, Y.H., Sarshar, N., and Roychowdhury, V.P. (2006) Assessment of long range correlation in time series: How to avoid pitfalls. *Phys. Rev. E* **73**, 016117.

158. Gao, J.B., Hwang, S.K, and Liu, J.M. (1999) When can noise induce chaos? *Phys. Rev. Lett.* **82**, 1132–1135.

159. Gao, J.B., Qi, Y., Cao, Y.H., and Tung, W.W. (2005) Protein coding sequence identification by simultaneously characterizing the periodic and random features of DNA sequences. *J. Biomed. Biotechnol.* **2**, 139–146.

160. Gao, J.B., Qi, Y., Cao, Y.H., Tung, W.W., and Roychowdhury, V.P. Deriving a novel codon index by combining period-3 and fractal features of DNA sequences, *Proc. Natl. Acad. Sci. USA*. In press.

161. Gao, J.B., Qi, Y., and Fortes, J.A.B. (2005) Bifurcations and fundamental error bounds for fault-tolerant computations. *IEEE Tran. Nanotech.* **4**, 395–402.

162. Gao, J.B. and Royshowdhury, V.P. (2000) Multifractal gene finder. Technical report, Electrical Engineering Department, UCLA.

163. Gao, J.B. and Rao, N.S.V. (2005) Complicated dynamics of Internet transport protocols. *IEEE Commun. Lett.* **9**, 4–6.

164. Gao, J.B., Rao, N.S.V., Hu, J., and Ai, J. (2005) Quasi-periodic route to chaos in the dynamics of Internet transport protocols. *Phys. Rev. Lett.* **94**, 198702.

165. Gao, J.B. and Rubin, I. (1999) Multiplicative multifractal modeling of long-range-dependent traffic. *Proceedings ICC'99*, Vancouver, Canada.

166. Gao, J.B. and Rubin, I. (1999) Multifractal modeling of counting processes of long-range-dependent network traffic. *Proceedings SCS Advanced Simulation Technologies Conference*, San Diego, CA.

167. Gao, J.B. and Rubin, I. (2000) Superposition of multiplicative multifractal traffic streams. *Proceedings ICC'2000*, New Orleans.

168. Gao, J.B. and Rubin, I. (2000) Statistical properties of multiplicative multifractal processes in modeling telecommunications traffic streams. *Electronics Lett.* **36**, 101–102.

169. Gao, J.B. and Rubin, I. (2000) Multifractal analysis and modeling of VBR video traffic. *Electronics Lett.* **36**, 278–279.

170. Gao, J.B. and Rubin, I. (2000) Superposition of multiplicative multifractal traffic processes. *Electronics Lett.* **36**, 761–762.

171. Gao, J.B. and Rubin, I. (2000) Multiplicative multifractal modeling of long-range-dependent (LRD) traffic in computer communications networks. *World Congress of Nonlinear Analysts*, July, Catonia, Sicily, Italy. Also *J. Nonlinear Analysis* **47**, 5765–5774 (2001).

172. Gao, J.B. and Rubin, I. (2001) Multifractal modeling of counting processes of long-range-dependent network traffic. *Comput. Commun.* **24**, 1400–1410.

173. Gao, J.B. and Rubin, I. (2001) Multiplicative multifractal modeling of long-range-dependent network traffic. *Int. J. Commun. Systems* **14**, 783–801.

174. Gao, J.B. and Rubin, I. (2001) Long-range-dependence properties and multifractal modeling of vBNS traffic. *Applied Telecommunications Symposium (ATS'01)*, Seattle, Washington, April.

175. Gao, J.B. and Tung, W.W. (2002) Pathological tremors as diffusional processes. *Biol. Cybern.* **86**, 263–270.

176. Gao, J.B., Tung, W.W., Cao, Y.H., Hu, J., and Qi, Y. (2005) Power-law sensitivity to initial conditions in a time series with applications to epileptic seizure detection. *Physica A* **353**, 613–624.

177. Gao, J.B., Tung, W.W., and Rao, N.S.V. (2002) Noise-induced Hopf-bifurcation-type sequence and transition to chaos in the Lorenz equations. *Phys. Rev. Lett.* **89**, 254101.

178. Gao, J.B. and Zheng, Z.M. (1993) Local exponential divergence plot and optimal embedding of a chaotic time series. *Phys. Lett. A* **181**, 153–158.

179. Gao, J.B. and Zheng, Z.M. (1994) Direct dynamical test for deterministic chaos and optimal embedding of a chaotic time series. *Phys. Rev. E* **49**, 3807–3814.

180. Gao, J.B. and Zheng, Z.M. (1994) Direct dynamical test for deterministic chaos. *Europhys. Lett.* **25**, 485–490.

181. Garfinkel, A., Spano, M.L., Ditto, W.L., and Weiss, J.N. (1992) Controlling cardiac chaos. *Science* **257**, 1230–1235.

182. Gaspard, P., Briggs, M.E., Francis, M.K., Sengers, J.V., Gammons, R.W., Dorfman, J.R., and Calabrese, R.V. (1998) Experimental evidence for microscopic chaos. *Nature* **394**, 865–868.

183. Gaspard, P. and Wang, X.J. (1993) Noise, chaos, and (ϵ, τ)-entropy per unit time. *Phys. Rep.* **235**, 291–343.

184. Geisel, T., Nierwetberg, J., and Zacherl, A. (1985) Accelerated diffusion in Josephson junctions and related chaotic systems. *Phys. Rev. Lett.* **54**, 616–620.

185. Gell-Mann, M. and Lloyd, S. (1996) Information measures, effective complexity, and total information. *Complexity* **2**, 44–52.

186. Georgopoulos, A.P., Schwartz, A.B., and Kettner, R.E. (1986) Neuronal population coding of movement direction. *Science* **233**, 1416–1419.

187. Gilbert, A.C., Willinger, W., and Feldmann, A. (1999) Scaling analysis of conservative cascades, with applications to network traffic. *IEEE Tran. Info. Theory* **45**, 971–991.

188. Gilden, D.L., Thornton, T., and Mallon, M.W. (1995) $1/f$ noise in human cognition. *Science* **267**, 1837–1839.

189. Gini, F. (2000) Performance analysis of two structured covariance matrix estimators in compound-Gaussian clutter. *Signal Processing* **80**, 365–371.

190. Gini, F., Montanari, M., and Verrazzani, L. (2000) Maximum likelihood, ESPRIT, and periodogram frequency estimation of radar signals in K-distributed clutter. *Signal Processing* **80**, 1115–1126.

191. Gkantsidis, C., Mihail, M., and Zegura, E. (2003) Spectral analysis of Internet topologies. *INFOCOM'03*.

192. Gleick, J. (1987) *Chaos*. Penguin Books.

193. Goh, K.I., Kahng, B., and Kim, D. (2001) Spectra and eigenvectors of scale-free networks. *Phys. Rev. E* **64**, 051903.

194. Goldberger, A.L., Rigney, D.R., Mietus, J., Antman, E.M., and Greenwald, S. (1988) Nonlinear dynamics in sudden cardiac death syndrome: Heart rate oscillations and bifurcations. *Experientia* **44**, 983–987.

195. Goldberger, A.L. and West, B.J. (1987) Applications of nonlinear dynamics to clinical cardiology. *Ann. NY Acad. Sci.* **504**, 155–212.

196. Golub, T.R., Slonim, D.K., Tamayo, P., Huard, C., Gaasenbeek, M., Mesirov, J.P., Coller, H., Loh, M.L., Downing, J.R., Caligiuri, M.A., Bloomfield, C.D., and Lander, E.S. (1999) Molecular classification of cancer: Class discovery and class prediction by gene expression monitoring. *Science* **286**, 531–537.

197. Gottwald, G.A. and Melbourne, I. (2004) A new test for chaos in deterministic systems. *Proc. R. Soc. Lond. Ser. A* **46**, 603–611.

198. Gouyet, J.F. (1995) *Physics and Fractal Structures*. Springer.

199. Granger, C.W.J. (1980) Long memory relationships and the aggregation of dynamic models. *J. Econ.* **14**, 227–238.

200. Granger, C.W.J. (1991) Developments in the nonlinear analysis of economic series. *Scand. J. Econ.* **93**, 263–276.

201. Granger, C.W.J. (1994) Is chaotic economic theory relevant for economics? A review article of Jess Benhabib: Cycles and chaos in economic equilibrium. *J. Int. Comp. Econ.* **3**, 139–145.

202. Grassberger, P. (1986) Toward a quantitative theory of self-generated complexity. *Int. J. Theo. Phys.* **25**, 907–938.

203. Grassberger, P. and Procaccia, I. (1983) Characterization of strange attractors. *Phys. Rev. Lett.* **50**, 346–349.

204. Grassberger, P. and Procaccia, I. (1983) Estimation of the Kolmogorov entropy from a chaotic signal. *Phys. Rev. A* **28**, 2591–2593.

205. Grimmett, G. and Stirzaker, D. (2001) *Probability and Random Processes*, 3rd ed. Oxford University Press.

206. Guckenheimer, J. and Holmes, P. (1990) *Nonlinear Oscillations, Dynamical Systems, and Bifurcations of Vector Fields*. Springer-Verlag.

207. Guerin, C.A. and Saillard, M. (2001) Electromagnetic scattering on fractional Brownian surfaces and estimation of the Hurst exponent. *Inverse Problems* **17**, 365–386.

208. Gusev, V.D., Nemytikova, L.A., and Chuzhanova, N.A. (1999) On the complexity measures of genetic sequences. *Bioinformatics* **15**, 994–999.

209. Haken, H. (1983) At least one Lyapunov exponent vanishes if the trajectory of an attractor does not contain a fixed point. *Phys. Lett. A* **94**, 71–72.

210. Hammer, P.W., Platt, N., Hammel, S.M., Heagy, J.F., et al. (1994) Experimental observation of on-off intermittency. *Phys. Rev. Lett.* **73**, 1095–1098.

211. Han, J., Gao, J.B., Qi, Y., Jonker, P., and Fortes, J.A.S. (2005) Towards hardware-redundant fault-tolerant logic for nanoelectronics. *IEEE Design and Test of Computers* (special issue), 328–339.

212. Hanson, J.E. and Crutchfield, J.P. (1997) Computational mechanics of cellular automata: An example. *Physica D* **103**, 169–189.

213. Haykin, S. (2001) *Adaptive Filter Theory*, 4th ed. Prentice Hall.

214. Haykin, S., Bakker, R., and Currie, B.W. (2002) Uncovering nonlinear dynamics — the case study of sea clutter. *Proc. IEEE* **90**, 860–881.

215. Heagy, J.F., Platt, N., and Hammel, S.M. (1994) Characterization of on-off intermittency. *Phys. Rev. E* **49**, 1140–1150.

216. Heath, D, Resnick, S., and Samorodnitsky, G. (1998) Heavy tails and long range dependence in ON/OFF processes and associated fluid models. *Math. Operations Res.* **23**, 145–165.

217. Helander P., Chapman S.C., Dendy R.O., Rowlands, G., and Watkins, N.W. (1999) Exactly solvable sandpile with fractal avalanching. *Phys. Rev. E* **59**, 6356–6360.

218. Henon, M. (1976) Two-dimensional mapping with a strange attractor. *Commun. Math. Phys.* **50**, 69–77.

219. Hentschel, H.G.E. and Procaccia, I. (1983) The infinite number of generalized diensions of fractals and strange attractors. *Physica D* **8**, 435–444.

220. Hively, L.M., Gailey, P.C., and Protopopescu, V.A. (1999) Detecting dynamical change in nonlinear time series. *Phys. Lett. A* **258**, 103–114.

221. Ho, K.K.L., Moody, G.B., Peng, C.K., Mietus, J.E., Larson, M.G., Levy, D., and Goldberger, A.L. (1997) Predicting survival in heart failure cases and controls using fully automated methods for deriving nonlinear and conventional indices of heart rate dynamics. *Circulation* **96**, 842–848.

222. Hommes, C.H. and Manzan, S. (2006) Testing for nonlinear structure and chaos in economic time series: A comment. *J. Macroecon.* **28**, 169–174.

223. Hon, E.H. and Lee, S.T. (1965) Electronic evaluations of the fetal heart rate patterns preceding fetal death: Further observations. *Am. J. Obstet. Gynecol.* **87**, 814–826.

224. Houghton, H.G. (1985) *Physical Meteorology*. MIT Press, pp. 272.

225. Hu, J., Gao, J.B., Cao, Y.H., Bottinger, E., and Zhang, W.J. (2007) Exploiting noise in array CGH data to improve detection of gene copy number change. *Nucl. Acids Res.* **35**, e35.

226. Hu, J., Gao, J.B., and Principe, J.C. (2006) Analysis of biomedical signals by the Lempel-Ziv complexity: The effect of finite data size. *IEEE Trans. Biomed. Eng.* **53**, 2606–2609.

227. Hu, K., Ivanov, P.C., Chen, Z., Carpena, P., and Stanley, H.E. (2001) Effect of trends on detrended fluctuation analysis. *Phys. Rev. E* **64**, 011114.

228. Hu, J., Tung, W.W., and Gao, J.B. (2006) Detection of low observable targets within sea clutter by structure function based multifractal analysis. *IEEE Trans. Antennas Prop.* **54**, 135–143.

229. Hu, J., Tung, W.W., Gao, J.B., and Cao, Y.H. (2005) Reliability of the 0-1 test for chaos. *Phys. Rev. E* **72**, 056207.

230. Huang, N.E., Shen, Z., and Long, S.R. (1999) A new view of nonlinear water waves: The Hilbert spectrum. *Annu. Rev. Fluid Mech.* **31**, 417–457.

231. Huberman, B.A. and Hogg, T. (1986) Complexity and adaptation. *Physica D* **22**, 376–384.

232. Huikuri, H.V., Seppanen, T., Koistinen, M.J., Airaksinen, K.E.J., Ikaheimo, M.J., Castellanos, A., and Myerburg, R.J. (1996) Abnormalities in beat-to-beat dynamics

of heart rate before the spontaneous onset of life-threatening ventricular tachyarrhythmias in patients with prior myocardial infarction. *Circulation* **93**, 1836–1844.

233. Hwa, R.C. and Ferree, T.C. (2002) Scaling properties of fluctuations in the human electroencephalogram. *Phys. Rev. E* **66**, 021901.

234. Hwang, S.K., Gao, J.B., and Liu, J.M. (2000) Noise-induced chaos in an optically injected semiconductor laser. *Phys. Rev. E* **61**, 5162–5170.

235. Ivanov, P.C., Amaral, L.A.N., Goldberger, A.L., Havlin, S., Rosenblum, M.G., Struzik, Z.R., and Stanley, H.E. (1999) Multifractality in human heartbeat dynamics. *Nature* **399**, 461–465.

236. Iyengar, N., Peng, C.K., Morin, R., Goldberger, A.L., and Lipsitz, L.A. (1996) Age-related alterations in the fractal scaling of cardiac interbeat interval dynamics. *Am. J. Physiol.* **271**, R1078–R1084.

237. Jaeger, H.M., Liu, C.H., and Nagel, S.R. (1989) Relaxation at the angle of repose. *Phys. Rev. Lett.* **62**, 40–43.

238. Jakeman, E. and Pusey, P.N. (1976) A model for non-Rayleigh sea echo. *IEEE Trans. Antennas Prop.* **24**, 806–814.

239. Janicki, A. and Weron, A. (1994) Can one see α-stable variables and processes? *Stat. Sci.* **9**, 109–126.

240. Jensen, H., Christensen, K., and Fogedby, H. (1989) $1/f$ noise, distribution of lifetimes, and a pile of sand. *Phys. Rev. B* **40**, 7425–7427.

241. Jeong, J., Kim, M.S., and Kim S.Y. (1999) Test for low-dimensional determinism in electroencephalograms. *Phys. Rev. E* **60**, 831–837.

242. Jerger, K.K., Netoff, T.I., Francis, J.T., Sauer, T., Pecora, L., Weinstein, S.L., and Schiff, S.J. (2001) Early seizure detection. *J. Clin. Neurophysiol.* **18**, 259–268.

243. Johnson, J.B. (1925) The Schottky effect in low frequency circuits. *Phys. Rev.* **26**, 71–85.

244. Kaneko, K. and Tsuda, I. (2000) *Complex Systems: Chaos and Beyond.* Springer, pp. 219–236.

245. Kantelhardt, J.W., Koscielny-Bunde, E., Rego, H.H.A., Havlin, S., and Bunde A. (2001) Detecting long-range correlations with detrended fluctuation analysis. *Physica A* **295**, 441–454.

246. Kantelhardt, J.W., Zschiegner, S.A., Bunde, E., Havlin, S., Bunde, A., and Stanley, H.E. (2002) Multifractal detrended fluctuation analysis of nonstationary time series. *Physica A* **316**, 87–114.

247. Kanter, M. (1975) Stable densities under change of scale and total variation inequalities. *Ann. Probability* **3**, 697–707.

248. Kantz, H. (1994) A robust method to estimate the maximal Lyapunov exponent of a time series. *Phys. Lett. A* **185**, 77–87.

249. Kantz, H. and Schreiber, T. (1997) *Nonlinear Time Series Analysis.* Cambridge University Press.

250. Kaplan, D. (1995) *Understanding Nonlinear Dynamics.* Springer-Verlag.

251. Kaplan, D.T. and Glass, L. (1992) Direct test for determinism in a time-series. *Phys. Rev. Lett.* **68**, 427–430.

252. Kaplan, D.T. and Goldberger, A.L. (1991) Chaos in cardiology. *J. Cardiovasc. Electrophysiol.* **2**, 342–354.

253. Kaulakys, B. and Meskauskas, T. (1998) Modeling $1/f$ noise. *Phys. Rev. E* **58**, 7013–7019.

254. Kennel, M.B. (1997) Statistical test for dynamical nonstationarity in observed time-series data. *Phys. Rev. E* **56**, 316–321.

255. Kennel, M.B., Brown, R., and Abarbanel, H.D.I. (1992) Determining embedding dimension for phase-space reconstruction using a geometrical construction. *Phys. Rev. A* **45**, 3403–3411.

256. Kennel, M.B. and Isabelle, S. (1992) Method to distinguish possible chaos from colored noise and to determine embedding parameters. *Phys. Rev. A* **46**, 3111–3118.

257. Khinchin, A.I. and Gnedenko, B.V. (1962) *An Elementary Introduction to the Theory of Probability*. Dover.

258. Kida, S. (1991) Log-stable distribution and intermittency of turbulence. *J Phys. Soc. Jpn.* **60**, 5–8.

259. Kim, S.P., Sanchez, J.C., Erdogmus, D., Rao, Y.N., Wessberg, J., Principe, J.C., and Nicolelis, M. (2003) Divide-and-conquer approach for brain-machine interfaces: Nonlinear mixture of competitive linear models. *Neural Networks* **16**, 865–871.

260. Klafter, J., Shlesinger, M.F., and Zumofen, G. (1996) Beyond Brownian motion. *Physics Today* **49**, 33–39.

261. Kleiger, R.E., Miller, J.P., Bigger, J.T., Moss, A.J., and the Multicenter Post-infarction Research Group (1987) Decreased heart rate variability and its association with increased mortality after myocardial infarction. *Am. J. Cardiol.* **59**, 256–262.

262. Kleinfelter, N., Moroni, M., and Cushman, J.H. (2005) Application of a finite-size Lyapunov exponent to particle tracking velocimetry in fluid mechanics experiments. *Phy. Rev. E* **72**, 056306.

263. Kleinrock, L. (1975) *Queueing Systems*. Wiley.

264. Kobayashi, M. and Musha, T. (1982) $1/f$ fluctuation of heartbeat period. *IEEE Trans. Biomed. Eng.* **29**, 456–457.

265. Kolmogorov, A.N. (1941) The local structure of turbulence in incompressible viscous fluid for very large Reynolds number. *Dokl. Akad. Nauk SSSR* **30**, 299–303.

266. Kolmogorov, A.N. (1962) A refinement of previous hypotheses concerning the local structure of turbulence in a viscous incompressible fluid at high reynolds number. *J. Fluid Mech.* **13**, 82–85.

267. Kolmogorov, A.N. (1965) Three approaches to the quantitative definition of "information." *Probl. Info. Transm.* **1**, 1–7.

268. Kolmogorov, A.N. (1968) Logical basis for information theory and probability theory. *IEEE Trans. Inform. Theory* **IT-14**, 662–664.

269. Kolmogorov, A.N. (1983) Combinatorial foundations of information theory and the calculus of probabilities. *Russ. Math. Surveys* **38**, 29–40.

270. Koppel, M. (1987) Complexity, depth and sophistication. *Complex Systems* **1**, 1087–1091.

271. Kore, L. and Bosman, G. (1999) Random telegraph signal of a multiple quantum well infrared photodetector. *J. Appl. Phys.* **86**, 6586–6589.

272. Kuusela, T.A., Jartti, T.T., Tahvanainen, K.U.O., and Kaila, T.J. (2002) Nonlinear methods of biosignal analysis in assessing terbutaline-induced heart rate and blood pressure changes. *Am. J. Physiol. — Heart and Circulatory Physiol.* **282**, H773–H783.

273. Kyrtsou, C. and Serletis, A. (2006) Univariate tests for nonlinear structure. *J. Macroecon.* **28**, 154–168.

274. Labovitz, C., Malan, G.R., and Jahanian, F. (1998) Internet routing instability. *IEEE-ACM Trans. Networking* **6**, 515–528.

275. Lai, Y.C., Harrison, M.A.F., Frei, M.G., and Osorio, I. (2003) Inability of Lyapunov exponents to predict epileptic seizures. *Phys. Rev. Lett.* **91**, 068102.

276. Latora, V., Baranger, M., Rapisarda, A., and Tsallis, C. (2000) The rate of entropy increase at the edge of chaos. *Phys. Lett. A* **273**, 97–103.

277. Latora, V., Rapisarda, A., and Tsallis, C. (2001) Non-Gaussian equilibrium in a long-range Hamiltonian system. *Phys. Rev. E* **64**, 056134.

278. Lavagno, A., Kaniadakis, G., Rego-Monteiro, M., Quarati, P., and Tsallis. C. (1998) Non-extensive thermostatistical approach of the peculiar velocity function of galaxy clusters. *Astrophys. Lett. Commun.* **35**, 449–455.

279. Lee, H.N. (2001) Impact of flow control windows in TCP on fractal scaling of traffic exiting from a server pool. *Proc. GlobeCom*, San Antonio, Texas, November.

280. Lee, J.M., Kim, D.J., Kim, I.Y., Park, K.S., and Kim, S.I. (2002) Detrended fluctuation analysis of EEG in sleep apnea using MIT/BIH polysomnography data. *Computers Biol. Med.* **32**, 37–47.

281. Leland, W.E., Taqqu, M.S., Willinger, W., and Wilson, D.V. (1994) On the self-similar nature of Ethernet traffic (extended version). *IEEE/ACM Trans. Networking* **2**, 1–15.

282. Lempel, A. Lempel and Ziv, J. (1976) On the complexity of finite sequences. *IEEE Trans. Info. Theory* **22**, 75–81.

283. Li, W. (1991) On the relationship between complexity and entropy for Markov chains and regular languages. *Complex Systems* **5**, 381–399.

284. Li, W. and Kaneko, K. (1992) Long-range correlation and partial $1/f$-alpha spectrum in a noncoding DNA-sequence. *Europhys. Lett.* **17**, 655–660.

285. Liebert, W., Pawelzik, K., and Schuster, H.G. (1991) Optimal embedding of chaotic attractors from topological considerations. *Europhys. Lett.* **14**, 521–526.

286. Lindgren, K. and Nordahl, M.G. (1988) Complexity measures and cellular automata. *Complex Systems* **2**, 409–440.

287. Lloyd, S. and Pagels, H. (1988) Complexity as thermodynamic depth. *Ann. Phys.* **188**, 186–213.

288. Lorenz, E.N. (1963) Deterministic nonperiodic flow. *J. Atmos. Sci.* **20**, 130–141.

289. Lyra, M.L. and Tsallis, C. (1998) Nonextensivity and multifractality in low-dimensional dissipative systems. *Phys. Rev. Lett.* **80**, 53–56.

290. Mackey, M.C. and Glass, L. (1977) Oscillation and chaos in physiological control systems. *Science* **197**, 287–288.

291. Majda, A.J. and Kramer, P.R. (1999) Simplified models for turbulent diffusion: Theory, numerical modeling, and physical phenomena. *Phys. Rep.* **314**, 238–574.

292. Malik, M., Farrell, T., Cripps, T.R., and Camm, A.J. (1989) Heart rate variability in relation to prognosis after myocardial infarction: Selection of optimal processing techniques. *Eur. Heart J* **10**, 1060–1074.

293. Mandelbrot, B.B. (1974) Intermittent turbulence in self-similar cascades: Divergence of high moments and dimension of carrier. *J. Fluid Mech.* **62**, 331–358.

294. Mandelbrot, B.B. (1982) *The Fractal Geometry of Nature*. San Francisco: Freeman.

295. Mandelbrot, B.B. (1982) Comment on computer rendering of fractal stochastic models. *Commun. ACM* **25**, 581–583.

296. Mandelbrot, B.B. (1997) *Fractals and Scaling in Finance*. New York: Springer.

297. Mandelbrot, B.B. and Ness, V. (1968) Fractional Brownian motions, fractional noises and applications. *SIAM Rev.* **10**, 422–437.

298. Mane, R. (1981) On the dimension of the compact invariant sets of certain nonlinear maps. In *Dynamical Systems and Turbulence*, edited by Rand, D.A. and Young, L.S., *Lecture Notes in Mathematics*, Vol. 898. Springer-Verlag, pp. 230–242.

299. Manuca, R (1996) Stationarity and nonstationarity in time series analysis. *Phys. D* **99**, 134–161.

300. Martinerie, J., Adam, C., Le Van Quyen, M., Baulac, M., Clemenceau, S., Renault, B., and Varela, F.J. (1998) Epileptic seizures can be anticipated by non-linear analysis. *Nat. Med.* **4**, 1173–1176.

301. Martin-Lof, P. (1966) The definition of random sequences. *Inform. Contr.* **9**, 602–619.

302. Martorella, M., Berizzi, F., and Mese, E.D. (2004) On the fractal dimension of sea surface backscattered signal at low grazing angle. *IEEE Trans. Antennas Prop.* **52**, 1193–1204.

303. Mayer-Kress, G., Yates, F.E., Benton, L., Keidel, M., Tirsch, W., Pappl, S.J., and Geist, K. (1988) Dimensional analysis of nonlinear oscillations in brain, heart, and muscle. *Math Biosci.* **90**, 155–182.

304. Meneveau, C. and Sreenivasan, K.R. (1987) Simple multifractal cascade model for fully-developed turbulence. *Phys. Rev. Lett.* **59**, 1424–1427.

305. Metzler, R. and Klafter, J. (2000) The random walk's guide to anomalous diffusion: A fractional dynamics approach. *Phys. Rep.* **339**, 1–77.

306. Metzler, R. and Klafter, J. (2004) The restaurant at the end of the random walk: Recent developments in fractional dynamics of anomalous transport processes. *J. Phys. A* **37**, R161–R208.

307. Meyer, Y., Sellan, F., and Taqqu, M.S. (1999) Wavelets, generalized white noise and fractional integration: The synthesis of fractional Brownian motion. *J. Fourier Anal. Appl.* **5**, 465–494.

308. Micolich, A.P., Taylor, R.P., Davies, A.G., Bird, J.P., Newbury, R., Fromhold, T.M., Ehlert, A., Linke, H., Macks, L.D., Tribe, W.R., Linfield, E.H., Ritchie, D.A., Cooper, J., Aoyagi, Y., and Wilkinson, P.B. (2001) The evolution of fractal patterns during a classical-quantum transition. *Phys. Rev. Lett.* **87**, 036802.

309. Mihail, M. and Papadimitriou, C. (2002) On the Eigenvalue power-law. *RANDOM 02*, Harvard, MA.

310. Milotti, E. (2002) $1/f$ noise: A pedagogical review (downloadable at http://arxiv.org/abs/physics/0204033).

311. Molina, M., Castelli, P., and Foddis, G. (2000) Web traffic modeling exploiting TCP connections' temporal clustering through HTML-REDUCE. *IEEE Network* **14**, 46–55.

312. Mondragon, R.J. (1999) A model of packet traffic using a random wall model. *Int. J. Bif. Chaos* **9**, 1381–1392.

313. Mondragon, R.J., Arrowsmith, D.K., and Pitts, J.M. (2001) Chaotic maps for traffic modelling and queueing performance analysis. *Performance Evaluation* **43**, 223–240.

314. Mondragon, R.J., Pitts, J.M., and Arrowsmith, D.K. (2000) Chaotic intermittency-sawtooth map model of aggregate self-similar traffic streams. *Electronics Lett.* **36**, 184–186.

315. Montroll, E.W. and Shlesinger, M.K. (1982) On $1/f$ noise and other distributions with long tails (log-normal distribution, Levy distribution, Pareto distribution, scale-invaraint process). *Proc. Natl. Acad. Sci. USA* **79**, 3380–3383.

316. Montroll, E.W. and Shlesinger, M.K. (1983) Maximum entropy formalism, fractals, scaling phenomena, and $1/f$ noise: A tale of tails. *J. Stat. Phys.* **32**, 209–230.

317. Nagarajan, R. (2002) Quantifying physiological data with Lempel-Ziv complexity: Certain issues. *IEEE Trans. Biomed. Eng.* **49**, 1371–1373.

318. Nathanson, F.E. (1969) *Radar Design Principles*, McGraw Hill, pp. 254–256.

319. Nicolelis, M.A.L., Ghazanfar, A.A., Faggin, B.M., Votaw, S., and Oliveira, L.M.O. (1997) Reconstructing the engram: Simultaneous, multisite, many single neuron recordings. *Neuron* **18**, 529–537.

320. Nohara, T. and Haykin, S. (1991) Canadian east coast radar trials and the K-distribution. *Proc. Inst. Elect. Eng.* **F138**, 80–88.

321. Norros, I. (1994) A storage model with self-similar input. *Queueing Systems* **16**, 387–396.

322. Norros, I. (1995) On the use of fractional Brownian motions in the theory of connectionless networks. *IEEE JSAC* **13**, 953–962.

323. Obukhov, A.M. (1962) Some specific features of atmospheric turbulence. *J. Fluid Mech.* **13**, 77–81.

324. Oppenheim, A.V., Schafer, R.W., and Buck, J.R. (1999) *Discrete-Time Signal Processing*, 2nd ed. Prentice Hall.

325. Orlov, Y.L. and Potapov, V.N. (2004) Complexity: An internet resource for analysis of DNA sequence complexity. *Nucl. Acids Res.* **32**, W628–W633.

326. Ortega, G.J. (1995) A new method to detect hidden frequencies in chaotic time series. *Phys. Lett. A* **209**, 351–355.

327. Ortega, G.J. (1996) Invariant measures as Lagrangian variables: Their application to time series analysis. *Phys. Rev. Lett.* **77**, 259–262.

328. Osborne, A.R. and Provenzale, A. (1989) Finite correlation dimension for stochastic systems with power-law spectra. *Physica D* **35**, 357–381.

329. Oswiecimka, P., Kwapien, J., and Drozdz, S. (2005) Multifractality in the stock market: Price increments versus waiting times. *Physica A* **347**, 626–638.

330. Ott, E. (2002) *Chaos in Dynamical Systems*. Cambridge University Press.

331. Ott, E., Sauer, T., and Yorke, J.A. (1994) *Coping with Chaos: Analysis of Chaotic Data and the Exploitation of Chaotic Systems*. Wiley.

332. Otu, H.H. and Sayood, K. (2003) A new sequence distance measure for phylogenetic tree construction. *Bioinformatics* **19**, 2122–2130.

333. Over, T.M. and Gupta, V.K. (1996) A space-time theory of mesoscale rainfall using random cascades. *J. Geophys. Res.* **101**, 26319–26331.

334. Packard, N.H., Crutchfield, J.P., Farmer, J.D., and Shaw, R.S. (1980) Geometry from a time series. *Phys. Rev. Lett.* **45**, 712–716.

335. Papoulis, A. and Unnikrishna Pillai, S. (2001) *Probability, Random Variables and Stochastic Processes with Errata Sheet*, 4th ed. McGraw-Hill.

336. Parisi, G. and Frisch, U. (1985) On the singularity structure of fully developed turbulence. In *Turbulence and Predictability in Geophysical Fluid Dynamics and Climate Dynamics*, edited by Ghil, M., Benzi, R., and Parisi, G. North-Holland, pp. 71–84.

337. Paxson, V. and Floyd, S. (1995) Wide area traffic — the failure of Poisson modeling. *IEEE/ACM Trans. Networking* **3**, 226–244.

338. Pecora, L.M. and Caroll, L.M. (1990) Synchronization in chaotic systems. *Phys. Rev. Lett.* **64**, 821–824.

339. Pei, X. and Moss, F. (1996) Characterization of low-dimensional dynamics in the crayfish caudal photoreceptor. *Nature* **379**, 618–621.

340. Peng, C.K., Buldyrev, S.V., Goldberger, A.L., Havlin, S., Sciortino, F., Simons, M., and Stanley, H.E. (1992) Long-range correlations in nucleotide sequences. *Nature* **356**, 168–170.

341. Peng, C.K., Buldyrev, S.V., Havlin, S., Simons, M., Stanley, H.E., and Goldberger, A.L. (1994) Mosaic organization of DNA nucleotides. *Phys. Rev. E* **49**, 1685–1689.

342. Peng, C.K., Havlin, S., Stanley, H.E., and Goldberger, A.L. (1995) Quantification of scaling exponents and crossover phenomena in nonstationary heartbeat time series. *Chaos* **5**, 82–87.

343. Peng, C.K., Mietus, J., Hausdorff, J.M., Havlin, S., Stanley, H.E., and Goldberger, A.L. (1993) Long-range anticorrelations and non-Gaussian behavior of the heartbeat. *Phys. Rev. Lett.* **70**, 1343–1346.

344. Pesin, Y.B. (1977) Characteristic Lyapunov exponents and smooth ergodic theory. *Russian Math. Survey* **32**, 55–114.

345. Pincus, S.M. (1991) Approximate entropy as a measure of system complexity. *Proc. Natl. Acad. Sci. USA* **88**, 2297–2301.

346. Pincus, S.M. and Viscarello, R.R. (1992) Approximate entropy: A regularity statistic for fetal heart rate analysis. *Obstet. Gynecol.* **79**, 249–255.

347. Pipiras, V. (2005) Wavelet-based simulation of fractional Brownian motion revisited. *Appl. Comput. Harmonic Anal.* **19**, 49–60.

348. Pippenger, N. (1988) Reliable computation by formulas in the presence of noise. *IEEE Tran. Inform. Theory* **34**, 194–197.

349. Plastino, A.R. and Plastino, A. (1993) Stellar polytropes and tsallis entropy. *Phys. Lett. A* **174**, 384–386.

350. Platt, N., Hammel, S.M., and Heagy, J.F. (1994) Effects of additive noise on on-off intermittency. *Phys. Rev. Lett.* **72**, 3498–3501.

351. Pomeau, Y. and Manneville, P. (1980) Intermittent transition to turbulence in dissipative dynamical systems. *Commun. Math. Phys.* **74**, 189–197.

352. Poon, C.S. and Barahona, M. (2001) Titration of chaos with added noise. *Proc. Natl. Acad. Sci. USA* **98**, 7107–7112.

353. Press, W.H. (1978) Flicker noises in astronomy and elsewhere. *Comments on Astrophysics* **7**, 103–119.

354. Pritchard, W.S., Duke, D.W., and Krieble, K.K. (1995) Dimentional analysis of resting human EEC .2. Surrogate-data testing indicates nonlinear but not low-dimensional chaos. *Psychophysiology* **32**, 486–491.

355. Proakis, J.G. and Manolakis, D.K. (2006) *Digital Signal Processing*, 4th ed. Prentice Hall.

356. Protopopescu, V.A., Hively, L.M., and Gailey, P.C. (2001) Epileptic event forewarning from scalp EEG. *J. Clin. Neurophysiol.* **18**, 223–245.

357. Provenzale, A., Osborne, A.R., and Soj, R. (1991) Convergence of the K2 entropy for random noises with power law spectra. *Physica D* **47**, 361–372.

358. Provenzale, A., Smith, L.A., Vio, R., and Murante, G. (1992) Distinguishing between low-dimensional dynamics and randomness in measured time-series. *Physica D* **58**, 31–49.

359. Qi, Y., Gao, J.B., and Fortes, J.A.B. (2005) Markov chain and probabilistic computation: A general framework for fault-tolerant system architectures for nanoelectronics. *IEEE Trans. Nanotech.* **4**, 194–205.

360. Radhakrishnan, N. and Gangadhar, B. (1998) Estimating regularity in epileptic seizure time-series data. *Eng. Med. Biol. Magazine, IEEE* **17**, 89–94.

361. Rambaldi, S. and Pinazza, O. (1994) An accurate fractional Brownian motion generator. *Physica A* **208**, 21–30.

362. Ramos-Fernandez, G., Mateos, J.L., Miramontes, O., Cocho, G., Larralde, H., and Ayala-Orozco, B. (2004) Levy walk patterns in the foraging movements of spider monkeys (*Ateles geoffroyi*). *Behav. Ecol. Sociobiol.* **55**, 223–230.

363. Rao, N.S.V., Gao, J.B., and Chua. L. O. (2004) On dynamics of transport protocols in wide-area Internet connections. In *Complex Dynamics in Communication Networks*, edited by Kocarevi, L. and Vattay, G. Springer-Verlag, pp. 69–101.

364. Rapp, P.E., Cellucci, C.J., Korslund, K.E., Watanabe, T.A.A., and Jiménez-Montaño, M.A. (2001) Effective normalization of complexity measurements for epoch length and sampling frequency. *Phys. Rev. E* **64**, 016209.

365. Rapp, P.E., Cellucci, C.J., Watanabe, T.A.A., and Albano, A.M. (2005) Quantitative characterization of tide complexity of multichannel human EEGs. *Int. J. Bifurcation Chaos* **15**, 1737–1744.

366. Resnick, S. and Samorodnitsky, G. (2000) Fluid queues, on/off processes, and teletraffic modeling with highly variable and correlated inputs. In *Self-Similar Traffic and Performance Evaluation*, edited by Park, K. and Willinger, W. Wiley, pp. 171–192.

367. Richman, J.S. and Moorman, J.R. (2000) Physiological time-series analysis using approximate entropy and sample entropy. *Am. J. Physiol. Heart Circ. Physiol.* **278**, H2039–H2049.

368. Riedi, R.H., Crouse, M.S., Ribeiro, V.J., and Baraniuk, R.G. (1999) A multifractal wavelet model with application to network traffic. *IEEE Trans. Info. Theory* **45**, 992–1018.

369. Rissanen, J. (1989) *Stochastic Complexity in Statistical Inquiry*. World Scientific.

370. Robinson, P.A. (2003) Interpretation of scaling properties of electroencephalographic fluctuations via spectral analysis and underlying physiology. *Phys. Rev. E* **67**, 032902.

371. Rombouts, S.A.R.B., Keunen, R.W.M., and Stam, C.J. (1995) Investigation of nonlinear structure in multichannel EEG. *Phys. Lett. A* **202**, 352–358.

372. Rosenstein, M.T., Collins, J.J., and De Luca, C.J. (1993) A practical method for calculating largest Lyapunov exponents from small data sets. *Physica D* **65**, 117–134.

373. Rosenstein, M.T., Collins, J.J., and De Luca, C.J. (1994) Reconstruction expansion as a geometry-based framework for choosing proper delay times. *Physica D* **73**, 82–98.

374. Ross, D.L. et al. (2000) Systematic variation in gene expression patterns in human cancer cell lines. *Nat. Genet.* **24**, 227–235.

375. Ross, S. (2002) *Introduction to Probability Models*. Harcourt Academic Press.

376. Rossler, O.E. (1979) Equation for hyperchaos. *Phys. Lett. A* **71**, 155–157.

377. Rothstein, J. (1979) Generalized entropy, boundary conditions, and biology. In *The Maximum Entropy Formalism*, edited by Levine, R.D. and Tribus, M. MIT Press, pp. 423–468.

378. Ruelle, D. (1978) *Thermodynamic Formalism*. Addison Wesley Longman.

379. Ruelle, D. and Takens, F. (1971) On the nature of turbulence. *Commun. Math. Phys.* **20**, 167.

380. Salvino, L.M.W. and Cawley, R. (1994) Smoothness implies determinism — a method to detect it in time series. *Phys. Rev. Lett.* **73**, 1091–1094.

381. Samorodnitsky, G. and Taqqu, M.S. (1994) *Stable Non-Gaussian Random Processes*. Chapman & Hall.

382. Sanchez, J.C., Carmena, J.M., Lebedev, M.A., Nicolelis, M.A.L., Harris, J.G., and Principe, J.C. (2004) Ascertaining the importance of neurons to develop better brain-machine interfaces. *IEEE Trans. Biomed. Eng.* **51**, 943–953.

383. Sanchez, J.C., Erdogmus, D., Nicolelis, M.A.L., Wessberg, J., and Principe, J.C. (2005) Interpreting spatial and temporal neural activity through a recurrent neural network brain-machine interface, *IEEE Trans. Neur. Sys. Reh.* **13**, 213–219.

384. Sano, M. and Sawada, Y. (1985) Measurement of the Lyapunov spectrum from a chaotic time series. *Phys. Rev. E* **55**, 1082–1085.

385. Sauer, T., Yorke, J.A., and Casdagli, M. (1991) Embedology. *J. Stat. Phys.* **65**, 579–616.

386. Saul, J.P., Albrecht, P., Berger, R.D., and Cohen, R.J. (1988) Analysis of long-term heart rate variability: Methods, $1/f$ scaling and implications. *Comput. Cardiol.* **14**, 419–422.

387. Saupe, D. (1988) Algorithms for random fractals. In *The Science of Fractal Images*, edited by Peitgen, H. and Saupe, D. Springer-Verlag, pp. 71–113.

388. Scafetta, N. and West, B.J. (2004) Multiscaling comparative analysis of time series and a discussion on "earthquake conversations" in California. *Phys. Rev. Lett.* **9213**, 8501–8504.

389. Scheinkman, J. and LeBaron, B. (1989) Nonlinear dynamics and stock returns. *J. Business* **62**, 311–337.

390. Schiff, S.J., Dunagan, B.K., and Worth, R.M. (2002) Failure of single-unit neuronal activity to differentiate globus pallidus internus and externus in Parkinson disease. *J. Neurosurg.* **97**, 119–128.

391. Schottky, W. (1926) Small-shot effect and flicker effect. *Phys. Rev.* **28**, 74–103.

392. Schreiber, T. (1997) Detecting and analysing nonstationarity in a time series using nonlinear cross predictions. *Phys. Rev. Lett.* **78**, 843–846.

393. Schwartz, A.B., Taylor, D.M., and Tillery, S.I.H. (2001) Extraction algorithms for cortical control of arm prosthetics. *Curr. Opin. Neurobiol.* **11**, 701–707.

394. Sellan, F. (1995) Wavelet transform based fractional brownian-motion synthesis. *Compt. Rend. Acad. Sci. Ser. I-Math.* **321**, 351–358.

395. Serletis, A. and Shintani, M. (2006) Chaotic monetary dynamics with confidence. *J. Macroecon.* **28**, 228–252.

396. Serruya, M.D., Hatsopoulos, N.G., Paninski, L., Fellows, M.R., and Donoghue, J.P. (2002) Instant neural control of a movement signal. *Nature* **416**, 141–142.

397. Shaw, R. (1984) *The Dripping Faucet as a Model Chaotic System*. Aerial Press.

398. She, Z.S., Jackson, E., and Orszag, S.A. (1991) Structure and dynamics of homogeneous turbulence: Models and simulations. *Proc. R. Soc. Lond. A* **434**, 101–124.

399. She, Z.S. and Leveque, E. (1994) Universal scaling laws in fully-developed turbulence. *Phys. Rev. Lett.* **72**, 336–339.

400. She, Z.S. and Waymire, E.C. (1995) Quantized energy cascade and log-Poisson statistics in fully-developed turbulence. *Phys. Rev. Lett.* **74**, 262–265.

401. Shintani, M. and Linton, O. (2003) Is there chaos in the world economy? A nonparametric test using consistent standard errors. *Int. Econ. Rev.* **44**, 331–358.

402. Shintani, M. and Linton, O. (2004) Nonparametric neural network estimation of Lyapunov exponents and a direct test for chaos. *J. Econometr.* **120**, 1–33.

403. Shlesinger, M.F., Zaslavsky, G.M., and Klafter, J. (1993) Strange kinetics. *Nature* **363**, 31–37.

404. Shuster, H.G. (1988) *Deterministic Chaos*. VCH.

405. Smith, L.A., Ziehmann, C., and Fraedrich, K. (1999) Uncertainty dynamics and predictability in chaotic systems. *Q. J. Roy. Meteorol. Soc.* **125**, 2855–2886.

406. Snow, E.S., Novak, J.P., Lay, M.D., and Perkins, F.K. (2004) $1/f$ noise in single-walled carbon nanotube devices. *Appl. Phys. Lett.* **85**, 4172–4174.

407. Solomon, T., Weeks, E., and Swinney, H. (1993) Observation of anomalous diffusion and Levy flights in a two dimensional rotating flow. *Phys. Rev. Lett.* **71**, 3975–3979.

408. Solomonoff, R.J. (1964) Formal theory of inductive inference (part I). *Inform. Contr.* **7**, 1–22.

409. Sprott, J.C. (2003) *Chaos and Time-Series Analysis*. Oxford University Press.

410. Stam, C.J., Pijn, J.P.M., Suffczynski, P., and daSilva, F.H.L. (1999) Dynamics of the human alpha rhythm: Evidence for non-linearity? *Clin. Neurophysiol.* **110**, 1801–1813.

411. Stapf, S., Kimmich, R., and Seitter, R. (1993) Proton and deuteron field-cycling NMR relaxometry of liquids in porous glasses: Evidence of Levy-walk statistics. *Phys. Rev. Lett.* **75**, 2855–2859.

412. Stern, L., Allison, L., Coppel, R.L., and Dix, T.I. (2001) Discovering patterns in *Plasmodium falciparum* genomic DNA. *Mol. Biochem. Parasitol.* **118**, 175–186.

413. Stoica, P. and Moses, R.L. (1997) *Spectral Analysis of Signals*. Prentice Hall.

414. Strang, G. and Nguyen, T. (1997) *Wavelet and Filter Banks*. Wellesley-Cambridge Press.

415. Strogatz, S.H. (2001) *Nonlinear Dynamics and Chaos: With Applications to Physics, Biology, Chemistry and Engineering*. Westview Press.

416. Sugihara, G. and May, R.M. (1990) Nonlinear forecasting as a way of distinguishing chaos from measurement error in time series. *Nature* **344**, 734–741.

417. Swinney, H.L. (1983) Observations of order and chaos in non-linear systems. *Physica D* **7**, 3–15.

418. Swinney, H.L. and Gollub, J.P. (1986) Characterization of hydrodynamic strange attractors. *Physica D* **18**, 448–454.

419. Szczepaski, J., Amigó, J.M., Wajnryb, E., and Sanchez-Vives, M.V. (2003) Application of Lempel-Ziv complexity to the analysis of neural discharges. *Network* **14**, 335–350.

420. Szepfalusy, P. and Gyorgyi, G. (1986) Entropy decay as a measure of stochasticity in chaotic systems. *Phys. Rev. A* **33**, 2852–2855.

421. Takens, F. (1981) Detecting strange attractors in turbulence. In *Dynamical Systems and Turbulence, Lecture Notes in Mathematics*, vol. 898, edited by Rand, D.A. and Young, L.S. Springer-Verlag, pp. 366.

422. Talkner, P. and Weber, R.O. (2000) Power spectrum and detrended fluctuation analysis: Application to daily temperatures. *Phys. Rev. E* **62**, 150–160.

423. Tang, Y. and Chen, S.G. (2005) Defending against Internet worms: A signature-based approach. *Proc. of IEEE INFOCOM'05*, Miami, Florida.

424. Taqqu, M.S., Teverovsky, V., and Willinger, W. (1995) Estimators for long-range dependence: An empirical study. *Fractals* **3**, 785–798.

425. Taqqu, M.S., Teverovsky, V., and Willinger, W. (1997) Is network traffic self-similar or multifractal? *Fractals* **5**, 63–73.

426. Taqqu, M.S., Willinger, W., and Sherman, R. (1997) Proof of a fundamental result in self-similar traffic modeling. *Comput. Commun. Rev.* **27**, 5–23.

427. Task Force of the European Society of Cardiology and the North American Society of Pacing and Electrophysiology (1996) Heart rate variability: Standards of measurement, physiological interpretation, and clinical use. *Circulation* **93**, 1043–1065.

428. Taylor, D.M., Tillery, S.I.H., and Schwartz, A.B. (2002) Direct cortical control of 3D neuroprosthetic devices. *Science* **296**, 1829–1832.

429. Taylor, R.P., Micolich, A.P., and Jonas, D. (1999) Fractal analysis of Pollock's drip paintings. *Nature* **399**, 422.

430. Theiler, J. (1986) Spurious dimension from correlation algorithms applied to limited time-series data. *Phys. Rev. A* **34**, 2427–2432.

431. Theiler, J. (1991) Some comments on the correlation dimension of $1/f$-alpha noise. *Phys. Lett. A* **155**, 480–493.

432. Theiler, J., Eubank, S., Longtin, A., Galdrikian, B., and Farmer. J.D. (1992) Testing for nonlinearity in time-series — the method of surrogate data. *Physica D* **58**, 77–94.

433. Theiler, J. and Rapp, P. (1996) Re-examination of the evidence for low-dimensional, nonlinear structure in the human electroencephalogram. *Electroencephalogr. Clin. Neurophysiol.* **98**, 213–222.

434. Thompson, D. (1961) *On Growth and Form*. Cambridge University Press.

435. Timmer, J., Haussler, S., Lauk, M., and Lucking, C.H. (2000) Pathological tremors: Deterministic chaos or nonlinear stochastic oscillators? *Chaos* **10**, 278–288.

436. Tirnakli, U. (2000) Asymmetric unimodal maps: Some results from q-generalized bit cumulants. *Phys. Rev. E* **62**, 7857–7860.

437. Tirnakli, U. (2002) Dissipative maps at the chaos threshold: Numerical results for the single-site map. *Physica A* **305**, 119–123.

438. Tirnakli, U. (2002) Two-dimensional maps at the edge of chaos: Numerical results for the Henon map. *Phys. Rev. E* **66**, 066212.

439. Tirnakli, U., Ananos, G.F.J., and Tsallis, C. (2001) Generalization of the Kolmogorov-Sinai entropy: Logistic-like and generalized cosine maps at the chaos threshold. *Phys. Lett. A* **289**, 51–58.

440. Tirnakli, U., Tsallis, C., and Lyra, M.L. (1999) Circular-like maps: Sensitivity to the initial conditions, multifractality and nonextensivity. *Eur. Phys. J. B* **11**, 309–315.

441. Tirnakli, U., Tsallis, C., and Lyra, M.L. (2002) Asymmetric unimodal maps at the edge of chaos. *Phys. Rev. E* **65**, 036207.

442. Tononi, G. and Edelman, G.M. (1998) Neuroscience — consciousness and complexity. *Science* **282**, 1846–1851.

443. Tononi, G., Edelman, G.M., and Sporns, O. (1998) Complexity and coherency: Integrating information in the brain. *Trends Cogni. Sci.* **2**, 474–484.

444. Tononi, G., McIntosh, A.R., Russell, D.P., and Edelman, G.M. (1998) Functional clustering: Identifying strongly interactive brain regions in neuroimaging data. *NeuroImage* **7**, 133–149.

445. Tononi, G., Sporns, O., and Edelman, G.M. (1994) A measure for brain complexity — relating functional segregation and integration in the nervous-system. *Proc. Natl. Acad. Sci. USA* **91**, 5033–5037.

446. Trulla, L.L., Giuliani, A., Zbilut, J.P., and Webber, C.L. (1996) Recurrence quantification analysis of the logistic equation with transients. *Phys. Lett. A* **223**, 255–260.

447. Trunk, G.V. and George, S.F. (1970) Detection of targets in non-Gaussian sea clutter. *IEEE Tran. Aero. Elec. Sys.* **6**, 620–628.

448. Tsallis, C. (1988) Possible generalization of Boltzmann-Gibbs statistics. *J. Stat. Phys.* **52**, 479–487.

449. Tsallis, C., Levy, S.V.F., Souza, A.M.C., and Maynard, R. (1995) Statistical-mechanical foundation of the ubiquity of Levy distributions in nature. *Phys. Rev. Lett.* **75**, 3589–3593.

450. Tsallis, C., Plastino, A.R., and Zheng, W.M. (1997) Power-law sensitivity to initial conditions — New entropic representation. *Chaos Solitons Fractals* **8**, 885–891.

451. Tsonis, A.A. and Elsner, J.B. (1992) Nonlinear prediction as a way of distinguishing chaos from random fractal sequences. *Nature* **358**, 217–220.

452. Tsybakov, B. and Georganas, N.D. (1997) On self-similar traffic in ATM queues: Definitions, overflow probability bound, and cell delay distribution. *IEEE/ACM Trans. Networking* **5**, 397–409.

453. Tulppo, M.P., Makikallio, T.H., Takala, T.E.S., Seppanen, T., and Huikuri, H.V. (1996) Quantitative beat-to-beat analysis of heart rate dynamics during exercise. *Am. J. Physiol.* **271**, H244–H252.

454. Tung, W.W., Moncrief, M.W., and Gao, J.B. (2004) A systemic view of the multiscale tropical deep convective variability over the tropical western Pacific warm pool. *J. Climate* **17**, 2736–2751.

455. Uchaikin, V.V. and Zolotarev, V.M. (1999) *Chance and Stability: Stable Distributions and Their Applications*. VSP BV.

456. van der Ziel, A. (1979) Flicker noise in electronic devices. *Adv. Electron. Phys.* **49**, 225–297.

457. Veres, A. and Boda, M. (2000) The chaotic nature of TCP congestion control. *Proc. IEEE INFOCOM 2000*, Piscataway, NJ, pp. 1715–1723.

458. Viswanathan, G.M., Afanasyev, V., Buldyrev, S.V., Murphy, E.J., Prince, P.A., and Stanley, H.E. (1996) Levy flight search patterns of wandering albatrosses. *Nature* **381**, 413–415.

459. Viswanathan, G.M., Buldyrev, S.V., Havlin, S., da Luz, M.G.E., Raposo, E.P., and Stanley, H.E. (1999) Optimizing the success of random searches. *Nature* **401**, 911–914.

460. Von Neumann J. (1956) Probabilistic logics and the synthesis of reliable organisms from unreliable components. In *Automata Studies*, edited by Shannon, C.E. and McCarthy, J. Princeton University Press, pp. 43–98.

461. Voss, R.F. (1985) Random fractal forgeries. In *Fundamental algorithms in Computer Graphics*, edited by Earnshaw, R.A. Springer-Verlag, pp. 805–835.

462. Voss, R.F. (1988) Fractals in nature: From characterization to simulation. In *The Science of Fractal Images*, edited by Peitgen, H. and Saupe, D. Springer-Verlag, pp. 21–70.

463. Voss, R.F. (1992) Evolution of long-range fractal correlations and $1/f$ noise in DNA-base sequences. *Phys. Rev. Lett.* **68**, 3805–3808.

464. Wackerbauer, R., Witt, A., Altmanspacher, H., Kurths, J., and Scheingraber, H. (1994) A comparative classification of complexity measures. *Chaos, Solitons Fractals* **4**, 133–173.

465. Wallace, C.S. and Boulton, D.M. (1968) An information measure for classification. *Comput. J.* **11**, 185–194.

466. Ward, K.D., Baker, C.J., and Watts, S. (1990) Maritime surveillance radar Part 1: Radar scattering from the ocean surface. *Proc. Inst. Elect. Eng.* **F137**, 51–62.

467. Watanabe, S. (1969) *Knowing and Guessing: A Quantitative Study of Inference and Information*. Wiley.

468. Watanabe, T.A.A., Cellucci, C.J., Kohegyi, E., Bashore, T.R., Josiassen, R.C., Greenbaun, N.N., and Rapp, P.E. (2003) The algorithmic complexity of multichannel EEGs is sensitive to changes in behavior. *Psychophysiology* **40**, 77–97.

469. Wayland, R., Bromley, D., Pickett, D., and Passamante, A. (1993) Recognizing determinism in a time series. *Phys. Rev. Lett.* **70**, 580–582.

470. Webber, C.L. and Zbilut, J.P. (1994) Dynamical assessment of physiological systems and states using recurrence plot strategies. *J. Appl. Physiol.* **76**, 965–973.

471. Wessberg, J., Stambaugh, C.R., Kralik, J.D., Beck, P.D., Laubach, M., Chapin, J.K., Kim, J., Biggs, J., Srinivasan, M.A., and Nicolelis, M.A.L. (2000) Real-time prediction of hand trajectory by ensembles of cortical neurons in primates. *Nature* **408**, 361–365.

472. West, B.J. (1996) Levy statistics of water wave forces. *Physica A* **230**, 359–363.

473. West, B.J. (1996) Extrema of fractal random water waves. *Int. J. Mod. Phys. B* **10**, 67–132.

474. West, B.J. (2006) Thoughts on modeling complexity. *Complexity* **11**, 33–43.

475. Wiggins, S. (2003) *Introduction to Applied Nonlinear Dynamical Systems and Chaos.* Springer.

476. Willinger, W., Taqqu, M.S., Sherman, M.S., and Wilson, D.V. (1997) Self-similarity through high-variability: Statistical analysis of ethernet LAN traffic at the source level. *IEEE/ACM Trans. Networking* **5**, 71–86.

477. Witt, A., Neiman, A., and Kurths, J. (1997) Characterizing the dynamics of stochastic bistable systems by measures of complexity. *Phys. Rev. E* **55**, 5050–5059.

478. Wolf, A., Swift, J.B., Swinney, H.L., and Vastano, J.A. (1985) Determining Lyapunov exponents from a time series. *Physica D* **16**, 285–317.

479. Wolf, M. (1997) $1/f$ noise in the distribution of prime numbers. *Physica A* **241**, 493–499.

480. Wolfram, S. (1984) Cellular automata as models of complexity. *Nature* **311**, 419–424.

481. Wolfram, S. (1984) Computation theory of cellular automata. *Commun. in Math. Phys.* **96**, 15–57.

482. Wolfram, S. (1984) Universality and complexity in cellular automata. *Physica D* **10**, 1–35.

483. Wornell, G.M. (1996) *Signal Processing with Fractals: A Wavelet-Based Approach.* Prentice Hall.

484. Wu, X. and Xu, J. (1991) Complexity and brain function. *Acta Biophys. Sinica* **7**, 103–106.

485. Xu, J., Liu, Z., Liu, R., and Yang, Q.F. (1997) Information transformation in human cerebral cortex. *Physica D* **106**, 363–374.

486. Yamamoto, Y. and Hughson, R.L. (1991) Coarse-graining spectral analysis: New method for studying heart rate variability. *J. Appl. Physiol.* **71**, 1143–1150.

487. Yano, J.-I., Fraedrich, K., and Blender, R. (2001) Tropical convective variability as $1/f$ noise. *J. Climate* **14**, 3608–3616.

488. Yin, Z.M. (1996) New methods for simulation of fractional brownian motion. *J. Comput. Phys.* **127**, 66–72.

489. Zhang, X.S. and Roy, R.J. (1999) Predicting movement during anaesthesia by complexity analysis of electroencephalograms. *Med. Biol. Eng. Comput.* **37**, 327–334.

490. Zhang, X.S. and Roy, R.J. (2001) Derived fuzzy knowledge model for estimating the depth of anesthesia. *IEEE Trans. Biomed. Eng.* **48**, 312–323.

491. Zhang, X.S., Roy, R.J., and Jensen, E.W. (2001) EEG complexity as a measure of depth of anesthesia for patients. *IEEE Trans. Biomed. Eng.* **48**, 1424–1433.

492. Zhang, X.S., Zhu, Y.S., Thakor, N.V., and Wang, Z.Z. (1999) Detecting ventricular tachycardia and fibrillation by complexity measure. *IEEE Trans. Biomed. Eng.* **46**, 548–555.

493. Zhang, X.S., Zhu, Y.S., and Wang, Z.M. (2000) Complexity measure and complexity rate information based detection of ventricular tachycardia and fibrillation. *Med. Biol. Eng. Comput.* **38**, 553–557.

494. Zhang, X.S., Zhu, Y.S., and Zhang, X.J. (1997) New approach to studies on ECG dynamics: Extraction and analyses of QRS complex irregularity time series. *Med. Biol. Eng. Comput.* **35**, 467–473.

495. Zheng, Y., Gao, J.B., Sanchez, J.C., Principe, J.C., and Okun, M.S. (2005) Multiplicative multifractal modeling and discrimination of human neuronal activity. *Phys. Lett. A* **344**, 253–264.

496. Zhou, Y.H., Gao, J.B., White, K.D., Merk, I., and Yao, K. (2004) Perceptual dominance time distributions in multistable visual perception. *Biol. Cybern.* **90**, 256–263.

497. Ziv, J. and Lempel, A. (1978) Compression of individual sequences via variable-rate coding. *IEEE Trans. Info. Theory* **24**, 530–536.

498. Zumbach, G. (2004) Volatility processes and volatility forecast with long memory. *Quant. Financ.* **4**, 70–86.

INDEX

$1/f$ processes, 8, 60, 81, 88–89, 97, 118, 195, 254, 264, 279, 315
  cascade representation of, 184–189
  modeled by ON/OFF trains, 205
  modeled by SOC, 206
  modeled by superposition of relaxation processes, 203
Absolute Values of the Aggregated Series Approach, 122–123
Acute lymphoblastic leukemia (ALL), 136
Acute myeloid leukemia (AML), 136
Adjacency matrix, 135
Approximate entropy, 253, 258, 304
Asset pricing model, 300
Attack signatures, 237
Autocorrelation, 42
Autoregressive (AR)
  model, 7, 126
  process, 9, 43, 57
Axioms of probability theory, 27
B-cell lymphoma, 136
Bernoulli shift, 20
Bernoulli trials, 36–37
Bernstein's inequality, 38
Bifurcation, 11, 13, 21, 213, 233
  diagram, 22–23, 215

  in continuous time systems, 213
  in discrete maps, 217
  in high-dimensional space, 218
  period-doubling, 21–22, 218, 220
  point, 21, 214
  saddle-node, 215, 233
  subcritical pitchfork, 216
  supercritical pitchfork, 215
  transcritical, 215, 234
Binary shift, 20
Bio-inspired computations, 11
Block entropy, 252
Bode plots, 60
Boltzmann-Gibbs entropy, 200
Box-Muller method, 41
Brain-machine interfaces (BMI), 11, 141, 150
Brownian motion (Bm), 81–83, 97, 108–110, 280, 282, 285
  box-counting dimension of, 87
  generation of, 82
Cantor set, 71–73, 77
  fractal dimension of, 71, 73
Cardiac chaos, 300, 305
Cardiovascular system, 8, 12, 299–300, 305
Cascade model, 157, 169, 179
Cell line, 136

Central limit theorem, 37, 99, 107, 210
Chaos, 2, 14, 18, 279
    low-dimensional, 10, 257, 274
    noise-induced, 274–275, 283, 285
    noisy, 256, 274, 285
Chaotic
    analysis, 7, 235
    attractor, 19, 137, 244
    dynamics, 247
Characteristic function, 33–34, 52
    for Cauchy distribution, 103
    for Levy distribution, 103
    for normal distribution, 103
    for stable distributions, 102
Characteristic scales, 280, 299
Characterization of chaos
    dimension, 244, 248
    D_q spectrum, 244–245
    entropy, 251, 295
    Lyapunov exponent, 10, 18, 89, 246, 248, 256,
        258, 282, 293, 301, 314
Chromosome, 149
Coefficient of variation, 31
Complementary cumulative distribution
        function (CCDF), 143, 197
Complexity, 313
Complexity measures, 3, 12, 291, 313
Conditional CDF, 30
Conditional PDF, 30
Convolution, 35, 101, 112
Correlation
    antipersistent, 86, 116, 126, 276
    dimension, 88, 244, 246, 253, 259, 293
    entropy, 259, 293
    integral, 88, 245–246, 253, 259
    long range, 2, 13, 97, 276
    persistent, 5, 86, 116, 276
Counting process, 49, 93, 144
Covariance, 42
Crack, 76
Cross-correlation, 43, 143
Cumulative distribution function (CDF), 28
Degree of freedom, 18, 261
DeMoivre-Laplace theorem, 38
Detrended Fano factor analysis, 138
Detrended fluctuation analysis (DFA), 125, 129,
        135
Device physics, 112
Devil's staircase, 73
Diffusion entropy analysis, 124
Diffusive behavior, 282
Dimension, 244, 248
    box-counting, 19, 88, 237, 244, 293

capacity, 244
    correlation, 88, 244, 246, 253, 259, 293
    fractal, 16, 70, 244
    graph, 87
    Hausdorff-Besicovitch, 70–71
    information, 244
    Kaplan-Yorke, 248
    Lyapunov, 248
    pointwise, 244
    topological, 16
Discrete Fourier transform (DFT), 58
Distinguishing chaos from noise, 258, 283–285,
        304
Distributed denial of service (DDoS), 237
Distribution, 34
    binomial, 37
    Cauchy, 38–39, 101, 109
    chi-squared, 37
    double exponential, 160
    Erlang, 36
    exponential, 34–35, 197
    Fischer-Tippet, 107
    Frechet, 106
    gamma, 36, 199
    geometrical, 36, 46
    heavy-tailed, 13, 39, 195
    K, 209
    Levy, 101, 111
    Log-normal, 37, 159
    normal, 36
    Pareto, 39, 199
    Poisson, 38
    power-law, 75
    Rayleigh, 37
    Zipf, 40
DNA sequences, 5, 11, 13, 61, 134, 147
    coding regions, 11, 61, 147
    noncoding regions, 11, 61, 147
    period-3 feature, 61, 147
DNA walk, 61, 134
Dynamic coalition of neurons, 145, 150
Economy, 10, 112, 290, 305
    dynamics, 300, 303
    time series analysis, 300
Edge of chaos, 206, 266–268
EEG, 12, 61, 291, 304
    alpha wave, 61, 297
    beta wave, 61, 297
    delta wave, 62, 297
    epileptic seizures, 12
    seizure detection, 292, 297–298
    seizure forewarning, 304
    theta wave, 61, 297

Eigen-decomposition, 135
Eigenvalue, 45, 137, 311
  spectrum, 134–135, 137
Eigenvector, 45, 137, 311
Embedding window, 240–241, 257, 273
Empirical orthogonal functions (EOFs), 134
Energy density spectrum, 55–56
Entropic index, 262
Entropy, 251, 295
ϵ-τ entropy, 253, 258, 271, 304
Equivalence relation, 100
Euclidean norm, 125
Euler constant, 107
Expectation, 30–31
  of multiple random variables, 32
Exponential sensitivity to initial conditions
    (ESIC), 18, 262
Extended entropy formalism, 200
Extended self-similarity, 165
Extreme events
  for exponential distribution, 106
  with power-law tail, 105
$f(\alpha)$ spectrum, 245
False nearest neighbor method, 241
Fano factor analysis, 122, 129
FARIMA, 92
Far-infrared laser data, 287
Fault-tolerant computations, 11, 218
  error bound, 218
Filaments, 170, 178
Finance, 111
Finite-size Lyapunov exponent (FSLE), 3, 254,
    271, 274, 304, 314–315
Fish transformation, 19
Fixed point, 19, 214
Fluctuation analysis (FA), 120, 129
Fluid flows, 258
Fluid mechanics, 112
Fluid motions, 6
Foraging movements of animals, 112
Foreign exchange rate, 303
Fourier analysis, 54, 58, 67
Fourier series representation, 54–55
Fourier transform, 54, 56, 280
Fractal, 2, 4, 12, 14–17, 69
  dimension, 16
  geometry, 15, 69, 76
  monofractal, 69, 88
  scaling, 2, 4, 121, 139, 145
  scaling break, 2, 4, 126
Fractional Brownian motion (fBm), 13, 84–86,
    109, 134, 264, 276, 286
  box-counting dimension of, 87

  latent dimension of, 87
  self-similar dimension of, 87
Fractional dynamics, 112
Fractional Gaussian noise (fGn), 85, 116
Fundamental frequency, 54
Gamma function, 36, 199
Gao and Zheng's method, 249–250
Gene, 11, 135, 147
Gene expression pattern, 136
Generalized central limit theorem, 108, 210
Gene transcriptional network, 136
Globus pallidus, 173, 309
  externa, 173, 176, 309
  interna, 173, 176, 309
Grassberger-Procaccia algorithm, 88, 245
Haar wavelet, 65
Harmonic oscillator, 117, 240
Heart rate variability (HRV), 8, 12, 290,
    298–299, 304
  congestive heart failure, 298–299
Heaviside step function, 88, 245
Henon map, 23
Hidden frequency phenomenon, 286, 305
Hidden Markov model, 147, 238
Higuchi's method, 122–123
Hurricane, 4–5
Hurst parameter, 5, 79, 84, 89, 116–117, 160,
    205, 211, 286, 293
  estimation of, 118–119, 121, 149, 286
Hysteresis, 217
Information theory, 2, 12, 291
Interarrival times, 49, 94, 196
Intermittency, 12, 164, 179
Internet traffic modeling, 189
Internet worm, 237, 258
Interspike interval, 49, 122, 142–143, 173, 309
Jacobian matrix, 302
Joint CDF, 29, 42
Joint PDF, 29
K_2 entropy, 253, 293
Karhunen-Loève (KL) expansion, 134, 312
Kolmogorov-Chaitin complexity, 292, 313, 315
Kolmogorov-Sinai (KS) entropy, 10, 251–252,
    262–263, 293, 313
  numerical calculations, 252
Kolmogorov-Smirnov (KS) test, 209
Kronecker delta function, 155
l_1 norm, 125
Laplace transform, 33, 197
  one-sided, 197
Large-deviation estimation, 38
Lempel-Ziv (LZ) complexity, 292, 313,
    315–317

encoding, 316
  normalization, 317
  parsing, 315
Levy flight, 104, 109–110, 112, 131, 277–279, 286, 315
Levy motions, 13, 108, 131, 265
  asymmetric, 109
  correlation structure of, 131
  graph dimension of, 109
  self-similarity dimension of, 109
  symmetric, 108, 265
Levy statistics, 109, 112
Levy walk, 109, 112, 131, 277
Limit cycle, 19, 220
Limiting scales, 280, 288, 290, 297
Logistic map, 21, 218, 266, 274–275
Longmemory process, 116
Long-range dependence, 7, 10, 116, 160, 207
Long-range interactions, 268
Lorenz attractor, 245, 275
Lyapunov exponent, 10, 18, 89, 246, 248, 256, 258, 282, 293, 301, 314
  numerical computations of, 248–249
  q-, 263
Machine learning, 2
Mackey-Glass delay differential system, 275
Mandelbrot set, 18, 23
Marginal density function, 29
Markov chain, 44, 197, 232
  homogeneous coutinuous-time, 46
  homogeneous discrete-time, 44
  stationary distribution of, 45
Markov process, 35, 43, 197
  homogeneous, 44
Mass function, 36
Mean-square error, 141
Memoryless property, 35–36
Microarray, 135–136, 151
Modified Bessel function, 196, 209
Molecular interaction network, 135
Moment, 31
  central, 31
Moment generating function, 33
Multifractal, 7, 10, 69, 120, 159, 268
Multifractal analysis
  DFA-based, 125, 149–150
  structure-function–based, 120
  wavelet-based, 125
Multipath propagation, 7, 137, 210
Multiplicative multifractal process, 153, 159–161, 171
  construction of, 154
  deterministic binomial, 155

D_q spectrum, 154, 179
  properties of, 157
  random, 156
  random binomial, 156
Multiplier, 154, 171, 180
Multiscale entropy analysis, 254, 258
Multiscale
  measure(s), 3
  phenomena, 4–8
  signal(s), 1–3
Multistable visual perception, 147
Multistage NAND multiplexing system, 228
Mutual information, 143, 241
Mutually exclusive events, 26
NAND gate, 219–220
NAND multiplexing unit, 226, 228
Network intrusion, 11, 237
Network traffic, 5, 10, 77, 93, 120–121, 149, 159, 176, 179, 205, 207
  data, 307–308
Neural information processing, 11
Neural network based Lyapunov exponent estimator, 10, 302
Neuronal firings, 11, 141–144, 173, 179
  data, 309
Noisy majority gate, 222
Nonextensive entropy formalism, 262
Nonextensive statistical mechanics, 268
Nonstationarity, 1, 12, 143, 209, 290–291, 305
Novikov's inequality, 167
Nucleotide, 5, 147
ON/OFF intermittency, 130, 206, 264, 277, 286
  correlation structure of, 130
ON/OFF train, 60, 130, 204–205, 278
Optimal embedding, 240–243, 258
Ordinary differential equation (ODE), 214, 236
Orthogonality, 54, 56
Packet loss probability, 94
Parsavel's theorem, 55–56
Pathologic tremors, 58, 280
  essential, 58
  Parkinson, 58
Pathway, 136
Permutation entropy, 292, 317
Phase diagram, 237, 239, 280
Phase space, 19, 236–237, 239–240, 262
  reconstruction, 236
Point process, 93
Poisson process, 48, 77, 183, 189, 197
Power-law, 2, 7–8, 16, 60, 69, 75, 77, 105, 134, 136, 145, 159, 205, 207, 262, 286
  networks, 50, 135

Power-law sensitivity to initial conditions (PSIC), 2, 13, 88, 262–263, 279
  characterization of $1/f$ processes, 264
  characterization of chaos, 263
  characterization of edge of chaos, 266–268
  characterization of Levy processes, 266
  computation of, 263
Power law
  through approximation by log-normal distribution, 196
  through maximization of Tsallis nonextensive entropy, 200
  through optimization, 202
  through queuing, 195
  through transformation of exponential distribution, 197
  truncation of, 2
Power spectral density (PSD), 55, 80, 128, 286
Principal component analysis (PCA), 132, 134–135, 311
Probability density function (PDF), 28
Probability generating function, 34
Probability measure, 26
Probability system, 27, 29
Probability theory, 25–26
Probability transition matrix, 233
Proper orthogonal decomposition (POD), 134
Protein-protein interaction, 135
Pseudorandom number generator, 285
Quasi-equilibrium, 164
Queueing system
  First-in-first-out (FIFO), 94, 178, 193
Queuing system
  M/M/1, 195, 209
Queuing theory, 94, 195, 210
Radar backscatter, 7, 137
Raindrops, 155, 178
Random telegraph signal, 60
Random variable, 27–28
  discrete, 34
  iid, 35
  independent, 29, 34
  simulation of, 40
Random walk, 27, 30, 83, 120, 149, 282
Random water waves, 112
Relative frequency, 26
Relaxation process, 203–204
Renyi entropy, 200–201
  order-q, 252
Reynolds number, 165
r-fold time, 314–315
Riemann-Stieltjes integral, 82

Rosenstein et al.'s and Kantz's algorithm, 249–250
Rossler attractor, 243
Rough surface, 97
Routes to chaos, 21–22
  intermittency, 22
  period-doubling, 21–22
  quasi-periodicity, 22
R/S statistic, 118, 129
Sample entropy, 253, 258, 304
Sampling theorem, 58, 290
Sand pile model, 207
Scale, 2–3, 7, 64, 258
Scale-dependent Lyapunov exponent (SDLE), 3, 13, 254, 271–272, 274
Scale-free, 2, 7
  networks, 50
Scaling law, 69, 106, 122, 185
Sea clutter, 7, 11, 137, 209
  data, 308
  target detection, 7, 137, 173, 258, 303
Self-affine process, 79
Self-organized criticality (SOC), 206
Self-similarity, 15, 23, 75, 265
  asymptotically second-order, 117
  exactly second-order, 117
  parameter, 79
  self-similar process, 79–80, 87–88, 117
Shannon entropy, 200, 292, 313
Shell, 242, 250, 257, 278
Short-range dependence, 116, 183
Sign function, 103
Similarity relation, 100
Singular value decomposition (SVD), 134, 312
Sojourn times, 47, 197
Spectral analysis, 2, 8
Stable distributions, 100–103, 108, 195, 265
  characteristic function for, 102
  simulation of, 109, 111
Stable laws, 14, 104, 111, 131
Stage-dependent multiplicative process, 181–183, 191
Standard deviation, 31
State space, 19
Statistical independence, 27, 30
Stieltjes integral, 31
Stochastic oscillation, 279, 286
Stochastic process, 41–42
  covariance stationary, 115
  stationary, 42
  wide-sense stationary, 42
Surrogate data, 254, 256, 258
Synthesis of fBm

Fast Fourier transform filtering, 91
Hierarchical Monte Carlo methods, 91
Random midpoint displacement method, 90
Successive random addition method, 91
Wavelet-based method, 92
Weierstrass-Mandelbrot function–based
    method, 92
Tail probability, 104–105, 197
Tangential motion, 245, 249
Taylor series expansion, 99, 214
Test for low-dimensional chaos, 254–256, 258
Theiler decorrelation time, 245, 250
Thermodynamics, 201
Throughput, 94
Time delay embedding, 236–237, 258, 262, 284
    delay time, 236
    embedding dimension, 236
Time-dependent exponent $\Lambda(k)$ curves, 242,
    254, 256, 259, 263, 272, 304
Topological entropy, 200
Torus, 259
Traffic model
    fBm, 93, 95
    LRD, 189
    Markovian, 93, 189
    Poisson, 93–94, 195
Tropical convection, 179
Tsallis distribution, 201–202, 209, 262
Tsallis entropy, 200–201, 262
Turbulence, 12, 268
    β-model, 168
    energy dissipation, 164
    energy-injection scale, 164
    inertial-range scale, 164

K41 theory, 164, 169
Kolmogorov energy spectrum, 116, 139, 164
    log-normal model, 167
    log-Poisson statistics, 169
    log-stable model, 168
    longitudinal velocity differences, 164
    molecular dissipation scale, 164
    p model, 169
    random β-model, 168
    SL model, 169
    universal scaling behavior, 163–164
Utilization level, 94, 196
Van der Pol's oscillator, 280
Variance, 31
Variance-time relation, 117, 122, 129, 160
Von Koch curve, 74
Von Neumann multiplexing system, 219, 226
Wavelet, 2, 4, 62
    approximation coefficient, 65
    detailed coefficient, 65
    Hurst parameter estimator, 119
    mother wavelet, 64
    multiresolution analysis, 64
    pyramidal structure, 64
    representation of fBm, 89
    scaling function, 64
Weierstrass-Mandelbrot function, 92, 97
Whittle's approximate maximum likelihood
    estimator, 119
Wiener-Khintchine theorem, 57, 254
Windowed Fourier transform, 62
Wolf et al.'s algorithm, 248, 271, 314
Yeast, 136
Z-transform, 34

Printed and bound by CPI Group (UK) Ltd, Croydon, CR0 4YY

16/04/2025

14658453-0003